ORGANIC CHEMISTRY: AN OVERVIEW

JAMES A. MOORE

University of Delaware

THOMAS J. BARTON

Iowa State University

SAUNDERS GOLDEN SUNBURST SERIES

W. B. SAUNDERS COMPANY • Philadelphia • London • Toronto

W. B. Saunders Company: West Washington Square
Philadelphia, PA 19105

1 St. Anne's Road
Eastbourne, East Sussex BN21 3UN, England

1 Goldthorne Avenue
Toronto, Ontario M8Z 5T9, Canada

Library of Congress Cataloging in Publication Data

Moore, James Alexander.

Organic chemistry.

(Saunders golden sunburst series)
Includes index.

1. Chemistry, Organic.   I. Barton, Thomas J., joint author.
   II. Title.

QD251.2.M664   547   77–72792

ISBN 0–7216–6516–0

Cover photograph of Colorado Independence Rock, Colorado National Monument, by Ed Cooper

Introduction to Organic Chemistry   ISBN 0–7216–6516–0

Last digit is the print number: 9   8   7   6   5   4   3

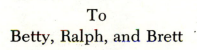
To
Betty, Ralph, and Brett

PREFACE

As our title states, this book is an overview, designed for use in a short course in organic chemistry for students in health sciences, agricultural sciences, and related curricula. One of our main concerns has been to keep the book to a reasonable length for such a course, and at the same time avoid superficiality. We do not subscribe to the idea that a quick glimpse at some structural formulas has significant educational value.

To meet these goals we have selected a central core of compounds and reactions, and have not hesitated to omit traditional but archaic topics such as synthetic methods for alkanes. By restricting the number of reactions covered, we have been able to provide the detailed explanations that are needed for students to really understand the process. We have hammered hard on the principles of bond-breaking and bond-making, and a second color has been used in structural formulas to help in following the bond changes that take place in reactions.

The organization and sequence of material are largely traditional. The first chapter contains the basic principles of atomic structure and a brief consideration of bond formation. Chapters on stereochemistry and structure determination by spectral methods are introduced early, at the points where they are needed and can be appreciated by the student. The amphoteric and nucleophilic properties of alcohols, amines, and thiols are presented in a unified treatment in Chapter 7; phenols and aromatic amines are treated together in a separate chapter to emphasize the chemistry common to both.

A very important part of organic chemistry, particularly in a non-major course, is the occurrence of organic compounds in nature and the uses of organic compounds as drugs, textiles, pesticides, food additives, and other products. These aspects are discussed and integrated with other material throughout the text as functional groups and structures are introduced. In addition, Special Topics sections distributed over the first two-thirds of the book highlight a number of relevant areas in greater depth.

An essential feature of any organic textbook is a generous supply of problems of various kinds and of carefully graded difficulty. We have made a special effort to meet this need. Problems are placed in the body of the text to provide immediate reinforcement of new concepts as they are introduced. Further problems are designed for drill and review, and to enable the student to apply and extend principles. Some chapters require more problems than others. We have attempted to provide as many problems as any student or instructor will be able to use. Moreover, we have avoided vague or open-ended questions that do not have a clearly defined answer based on material in this book. Solutions to all problems, with explanatory discussions of how to approach the problem, are given in the paperback text *Solutions to Problems in Organic Chemistry—an Overview.*

The text was reviewed in detail and criticized by several colleagues whose efforts are greatly appreciated: Dr. James Deyrup, University of Florida, Dr. Arnold Krubsack, University of Southern Mississippi, Dr. Manfred Reinecke, Texas Christian University, and Dr. Andrew Ternay, University of Texas, Arlington. We also thank Dr. Richard Bozak, California State University, Dr. Norbert Goeckner, Western Illinois University, and Dr. Robert Grubbs, Michigan State University, for their helpful comments on the manuscript. The expert ministrations of Mr. Jay Freedman of W. B. Saunders Company and of Mrs. Mary Ann Gregson and Miss Linda Charles are most gratefully acknowledged.

JAMES A. MOORE

THOMAS J. BARTON

A NOTE TO THE STUDENT

From the experience of a number of years in the classroom, we have found that a few comments about studying organic chemistry are helpful in getting students off to a good start in the course and making successful progress.

Why study organic chemistry? Most students are taking a short organic course because the people in charge of their curriculum have determined that organic chemistry is important for an understanding of their major field. It is generally agreed that a basic knowledge of organic compounds and their behavior is essential in such areas as agriculture, food science, home economics, nursing, and nutrition. Since our food, clothes, books, medicines, and bodies are made up of organic compounds, anyone concerned with the world around him must have an acquaintance with organic chemistry.

What does the course cover? As we will see in Chapter One, organic chemistry deals with compounds of just a few elements—mainly carbon, hydrogen, and oxygen—held together by covalent bonds. After reviewing some basic concepts from general chemistry about atomic bonding, we will look at how covalent bonds are formed and broken in reactions, and learn how to make predictions as to when and how bond changes will take place. Very early we will begin to study the shapes and structures of organic molecules, and how these factors control the course of reactions. Throughout the course, we will see how these reactions are applied to molecules of environmental, medical, or industrial importance.

What about questions and problems? Organic chemistry is a problem-solving course. You will find numerous problems within the text of most chapters; these follow immediately after the material on which the problems are based, and will give you a chance to check your understanding of the point just covered. We urge you to work these problems as you come to them. In addition, there are many problems at the end of each chapter. Study with pencil and paper at hand, and do as many problems as your schedule permits. Solutions to all problems in the book are provided in the supplementary paperback.

How to study? There is no magic formula for studying organic chemistry that will suit every individual. There is a good bit of material to be committed to memory, as in any beginning course. How much you must memorize will depend on how well you understand the basic principles, and learn to apply them. Organic chemistry is a "building" course. What you learn at one point can and usually will come up later in a new and slightly different situation. You cannot afford to cram for a test and quickly forget, since your need for that material will reappear throughout the course.

CONTENTS

5

6

7

8

ALDEHYDES AND KETONES 172

9

SPECTRAL IDENTIFICATION AND STRUCTURE DETERMINATION OF ORGANIC COMPOUNDS 204

1

ATOMS, MOLECULES AND BONDS

Historically, organic compounds were considered to be those produced by living systems, either plants or animals. These compounds were thought to be fundamentally different from the oxides and salts derived from minerals. Once it was discovered that "organic" compounds could be synthesized in the laboratory, sometimes with "inorganic" compounds as the starting materials, the distinction began to fade. Today the field of organic chemistry has broadened into the study of compounds which always contain carbon, usually contain hydrogen, and may also contain groups made up from other elements—most commonly oxygen, nitrogen, and the halogens. The old method of classification no longer holds, since every day many organic compounds are synthesized in the laboratory which have no counterparts (to our knowledge!) in nature. Since carbon can form bonds with a variety of metals, there is no strict line which differentiates organic from inorganic compounds. Thus, other than restricting ourselves to a study of the compounds of carbon, there is no reason for us to even try to strictly define organic chemistry. The same principles of atomic structure and bonding apply to organic compounds as to others, and we will begin with a brief review of these basic concepts.

1.1   Atoms, Ions, and Molecules

An atom is composed of a dense nucleus, containing neutrons and positively charged protons, surrounded by a number of negatively charged electrons sufficient to balance the charge. Familiar examples are the hydrogen atom, H, which has one positively charged proton in the nucleus and a single electron, and the carbon atom, C, which has six protons in its nucleus and six electrons to balance the charge. The atomic structures of the first ten elements are shown in Figure 1.1. As indicated in the figure, the electrons are

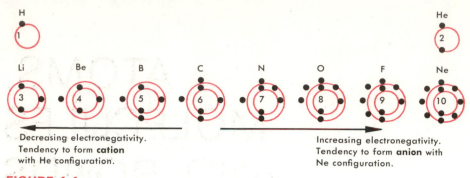

Decreasing electronegativity. Tendency to form **cation** with He configuration.

Increasing electronegativity. Tendency to form **anion** with Ne configuration.

FIGURE 1.1

Atomic structures of the first ten elements.

arranged in shells, the first containing two electrons and the second containing eight electrons.

An important principle of atomic structure is that a completely filled outer shell of electrons, as in the inert gases helium (atomic number 2) and neon (atomic number 10), is a particularly stable arrangement. Atoms may *gain, lose,* or *share* electrons to achieve a completely filled outer shell. If electrons are gained or lost, the neutral atom becomes electrically charged and is called an **ion.** Electron shells can also be completed by sharing pairs of electrons in neutral **molecules.** Both ions and molecules can be conveniently pictured by "electron dot" formulas. The symbol of the element represents the atomic nucleus plus all complete inner shells of electrons. Electrons in the *outer shell* are indicated by dots placed around the symbol. A pair of electrons that is shared between atoms is usually written as a dash.

Formation of ions; gain or loss of electrons

$$\text{H} \cdot \xrightarrow[e^-]{\text{lose}} \text{H}^+ + e^-$$

$$\cdot \ddot{\text{O}} \cdot + 2\,e^- \xrightarrow[2\,e^-]{\text{gain}} : \ddot{\ddot{\text{O}}} :^{2-}$$

Formation of molecules; sharing electrons

$$\text{H} \cdot + \text{H} \cdot \longrightarrow \text{H—H}$$

$$2\,\text{H} \cdot + \cdot \ddot{\text{O}} \cdot \longrightarrow \text{H—}\ddot{\text{O}}\text{—H}$$

Looking at Figure 1.1, we see that as we move to the right-hand side of the periodic chart, elements become increasingly **electronegative,** with a tendency to gain additional electrons and become negative ions, or **anions,** in order to complete the outer electron shell. When we approach the left side of the chart, the elements decrease in electronegativity and prefer to lose electrons, forming positively charged **cations** with the inert gas configuration. Thus, elements from opposite sides of the periodic table can transfer one or more electrons so that each can adopt the nearest inert gas configuration. For example, lithium fluoride is formed by the complete transfer of an electron from the electropositive lithium atom to the very electronegative fluorine atom. The resulting compound is a salt, and the attractive force between the two oppositely charged ions is called an **ionic bond.**

$$\text{Li} \cdot \xrightarrow[\text{loss}]{\text{electron}} e^- + \text{Li}^+ \qquad \textit{cation; He configuration}$$

$$:\!\ddot{\text{F}}\cdot \; + \; e^- \xrightarrow[\text{gain}]{\text{electron}} :\!\ddot{\text{F}}\!:^- \qquad \textit{anion; Ne configuration}$$

$$\text{Li} \cdot \; + \; :\!\ddot{\text{F}}\cdot \xrightarrow[\text{transfer}]{\text{electron}} \text{Li}^+ :\!\ddot{\text{F}}\!:^-$$

Problem 1.1   Write electron dot formulas showing the reaction of lithium and oxygen to form the compound lithium oxide, Li_2O.

In the center of the periodic table, the elements boron, carbon, and nitrogen have little tendency to form ions by gain or loss of electrons. For these elements, a completely filled outer shell would require formation of an ion with three or four unit charges, such as C^{4+} or N^{3-}, and the accumulation of like charges would lead to very large repulsive forces. For this reason, B, C, and N achieve an inert gas configuration by **sharing electrons.**

Thus carbon, for example, forms molecules by completing its electron shell with four electrons from other atoms. One, two, or even three pairs of electrons can be shared between two atoms. In writing these formulas, the electrons are combined to give an arrangement in which each atom is surrounded as far as possible by a completed outer shell—two electrons around hydrogen and eight around second row elements.

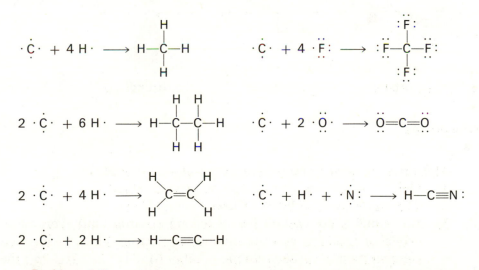

Problem 1.2   Write electron dot structures for a) CH_2F_2, b) C_2H_3F, c) CH_2O

These electron dot structures actually contain little structural information. They tell us nothing about the nature of the bonds between atoms, or the shapes of molecules, or the differences between single, double, and triple bonds. However, these diagrams are useful in helping to keep track of electrons, and are a point of departure for developing a picture of chemical bonding.

1.2 Atomic Orbitals

The picture of atoms that we have seen so far describes the arrangement of electrons in shells. We now need to look more closely at the distribution of electrons within these shells. The electrons are not floating randomly around the nucleus, but are located in **atomic orbitals,** which are specific regions in space where there is the highest probability of finding a maximum of two electrons having a particular energy.

The atoms that we will encounter most often in organic compounds have only two kinds of atomic orbitals, s and **p.** The s orbitals can be represented as spheres, in which the s electrons are most often located, with the nucleus at the center. The p orbitals are dumbell-shaped, with the nucleus between the two lobes. The energies of the atomic orbitals (actually the energies of the electrons in these orbitals) increase in the order 1s, 2s, 2p. There are three 2p orbitals—$2p_x$, $2p_y$, and $2p_z$—which are oriented perpendicular to one another in space and are equal in energy. The subscripts refer to the normal x, y, and z coordinates, and the numerals indicate the shells. The shapes and directional properties of the s and p orbitals are illustrated in Figure 1.2.

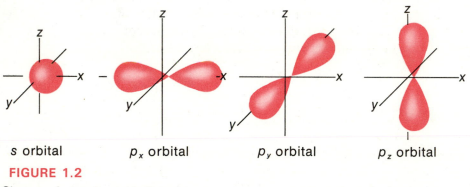

s orbital p_x orbital p_y orbital p_z orbital

FIGURE 1.2

Shapes of s and p orbitals.

The rules we need to remember about atomic orbitals are:
1) Only two electrons can occupy an atomic orbital.
2) The electrons fill orbitals of lower energy first.
3) An orbital is not occupied by a second electron until every other orbital of equal energy also contains one electron. Thus, the electron configuration of carbon (atomic number 6) is $1s^2$, $2s^2$, $2p_x$, $2p_y$; the superscript indicates the number of electrons in that orbital.

1.3 The Covalent Bond

Earlier we saw that molecules are formed by the sharing of electron pairs. With our model of atomic structure in terms of atomic orbitals we can describe electron sharing between atoms in more precise terms, namely as the formation of **covalent bonds.** Covalent bonds are formed by the overlap of two atomic orbitals, one from each bonding atom, to form a **molecular orbital** which contains the two bonding electrons. Thus, a covalent bond is one where two atoms *share* a pair of electrons. The basic rules for molecular orbitals are the same as were earlier presented for atomic orbitals. In the simplest possible case, we could imagine the molecule H_2 as being formed from the combination of two hydrogen atoms, $H\cdot$, to share the two electrons in a covalent bond.

$$H\cdot + \cdot H \longrightarrow H:H$$

This simple molecule can be written as H_2, $H:H$, or $H—H$, where the dash in the last structure represents a shared electron pair. For a more complete picture, we can expand to say that when two hydrogen atoms (each with one electron in a $1s$ atomic orbital) combine, the electrons are now found in a molecular orbital which is a product of the combination of the two $1s$ atomic orbitals. Such a bonding molecular orbital encompasses both hydrogens, has *cylindrical symmetry,* and is called a **sigma (σ) bond.** The σ bond can be visualized as an area shaped like a frankfurter, encompassing both bonded nuclei, where the two σ bonding electrons are located (Figure 1.3).

FIGURE 1.3

Formation of H_2.

The purpose of any theory or model of chemical bonding is to provide understanding of the fact that *atoms have a strong tendency to combine into molecules.* Since H_2 is a more stable arrangement than two $H\cdot$ atoms, energy is released in the bonding process. The energy released is called the *heat of formation* or *bond energy,* and for H_2 it is 104 kcal per mole. This means that we would have to add 104 kcal/mole to break the $H—H$ bond and form two hydrogen atoms.

$$H\cdot + H\cdot \underset{\text{addition of energy}}{\overset{\text{release of energy}}{\rightleftharpoons}} H_2 + 104 \text{ kcal/mole}$$

The simplest molecule in organic chemistry, **methane** or CH_4, can likewise be considered as arising from the combination of one carbon and four hydrogen atoms. (We are not concerned here with the actual preparation of methane, which is obtained from natural gas.) To put the methane molecule together, we can form four covalent bonds from the four electrons of the hydrogen atoms and the four electrons of carbon (recall that only the outer shell electrons are involved in bonding). We thus achieve filled outer shells, with the rare gas configuration of two electrons for each hydrogen and eight outer shell electrons for carbon. These shared electrons were formerly in atomic orbitals and are now in σ molecular orbitals.

Let us now look at the construction of methane more closely, and as a first possibility, consider that we combine the four hydrogen atoms with the 2s and 2p atomic orbitals of carbon. If this were done, we would expect three of the C—H bonds to lie at right angles to one another as a result of overlapping hydrogen 1s orbitals and the $2p_x$, $2p_y$, and $2p_z$ atomic orbitals, with the fourth bond formed from hydrogen 1s and carbon 2s overlap. However, it would be surprising if this actually were the structure of methane. Bonds are composed of electrons, and the negative charges should want to be as far away from each other as possible. The structure that would give maximum separation of the four bonds in space is a regular **tetrahedron,** or three-sided pyramid, with carbon at the center and each hydrogen at an apex. With tetrahedral arrangement, each H—C—H bond angle is 109.5°. As it turns out, many types of evidence show that methane does in fact have exactly this tetrahedral structure.

methane, a regular tetrahedron

Dashed lines --- are bonds directed behind the plane of the page, wedge lines ◄— are directed in front of this plane, and solid lines —— are bonds in the plane of the page.

1.4   Hybrid Orbitals

Since the tetrahedral structure of methane doesn't fit the geometry of our simple picture of overlapping four hydrogen 1s atomic orbitals with the 2s and 2p atomic orbitals of carbon, we must change that picture. One way of

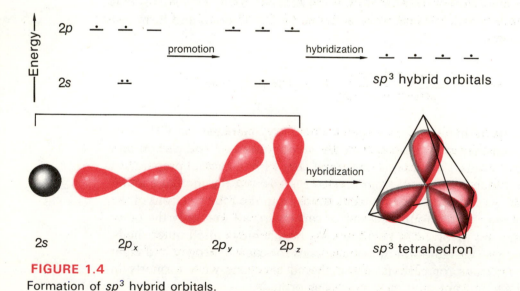

FIGURE 1.4
Formation of sp^3 hybrid orbitals.

looking at the process of methane formation from carbon and hydrogen atoms is to promote one of the $2s$ electrons of carbon into the vacant $2p$ orbital and then mix these four carbon orbitals together to obtain four new, entirely equal, tetrahedrally oriented orbitals. This "mixing" process is called **hybridization,** and the resulting orbitals are called **hybrid orbitals.** In this case they are sp^3 hybrid orbitals (Figure 1.4), since one s and three p orbitals were used to make the four new orbitals.

Now it is easy to see that overlap of the $1s$ orbitals of four hydrogen atoms with the four sp^3 hybrid orbitals of carbon will produce a tetrahedral molecule of methane (Figure 1.5). As in the bond in H_2, the two electrons in each bond are distributed symmetrically around the axis between the two atoms. This is, therefore, another example of a σ bond. The energy of each C—H bond is 99 kcal per mole; therefore, CH_4 is 396 kcal more stable than four hydrogen atoms and one carbon atom before bonding takes place.

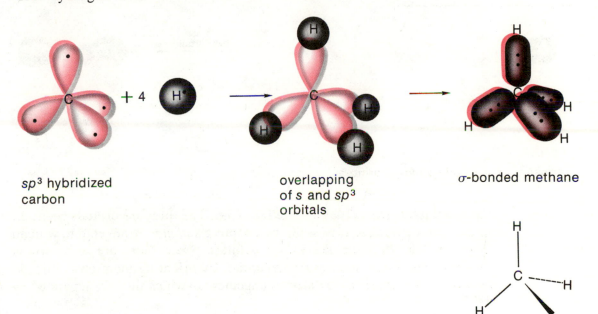

sp^3 hybridized carbon

overlapping of s and sp^3 orbitals

σ-bonded methane

FIGURE 1.5

Orbital construction of methane.

1.5 sp^2 and sp Hybridization

Hybridization of s and p orbitals can occur by using all three (as above), two, or only one of the p orbitals to produce **four sp^3, three sp^2,** or **two sp** hybrid orbitals, respectively (Figure 1.6). In each case the hybrid orbitals are directed in space to provide the maximum separation and therefore minimize electron-electron repulsions.

As we have seen in the case of methane, maximum separation of sp^3 orbitals, and the bonds formed by overlap with them, are obtained by directing

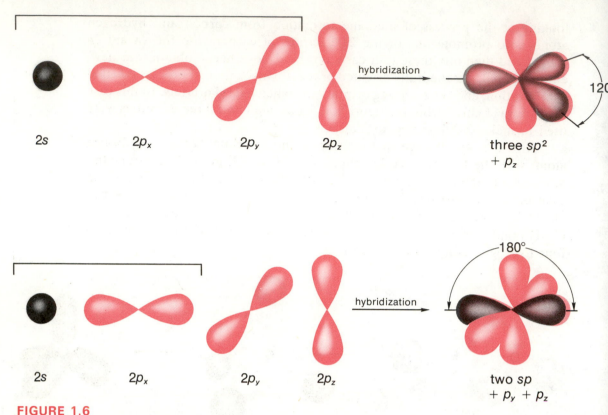

FIGURE 1.6

Formation of sp^2 and sp hybrid orbitals.

these orbitals to the corners of a tetrahedron. The three sp^2 orbitals get as far away from each other as possible by adopting a *planar* arrangement, with an angle of 120° between the hybrid orbitals. Since there are only two sp orbitals, they obtain maximum separation by orienting themselves directly away from one another—a *linear* arrangement in which they are separated by 180°.

For sp^2 and sp hybridized carbon, the remaining p orbitals, which are not involved in hybridization, are directed perpendicular to the plane or axis of the hybrid orbitals, respectively.

1.6 Carbon–Carbon Bonds

The central role of carbon in organic chemistry and in living matter depends upon the fact that orbitals of two carbon atoms overlap to form strong **C—C** bonds. Thus, a chain of virtually unlimited length can be built up from sp^3 hybridized carbon atoms linked together by **C—C σ bonds.*** The

---

*The ability to form a stable chain containing atoms of only one element is unique to carbon. Because of the much lower bond energies, chains of only a few atoms can be formed with other elements; for neighboring elements, the bond energies are C—C, 83.1 kcal; N—N, 38.4 kcal; Si—Si, 42.2 kcal.

remaining carbon orbitals are used to form σ bonds to hydrogen or other elements.

The simplest example of a molecule with a C—C σ bond is **ethane,** which has the molecular formula C_2H_6. In our construction of methane, if we had added only three hydrogen atoms to sp^3 hybridized carbon we would have one sp^3 orbital, containing one electron, left over. This $H_3C\cdot$ unit is called the **methyl free radical** or simply the methyl radical. Now we can combine two of these $H_3C\cdot$ units by overlapping their partially filled sp^3 orbitals to form a C—C σ bond (Figure 1.7).

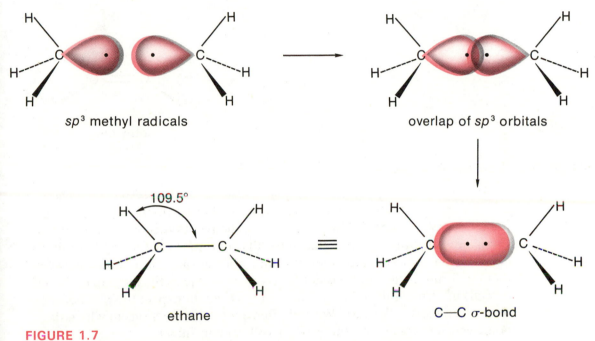

sp^3 methyl radicals   overlap of sp^3 orbitals

109.5°

ethane   C—C σ-bond

FIGURE 1.7

Orbital construction of ethane.

A second type of carbon–carbon bond is found in the molecule C_2H_4, **ethylene.** Recall that when we formed the three sp^2 hybrid orbitals on carbon (Figure 1.6), a $2p$ orbital was left over. If we now add hydrogen atoms to two of the sp^2 orbitals and couple two of the resulting $H_2C:$ units through the remaining sp^2 orbitals, we obtain $H_2\dot{C}$—$\dot{C}H_2$. The carbons are connected by a σ bond (sp^2-sp^2) and each carbon also has a p orbital containing one electron. The leftover p orbitals, directed above and below the plane of the six atoms, can now overlap sidewise, and a new carbon–carbon bond is formed to give the compound $H_2C{=}CH_2$, called ethylene.

This bond is different from our now familiar σ bond (in which all electron density is located around the axis between the two bonded atoms) because here the electron density is located above and below (but not in) the plane of the molecule. The bond formed in this sidewise overlap is called a **π (pi) bond** (Figure 1.8). Since the two carbons of ethylene are connected by both a σ bond (from sp^2-sp^2 overlap) and a π bond (from p-p overlap), this combination is known as a **double bond** and is written as C=C.

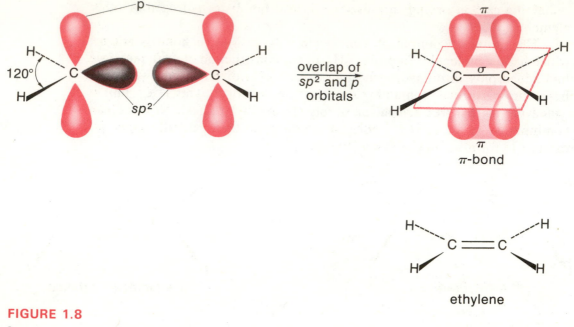

FIGURE 1.8

Construction of the double bond.

The C—C double bond is significantly shorter than the C—C single bond—1.35 Å *vs.* 1.54 Å (Table 1.1). Also, the two bonds which we write as C=C are not equal in strength. Since the *p* orbitals on the *sp²* hybridized carbons can only overlap sideways, and not end-on, the overlap between them in forming the π bond is not nearly as complete as the overlap of the *sp²* orbitals in forming the σ bond. Therefore, we would expect the π bond to be weaker, and this is the case. We will often encounter reactions in which the π bond will be broken but the σ bond will remain intact.

In a manner similar to our construction of $H_2C{=}CH_2$ from *sp²* hybridized carbons, we can prepare **HC≡CH,** called **acetylene,** from *sp* hybridized carbons. We simply overlap two *sp* atomic orbitals to form a σ bond between the two carbon atoms. This leaves, on each carbon atom, an *sp* orbital and two *p* orbitals, each with one electron. When both *p* orbitals on each carbon are allowed to overlap, two π bonds are formed, which are perpendicular to one another. This bonding situation for acetylene is called a **triple bond** and is conveniently drawn as C≡C. Again we must remember that these three lines stand for one σ bond (*sp-sp*) and two π bonds. As we would expect, the

TABLE 1.1   BOND LENGTHS AND ENERGIES

Bond	Length (Å)	Total Energy $\left(\dfrac{kcal}{mole}\right)$
C—C	1.54	83
C=C	1.35	146
C≡C	1.20	200

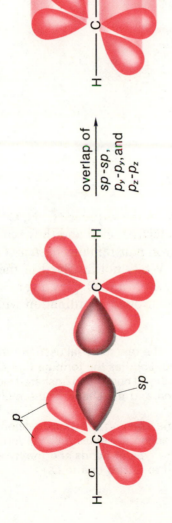

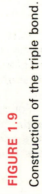

H—C≡C—H
acetylene

overlap of
sp-sp,
p_y-p_y, and
p_z-p_z

FIGURE 1.9

Construction of the triple bond.

carbon atoms are held closer together by three bonds than by two or one, and the carbon–carbon bond length in acetylene is found to be only 1.20 Å. The acetylene molecule is completed by combining the remaining sp orbital on each carbon with a hydrogen atom to form two C—H σ bonds. Our final result is a *linear* molecule of acetylene (Figure 1.9).

Carbon forms covalent bonds with many elements besides hydrogen and carbon. The most common are the halogens (F, Cl, Br, I), oxygen, and nitrogen. The halogens form σ bonds with carbon in the same way as we have seen for hydrogen—through overlap with sp^3, sp^2, or sp orbitals on carbon. Oxygen can form either single or double bonds to carbon, while nitrogen may form single, double, or triple bonds to carbon (Table 1.2).

TABLE 1.2   COMPOUNDS OF CARBON WITH N, O, AND F

Since each successive element in the sequence C, N, O, X* requires one fewer bond to complete its outer electron shell and thus achieve an inert gas configuration, an additional electron pair remains *unshared* (not involved in bonding), as shown in Table 1.2. While not involved in the bonding, these unshared electron pairs, especially on N and O, play a very important role in the chemistry of such compounds through coordination with various Lewis acids.

Problem 1.3   Vinyl chloride, the raw material used in manufacturing most "vinyl" plastics (Section 3.15), has the formula C_2H_3Cl. Write the structure of vinyl chloride, showing all bonds and unshared electrons, and state the hybridization of carbon and the shape of the molecule.

Problem 1.4   Formic acid, the irritating substance responsible for the sting of ants and the blistering action of nettles, has the formula CH_2O_2. Write the structure and state the type of bonds and the hybridization of carbon. (Hint: one H is bonded to C and one to O.)

1.7   Isomerism

The formulas CH_4, **methane**, and C_2H_6, **ethane**, represent the two simplest compounds of sp^3 carbon and hydrogen. Their structures follow

---

*The symbol X is used to indicate an unspecified halogen. Other such generic symbols will be introduced as needed.

directly from the bonding picture given above; the atoms can be connected in only one way. In building the next compound, C_3H_8, **propane,** another —CH_2— unit can be inserted into any of the six equivalent C—H bonds in ethane or into the C—C bond, and there will be only *one compound,* C_3H_8, no matter how we choose to write it.

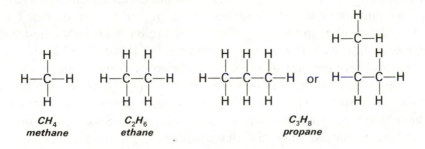

By the same reasoning, there is only *one* compound of the formula CH_2Cl_2. Replacement of *any* two of the four equivalent hydrogens in methane by two chlorine atoms gives exactly the same structure. The existence of only one compound of CH_2Cl_2 or of C_3H_8 depends upon the tetrahedral geometry of the sp^3 carbon (Figure 1.10A). For example, if the four covalent bonds around the carbon were in a planar arrangement, as in Figure 1.10B, it would be possible to have *two* different compounds with the molecular formula CH_2Cl_2. The fact that two compounds of the CX_2Y_2 type with the

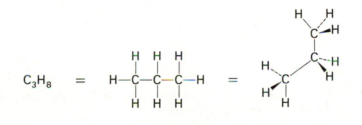

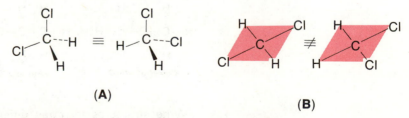

(A)

(B)

FIGURE 1.10

Possible structures of CH_2Cl_2. A. Tetrahedral structure. B. Incorrect planar structure.

same molecular formula have never been observed was one of the points that led to the original proposal of the tetrahedral structure of carbon in organic chemistry. No matter how we arrange tetrahedral CX_2Y_2, each X and each Y will have exactly the same geometrical relationship with every other atom in the molecule.

Continuing on to the next higher compound of carbon and hydrogen, C_4H_{10}, we encounter a new situation: the four carbon atoms can be connected to one another in two ways. One structure has a *straight chain* and the other is *branched,* with the central carbon having three C—C bonds. These two C_4H_{10} compounds are **isomers.** Isomers are *different compounds* which have exactly the same number and type of atoms (same molecular formula). Our two isomers of C_4H_{10} are called **structural isomers,** which means that they have different atom-to-atom bonding sequences. Since these molecules have different shapes, they also have different physical properties—*e.g.,* different boiling points, melting points, and solubilities.

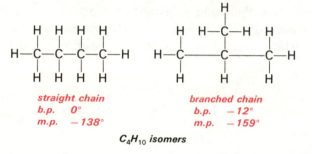

straight chain
b.p.   0°
m.p.   −138°

branched chain
b.p.   −12°
m.p.   −159°

C_4H_{10} isomers

Problem 1.5   There are three isomers with the formula C_5H_{12}. Write structures for these three compounds.

Inclusion of other elements besides carbon and hydrogen introduces further possibilities for isomerism. For example, two compounds of formula C_2H_6O exist, one containing two C—O bonds and the other containing one C—O and one O—H bond. These isomers have very different chemical properties, one being an unreactive ether and the other a highly reactive alcohol. The ether is almost totally insoluble in water, while the alcohol is infinitely soluble in water.

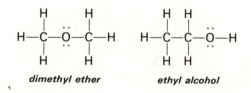

dimethyl ether   ethyl alcohol

Problem 1.6   Write structures for three different compounds with molecular formula C_3H_8O.

Problem 1.7   Write structures for all isomers of a) C_2H_5Cl, b) C_3H_7Cl.

An additional possibility for isomerism arises from the fact that the ends of a carbon chain can be joined to form a **cyclic** structure. The formula C_5H_{10} can represent a five-carbon chain, straight or branched, containing one double bond, or a **ring** connected with single bonds.

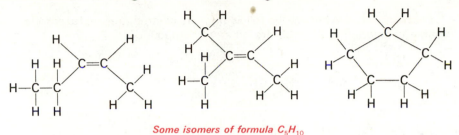

Some isomers of formula C_5H_{10}

1.8 Simplified Structural Formulas

The structures that we have been drawing thus far have been drawn to show all bonds in the molecules. These expanded formulas are useful in visualizing structures, but they become cumbersome when used for larger molecules. Examples of several condensed or abbreviated ways of writing organic structures are shown in Figure 1.11, and these will generally be used in the remainder of the text. In these condensed formulas, bonds and also unshared electron pairs on N, O, and halogen are omitted unless needed for clarity or to describe how a reaction works. Rings are often indicated by showing the bonding framework, with a CH_2 group understood to be at each

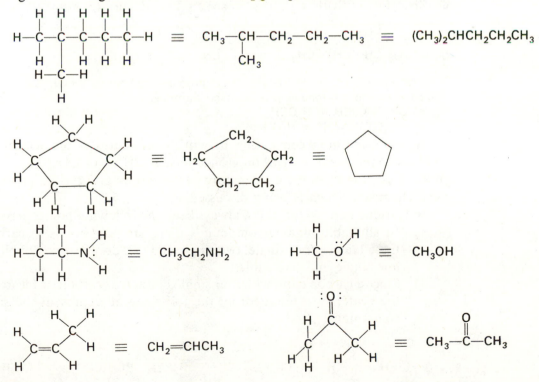

FIGURE 1.11

Extended and condensed structures.

angle. Double or triple bonds are clearly shown and charges on atoms are always included. It is important to remember that these abbreviated structural formulas have the same meaning as the three-dimensional perspective formulas.

Problem 1.8 Which of the following pairs of structures represent the same compound and which are different compounds?

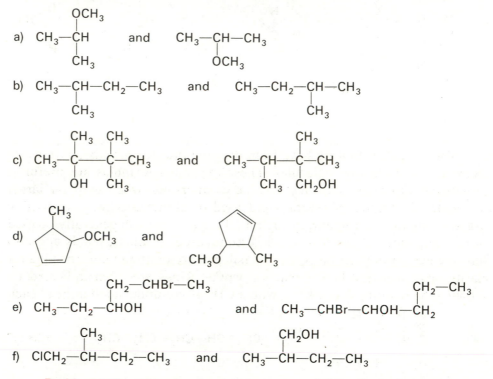

Problem 1.9 Write the expanded structures (all bonds shown) corresponding to the following condensed formulas:
a) $CH_3CHCl(CH_2)_3CH_2OCH_3$,
b) $CH_3C(CH_3)_2CHOHCH_2OH$.

Another common convention in writing structures is the use of the symbol R to represent *any* carbon group, such as CH_3— or CH_3CH_2—. The symbol R is used when the exact nature of the carbon group is unimportant to the chemistry which is being discussed.

In learning to draw formulas, always keep the following points in mind:
1) In all stable organic compounds there are *four* bonds to carbon; these may be four single, two single and one double, two double, or one single and one triple.
2) Since a carbon atom with four single bonds (sp^3 carbon) is located at the center of a tetrahedron, the following all represent the same compounds:

3) If the three-dimensional structure of a compound is to be written, bonds in the plane of the paper are shown as normal solid lines, bonds going behind the paper are dashed, and bonds coming out in front of the paper are shown as heavy wedges.

$$
\begin{array}{c}
\text{H} \\
| \\
\text{R}-\text{C}-\text{H} \\
| \\
\text{Cl}
\end{array}
\equiv
\begin{array}{c}
\text{H} \\
| \\
\text{R}--\text{C} \\
\diagup \quad \text{H} \\
\text{Cl}
\end{array}
$$

1.9 Electronegativity, Bond Polarity, and Bond Breaking

The *electronegativity* (Section 1.1) of an element is a measure of its ability to attract electrons to itself. Although bonds from carbon to other elements involve the same kind of molecular orbital picture described for C—C bonds, the electrons making up the bond are not symmetrically distributed in the space between the bonded atoms. As we proceed across the periodic chart in the order C—C, C—N, C—O, C—F, the electron density of the bond is increasingly shifted to the more electronegative atom. Table 1.3 gives the relative electronegativities of several elements commonly found in organic compounds.

TABLE 1.3	ELECTRONEGATIVITIES OF ELEMENTS			
H 2.2	C 2.5	N 3.0	O 3.5	F 4.0
	Si 1.9	P 2.2	S 2.5	Cl 3.0

A covalent bond between elements of different electronegativities is **polar.** The electrons in the bond are unevenly distributed, and the bond therefore has a negative and a positive end. This situation can be indicated by the symbol $\longmapsto$ with the arrow pointing toward the more negative end, as in C $\longmapsto$ Cl or C $\longleftarrow\!\!+$ Li. Another way of representing bond polarity is to show partial charges, $\delta+$ or $\delta-$, on the atoms. Knowing the polarity of a particular bond will help us predict how that bond will break in a chemical reaction.

$$
\begin{array}{lll}
\text{H}_3\text{C} \longmapsto \text{Cl} & \text{or} & \overset{\delta+}{\text{H}_3\text{C}}-\overset{\delta-}{\text{Cl}} \\
\text{H}_3\text{C} \longmapsto \text{OH} & \text{or} & \overset{\delta+}{\text{H}_3\text{C}}-\overset{\delta-}{\text{OH}} \\
\text{H}_3\text{C} \longleftarrow\!\!+ \text{MgCl} & \text{or} & \overset{\delta-}{\text{H}_3\text{C}}-\overset{\delta+}{\text{MgCl}}
\end{array}
$$

A convenient way to show how a bond breaks is the **curved arrow** notation. In reactions of a molecule A $\longrightarrow$ B (note that B is the more electronegative atom), the A—B bond will generally break so that atom B keeps both of the bonding electrons. The movement or redistribution of this pair of electrons to B is indicated by the direction of the curved arrow.

$$\overset{\delta^+}{A}\underset{}{-}\overset{\delta^-}{B} \longrightarrow A^+ :B^-$$

The curved arrow means specifically the *movement of an electron pair*. The atom to which the electrons move will always have one more negative charge than before the electron shift, and the atom from which the arrow comes will have one more positive charge. It is important to understand why these charges appear or disappear. We could have formed the molecule A—B by combining two neutral (uncharged) atoms A· and B·. If we now break the bond as shown above so that B gets both of the electrons, B will have one more negative charge than can be balanced by the positive charges of its nucleus. Thus B will become an anion, :B⁻, possessing a net negative charge of one. The opposite is true for atom A, which has lost an electron in the process and now is the cation, A⁺, possessing a net positive charge of one.

Curved arrows to represent **electron pair movement** will be used throughout this text. The following examples illustrate the use of this notation in several typical reactions. Note that only the unshared pairs that are involved in the reaction are actually shown.

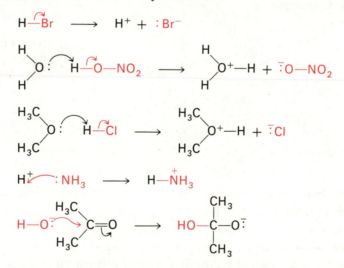

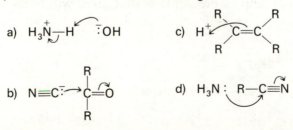

Problem 1.10   Write the products that would result from the indicated electron pair shifts. Be sure to show charges.

It is also possible to break a covalent bond so that each atom receives one of the two electrons which made up the bond. In this type of bond cleavage, which is indicated by single-barbed "fishhooks," neither fragment will be charged if the starting molecule is neutral. The resulting odd-electron species, each with an *unpaired electron,* are called **free radicals,** and are usually highly reactive species.

$$A{-}B \longrightarrow A\cdot \; + B\cdot$$

free radicals

In a compound R_3CX, the C—X bond can be broken in three ways, depending on the nature of the group X:

1) The bonding electron pair can remain on carbon to give a negatively charged carbon or **carbanion.**
2) The electron pair of the C—X bond can remain with X to leave a positively charged carbon called a **carbocation.**
3) The bond may undergo cleavage with carbon keeping *one* electron to become a **carbon free radical.**

One of these three types of bond cleavage will be involved in most of the reactions that we will see in organic chemistry, and it is important that you understand fully how the covalent bond to carbon can break.

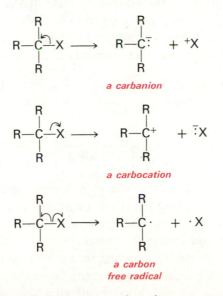

a carbanion

a carbocation

a carbon
free radical

The compound R_3CX has a neutral carbon atom, with one electron in each of the four covalent bonds plus the two $1s$ electrons of carbon balancing the nuclear charge of six protons. If X^+ is removed, with both electrons in the C—X bond remaining on carbon, one extra electron is associated with the carbon atom to form the carbanion ($R_3C\bar{:}$). If both electrons are removed with the group $X\bar{:}$, only three electrons in the outer shell plus the two $1s$ electrons are available to offset the nuclear charge, and carbon has a positive charge. In the radical there are three electrons in the three bonds plus one unpaired electron, or a total of four in the outer shell, plus the two $1s$ electrons, and carbon is neutral.

Problem 1.11 In the following equations, the formula R_3C for the product is incomplete in each case. Complete the formulas with electron dots, charges, or both.

a) $R_3\overset{\frown}{C}\cdot H \overset{\frown}{\cdot} Cl \longrightarrow R_3C + H{-}Cl$

b) $R_3\overset{\frown}{C}{-}H \overset{\frown}{\cdot} :B \longrightarrow R_3C + H{-}B$

c) $R_3C\overset{\frown}{-}Cl \longrightarrow R_3C + :Cl$

1.10 Acids and Bases

Many reactions that we will see in this book will depend upon the behavior of organic compounds as **acids** or **bases,** and we need to review and expand the meanings of these terms. The definitions that are most common in general chemistry courses are those of the Brönsted-Lowry concept, in which acids are defined as compounds such as HCl or H_2SO_4 that **donate a proton** to a base. Bases are species such as ^-OH or H_2O which can **accept a proton.**

Transfer of a proton from an acid HA to a base $B:$ leads to an equilibrium with $A:$ and HB. $A:$ is the conjugate base of the acid HA, and HB is the conjugate acid of $B:$. The relative strengths of two acids can be compared by determining the **equilibrium constant** (K_{eq}) for the reaction of each acid with a single base. The greater the strength of an acid, the larger the K_{eq}.

$$\underset{\substack{acid}}{HA} + \underset{\substack{base}}{B:} \rightleftharpoons \underset{\substack{conjugate \\ base\ of\ HA}}{A:} + \underset{\substack{conjugate \\ acid\ of\ B:}}{^+B{-}H}$$

$$K_{eq} = \frac{[A^-][^+BH]}{[HA][B:]}$$

The acidity of hydrogen attached to first-row elements increases enormously (by *ca.* 10^{40}!) on going from the center to the right in the periodic table. This is in keeping with the greater electronegativity of the atom attached to hydrogen and hence the greater ability of the conjugate base to exist as an anion. Of course, this means that the basicity of first row anions (the conjugate bases) must decrease in going from the center to the right of the periodic chart. We should also recall that the polarity of bonds to hydrogen also increases in moving to the right side of the chart.

$$H{-}CH_3 \ll H{-}NH_2 \ll H{-}OH \ll H{-}F$$

increasing acidity ⟶
increasing bond polarity

$$:CH_3 \gg :NH_2 \gg :OH \gg :F$$

⟵ increasing basicity

A more general definition of acids and bases comes from the **Lewis concept.** In this system, acids are defined as compounds which **accept an electron pair** from a base, and bases are defined as compounds which can **donate an electron pair** to an acid. We should note that all Brönsted-Lowry acids and bases are still classified as acids or bases under the Lewis system. However, compounds such as BF_3 and $AlCl_3$, which require an electron pair to complete the inert gas configuration about the central atom, will now be classified as (Lewis) acids. The reactions

$$F_3B \; + \; :NH_3 \longrightarrow F_3\bar{B}\!-\!\overset{+}{N}H_3$$

Lewis acid Lewis base

$$Cl_3Al + :Cl^- \longrightarrow Cl_3\bar{Al}\!-\!Cl \equiv {}^-AlCl_4$$

$$H^+ + H_2C\!=\!CH_2 \longrightarrow H_3C\!-\!\overset{+}{C}H_2$$

$$H^+ + :\!O\!\!<\!\!\begin{smallmatrix}CH_3\\CH_3\end{smallmatrix} \longrightarrow H\!-\!\overset{+}{O}\!\!<\!\!\begin{smallmatrix}CH_3\\CH_3\end{smallmatrix}$$

are all typical acid-base reactions by the Lewis definition.

Two other terms that are used extensively in organic chemistry are **electrophile** (electron-seeking species) and **nucleophile** (nucleus-seeking, or electron-donating). For our purposes these terms have essentially the same meaning as Lewis acid and base, respectively, but in organic reactions they are used to refer specifically to reactions occurring at carbon. The use of these terms is illustrated by the following example.

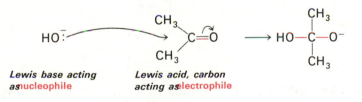

Lewis base acting as nucleophile *Lewis acid, carbon acting as electrophile*

1.11 Reactive Intermediates: Properties of Carbocations, Carbanions, and Carbon Free Radicals

Carbocations,* R_3C^+, can be formed by removal of an electron pair from a σ bond or a π bond to carbon. The electron-deficient carbon in a carbocation uses sp^2 hybrid orbitals for the three σ bonds and is therefore planar, with *ca.* 120° bond angles like a carbon in $H_2C\!=\!CH_2$. The remaining p orbital is vacant and is perpendicular to this plane. As we would expect, carbocations are extremely electrophilic and are very strong Lewis acids. For

---

*Such compounds are often referred to as carbonium ions and more recently as carbenium ions.

example, carbocations react immediately with water, a relatively weak base, to form a C—O σ bond. In this reaction, oxygen donates an electron pair and thus acquires a positive charge.

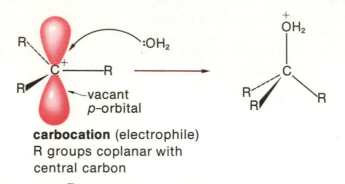

carbocation (electrophile)
R groups coplanar with
central carbon

Carbanions, $R_3C\colon^-$, are very powerful bases and nucleophiles. They are such strong bases that compounds which we would not normally think of as acids, for example NH_3, will behave as acids when in the presence of a carbanion. Carbanions of the type $R_3C\colon^-$ have tetrahedral geometry, with the unshared electron pair occupying an sp^3 hybrid orbital.

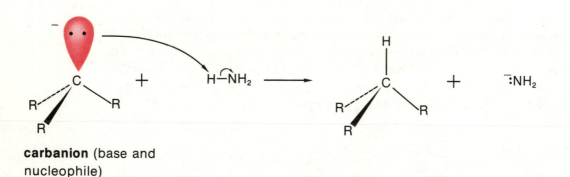

carbanion (base and
nucleophile)

Carbon free radicals, $R_3C\cdot$, formed by single-electron cleavage of a σ or π bond to carbon, are highly reactive. Since a group that donates only one electron is required in order to complete normal bonding for the carbon radical, the characteristic reactions of free radicals are *abstraction of an atom or group with a single bonding electron* from another molecule, and combination with another free radical.

$$R_3C\cdot + Cl—Cl \longrightarrow R_3C—Cl + Cl\cdot$$

*carbon
radical*

$$R_3C\cdot \quad \cdot CR_3 \longrightarrow R_3C—CR_3$$

In this brief preview we have seen how carbocations react with nucleophiles, and carbanions react with electrophiles. There are also many reactions which occur by attack of a nucleophile at a *partially electron-deficient* carbon, or by attack of an electrophile at the electron-rich end of a bond. From our understanding of bond polarities, we can often predict whether a carbon in a particular molecule will be susceptible to attack by electrophiles or nucleophiles. For example, it will not surprise us to find that the carbon atom in H_3C—Br is readily attacked by nucleophiles, since the greater electronegativity of bromine polarizes the C $\longleftrightarrow$ Br bond so as to leave carbon with a partial positive charge. Thus, we might predict that ^-OH would react with CH_3Br, and indeed it does.

$$HO\overset{..}{:} \quad H_3\overset{\delta^+}{C} \text{—} \overset{\delta^-}{Br} \quad \longrightarrow \quad HO\text{—}CH_3 + \overset{..}{:}Br$$

UPON COMPLETING CHAPTER 1

Since you are reading this section, we hope that means you have now read all of Chapter 1 and worked problems 1.1 through 1.11. You should now work all of the Additional Problems and, if you encounter trouble with any of them, go back and carefully reread the appropriate section. While all of the material in this book is considered important, a firm understanding of the material in Chapter 1 is crucial to your understanding of all of the material yet to come. There are two ways to check to see if you really understood this material: (1) by trying to work all of the problems, or (2) by waiting until the first examination. The former method will be less painful in the long run.

SUMMARY AND HIGHLIGHTS

1. **Ionic** (electrostatic) bonds as in Na^+Cl^- are formed by electron transfer between elements of greatly different electronegativity. **Covalent** bonds are formed by the *sharing* of two electrons between elements of similar electronegativities such as C and H.

2. **Bond formation** is a heat releasing (exothermic) process, and atoms have a strong tendency to combine into molecules, thus completing their inert-gas electronic configurations.

3. An **orbital** is a region in space where up to two electrons of a particular energy may be found. Overlap of atomic orbitals leads to molecular orbitals. A covalent bond has two electrons in a molecular orbital associated with both of the bonding atoms.

4. A σ **bond** is a covalent bond with cylindrical symmetry (like a hot dog).

5. A π **bond** is a covalent bond formed by overlapping parallel p orbitals on adjacent atoms. It has one lobe above and one below the plane formed by the surrounding σ bonds of the molecule.

6. The combination of one σ bond and one π bond connecting two atoms is called a **double bond.** One σ and *two* π bonds constitute a **triple bond.**

7. If carbon is bonded to four atoms (requiring that the carbon be sp^3 hybridized), it will be at the center of a **tetrahedron. Planar** molecules result

from sp^2 hybridized carbon, while sp hybridized carbon gives **linear** molecules.

8. **Structural isomers** are different compounds with the same molecular formula but with different atom-to-atom bonding sequences.

9. A covalent bond between two atoms of different electronegativities has an unequal distribution of bonding electrons and thus is **polar.**

10. Reactions occur by breaking and making bonds. A molecule A—B can break to give ions $A^+ + {:}B$, or $A{:} + B^+$, or radicals $A\cdot + B\cdot$.

11. A **curved arrow** $\frown$ is used to designate movement of a pair of electrons. A single-barbed arrow $\frown$ indicates movement of a single electron.

12. **Lewis acids** are species which *accept* an electron pair; they are **electrophiles. Lewis bases** donate an electron pair, and are often referred to as **nucleophiles** when they donate electrons to carbon in a chemical reaction.

13. Breaking of bonds (σ or π) to carbon can give **carbocations,** R_3C^+, which are electrophiles; **carbanions,** R_3C^-, which are nucleophiles; or **carbon free radicals,** $R_3C\cdot$.

14. Except for the three reactive intermediates mentioned in the preceding paragraph (plus one we shall not encounter), carbon in organic compounds will always have **four bonds.**

ADDITIONAL PROBLEMS

1.12 Showing all bonds, unshared electrons, and charges on atoms, write an electron dot structure for each of the following molecules and ions:

    a) C_2H_4 b) CH_2O c) N_2H_4

    d) CN^- e) C_2H_2 f) $\cdot CH_2Cl$

1.13 Indicate the types of hybrid orbitals used by carbon in the following compounds and ions, and state the geometry of the bonding arrangement (tetrahedral, planar, or linear) around carbon.

    a) $H_2C{=}O$ b) $H_3C{:}^-$ c) $HC{\equiv}N$

    d) H_3CCl e) CO_2

1.14 Write structures (showing all bonds, unshared valence electrons, and charges) for a carbocation, a carbanion, and a carbon free radical, each having the formula C_2H_5.

1.15 Draw wedge–dashed-line three-dimensional structures for CCl_4, H_3CCH_2Br, $H_3C{-}O{-}CH_3$, and H_3CCH_2OH. Assume that the bond angles for C—O—C and C—O—H are the same as for C—C—H.

1.16 Since the O atom in water is bonded to hydrogen and also has unshared electron pairs, H_2O may function either as an acid or as a base. Write equations for the expected reactions of the following compounds with water, showing electron pair movement with curved arrows.

    a) H—Cl   b) H—Li   c) $H_3C^-Na^+$

    d) Na_2O e) $Na^+NH_2^-$

1.17 Classify the following as likely electrophiles or nucleophiles:

    a) H^+ b) $:\ddot{Cl}^+$ c) $CH_3{-}\overset{\cdot\cdot}{N}{-}H$ d) $CH_3CH_2^+$

                                                             H

1.18   Identify the electrophilic Lewis acid and the nucleophilic base in the following reactions:

a) $H_3\overset{+}{C} + H_2C{=}CH_2 \longrightarrow H_3C{-}CH_2{-}\overset{+}{C}H_2$

b) $CH_3\ddot{O}H + R_3\overset{+}{C} \longrightarrow CH_3\overset{+}{\underset{H}{\ddot{O}}}{-}CR_3$

1.19   Suggest an explanation for the fact that a solution of HCl in benzene (C_6H_6) does not conduct an electric current, but after addition of $H_3C{-}O{-}CH_3$ the solution has significant conductivity.

1.20   Show the products that would be formed by the following reactions:

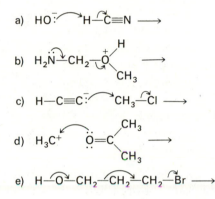

a) $HO\ddot{:}^- \quad H{-}C{\equiv}N \longrightarrow$

b) $H_2\ddot{N}{-}CH_2{-}\overset{+}{O}\overset{H}{\diagdown}_{CH_3} \longrightarrow$

c) $H{-}C{\equiv}C\ddot{:}^- \quad CH_3{-}Cl \longrightarrow$

d) $H_3C^+ \quad \ddot{O}{=}C\overset{CH_3}{\underset{CH_3}{\diagup}} \longrightarrow$

e) $H{-}O{-}CH_2{-}CH_2{-}CH_2{-}Br \longrightarrow$

1.21   On an exam the following structures appeared in student answers. Criticize them.

a) H_2CBr_2Cl

b) $CH_3{=}CBr_2$

c) $H_3C{-}CH_2 \longleftarrow + Cl$

d) $Cl_2C{=}H$

e)

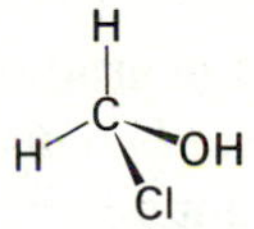

$H{-}\underset{Cl}{\overset{H}{\underset{|}{\overset{|}{C}}}}{-}OH$

f) 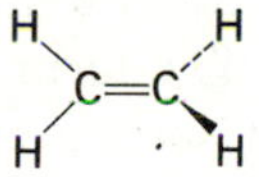 $\overset{H}{\underset{H}{\diagdown}}C{=}C\overset{H}{\underset{H}{\diagup}}$

2
SATURATED HYDROCARBONS

Hydrocarbons are compounds containing only carbon and hydrogen, and they are the starting point for organizing the vast array of more complex organic compounds. The classification scheme used for organic compounds is based upon the carbon skeleton of a parent hydrocarbon plus **functional groups.** A functional group is a portion of the molecule such as $\diagup C{=}C\diagdown$, —OH, or —NH_2, which undergoes a characteristic set of reactions.

Alkane is the general name for hydrocarbons containing only C—C and C—H single bonds. The term **saturated hydrocarbon** is often used instead of alkane; in this sense, saturated means *no double bonds* or maximum number of hydrogens. The alkanes form a **homologous series** with the general formula C_nH_{2n+2}. The term "homolog" refers to a series of compounds in which each member differs from the preceding member by one —CH_2— **(methylene)** group; alkanes are only the first example of homologous series that we shall encounter. The first member of the alkanes is methane, CH_4, and the next member of the series therefore has the formula C_2H_6 ($n{=}2$). The

TABLE 2.1 **STRUCTURAL ISOMERS OF C_nH_{2n+2}**

Formula	Possible Isomers	Formula	Possible Isomers
C_3H_8	1	C_8H_{18}	18
C_4H_{10}	2	C_9H_{20}	35
C_5H_{12}	3	$C_{10}H_{22}$	75
C_6H_{14}	5	$C_{20}H_{42}$	366,319
C_7H_{16}	9		

number of possible isomers multiplies rapidly as the number of carbons increases, as seen in Table 2.1.

2.1 Nomenclature

The naming of organic compounds is an essential part of organic chemistry. A structural formula is a clear representation of a molecule, but it cannot be pronounced or alphabetically indexed; names are needed for compounds just as for other objects. In the early days of organic chemistry, nomenclature was quite haphazard. Names for compounds might be based on their source, their taste, or their smell. As the number and diversity of compounds grew, this informal nomenclature became chaotic, and in 1892 the International Union of Pure and Applied Chemistry, or **IUPAC,** drafted a set of rules for the systematic naming of organic compounds. In this book we will emphasize these **systematic names,** but many simple compounds are still best known by older "trivial" names, and these will also be encountered. It should be emphasized that these trivial names are not "wrong"; for many compounds, two or more names are approved by the IUPAC rules.

Learning to name organic compounds is hardly an exciting task, but it will be difficult to make further progress unless a few basic rules of nomenclature are now committed to memory. The IUPAC system begins with names of the "straight-chain" alkanes listed in Table 2.2. You should memorize at least the first ten of these.

TABLE 2.2 *n*-ALKANES

Name	Formula	Boiling Point, °C	Density, g/ml	Name	Formula	Boiling Point, °C	Density, g/ml
Methane	CH_4	−162		Octane	C_8H_{18}	126	0.70
Ethane	C_2H_6	−88		Nonane	C_9H_{20}	151	0.72
Propane	C_3H_8	−42		Decane	$C_{10}H_{22}$	174	0.73
Butane	C_4H_{10}	0		Undecane	$C_{11}H_{24}$	196	0.74
Pentane	C_5H_{12}	36	0.63	Dodecane	$C_{12}H_{26}$	216	0.75
Hexane	C_6H_{14}	69	0.66	Tridecane	$C_{13}H_{28}$	234	0.76
Heptane	C_7H_{16}	98	0.68	Eicosane	$C_{20}H_{42}$		

The names in Table 2.2 are those of the *normal* alkanes (*n*-alkanes), that is, those that are "straight-chained" or "unbranched." The names of branched alkanes are formed by naming the **alkyl** substituent which has replaced a hydrogen on the parent alkane. These alkyl substituents, which correspond to an alkane minus one hydrogen, are named by converting the **-ane** ending of the corresponding alkane to **-yl.** For example, from methane, CH_4, we obtain *methyl,* —CH_3. Table 2.3 lists the most important alkyl groups which you must learn.

TABLE 2.3 IMPORTANT ALKYL SUBSTITUENTS

Alkyl Group	Structure
Methyl	CH_3-
Ethyl	CH_3CH_2-
n-Propyl (1-Propyl)	$CH_3CH_2CH_2-$
Isopropyl (2-Propyl)	$CH_3\overset{\mid}{C}HCH_3$
n-Butyl (1-Butyl)	$CH_3CH_2CH_2CH_2-$
sec-Butyl (2-Butyl)	$CH_3CH_2\overset{\mid}{C}HCH_3$
t-Butyl	$CH_3-\overset{\mid}{\underset{\underset{CH_3}{\mid}}{C}}-CH_3$

IUPAC Rules for Systematic Nomenclature. To name a compound with a branched or otherwise substituted chain, these steps are followed:

1) Identify the *longest straight chain* of carbon atoms and assign the parent name of the compound from the number of carbons (Table 2.2).

$$CH_3CHCH_2CH_2CHCH_2CH_3$$
$$\overset{\mid}{CH_3} \qquad \overset{\mid}{CH_2CH_3}$$

longest chain = 7 carbons
parent name = heptane

2) Number the parent chain beginning at the end which will result in the *lowest possible numbers* for the substituents attached to the chain.

$$\overset{1}{CH_3}\overset{2}{C}H\overset{3}{C}H_2\overset{4}{C}H_2\overset{5}{C}H\overset{6}{C}H_2\overset{7}{C}H_3$$
$$\overset{\mid}{CH_3} \qquad \overset{\mid}{CH_2CH_3}$$

alkyl groups at positions 2 and 5,
not 3 and 6

3) Name each substituent and indicate its position by the number of the carbon to which it is attached.

$$CH_3CHCH_2CH_2CHCH_2CH_3$$
$$\overset{\mid}{CH_3} \qquad \overset{\mid}{CH_2CH_3}$$

2-methyl-5-ethylheptane

4) If the parent chain is substituted by the same group more than once, the number of these identical groups is indicated by the prefixes *di-, tri-, tetra,* and so forth.

5) If there are two or more substituents, they may appear either in a) order of increasing complexity or b) alphabetic order.

$$CH_3CH_2CHCl_2 \qquad \textit{1,1-dichloropropane}$$

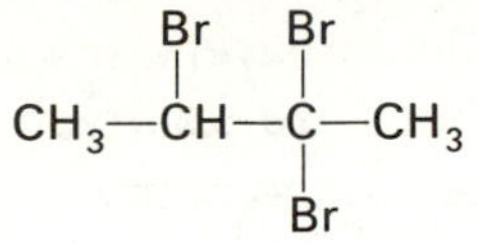

$$\begin{array}{cc} & \overset{\displaystyle Br}{|} \quad \overset{\displaystyle Br}{|} \\ CH_3—CH—C—CH_3 & \textit{2,2,3-tribromobutane} \\ & \underset{\displaystyle Br}{|} \end{array}$$

Examples.

$$\begin{array}{c} \overset{\displaystyle Cl}{|} \\ CH_3CHCHCH_3 \\ \underset{\displaystyle CH_2CH_3}{|} \end{array}$$

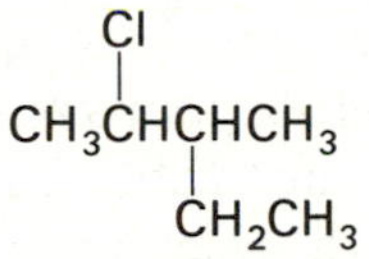

2-chloro-3-methylpentane
**(Note that the longest
chain may not be written
horizontally.)**

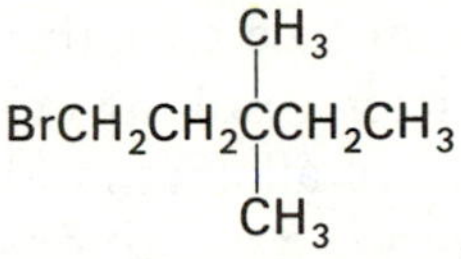

$$\begin{array}{c} \overset{\displaystyle CH_3}{|} \\ BrCH_2CH_2CCH_2CH_3 \\ \underset{\displaystyle CH_3}{|} \end{array}$$

1-bromo-3,3-dimethylpentane

$$CH_3CCl_2(CH_2)_3CH(CH_3)_2$$

2,2-dichloro-6-methylheptane

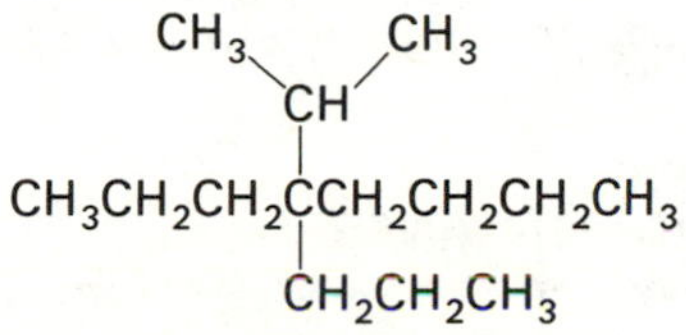

$$\begin{array}{c} CH_3 \quad\quad CH_3 \\ \diagdown \;\; CH \;\; \diagup \\ CH_3CH_2CH_2CCH_2CH_2CH_2CH_3 \\ \underset{\displaystyle CH_2CH_2CH_3}{|} \end{array}$$

4-isopropyl-4-n-propyloctane

Problem 2.1   Give names for the following compounds:

a) 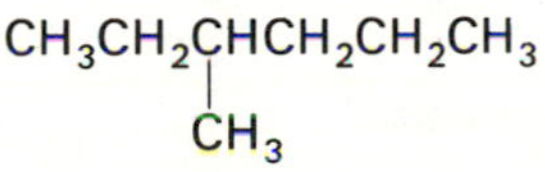 $\begin{array}{c} CH_3CH_2CHCH_2CH_2CH_3 \\ \underset{\displaystyle CH_3}{|} \end{array}$ b) $(CH_3)_2CHCH(CH_3)_2$

c) 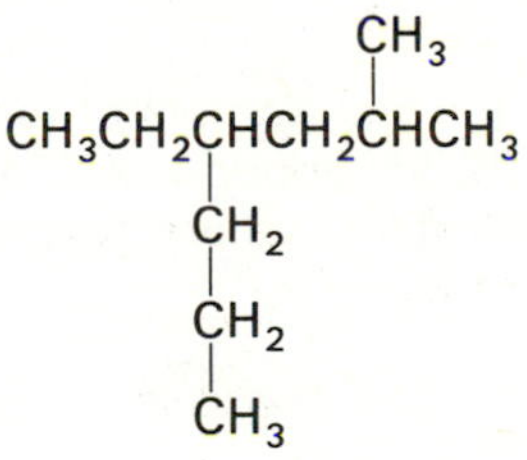 $\begin{array}{c} \overset{\displaystyle CH_3}{|} \\ CH_3CH_2CHCH_2CHCH_3 \\ \underset{\displaystyle CH_2}{|} \\ \underset{\displaystyle CH_2}{|} \\ \underset{\displaystyle CH_3}{|} \end{array}$ d) $CH_3CH_2CHClCHClCH_3$

Problem 2.2   Write structures for the following compounds:

a)   3-chloropentane   b)   2,2,4-trimethyloctane
c)   3-ethylheptane   d)   4-isopropyldecane
e)   2,2,3,3-tetramethylbutane

2.2   Properties of Alkanes

The first four members of the *n*-alkane series are gases at room temperature and atmospheric pressure. From pentane up to $C_{20}H_{42}$ they are colorless liquids with progressively increasing boiling points. Above C_{20} the homologs are solids at room temperature; these are the familiar **paraffin**

waxes. Since carbon and hydrogen have similar electronegativities, there are no polar bonds in alkanes; and in keeping with the rule of thumb "like dissolves like," these nonpolar molecules are very insoluble in the polar solvent water. Pure saturated hydrocarbons are relatively nontoxic and are reasonably pleasant-smelling. A mixture of long-chain *n*-alkanes called Nujol is a traditional intestinal lubricant.

Physical properties, such as boiling point, melting point, and density, reflect the forces between molecules in the liquid or solid state. Owing to the greater molecular surface area, the attractive forces in a liquid increase as the chain becomes longer, resulting in higher boiling points and greater densities. Since the densities of alkanes are less than one (Table 2.2), they all float on water—something we have all observed with gasoline or oil.

Within a group of compounds of the same chemical formula (*e.g.*, the isomeric pentanes), physical properties vary with extent of branching, as seen in Table 2.4. 2,2-Dimethylpropane is a compact, nearly spherical molecule, and intermolecular attractive forces are therefore considerably reduced compared to *n*-pentane, in which a much larger molecular surface is exposed.

TABLE 2.4   ISOMERIC PENTANES

Name	Structure	B.P., °C
n-Pentane	$CH_3CH_2CH_2CH_2CH_3$	36
2-Methylbutane (isopentane, *trivial*)	$(CH_3)_2CHCH_2CH_3$	28
2,2-Dimethylpropane (neopentane, *trivial*)	$CH_3C(CH_3)_2CH_3$	9

Problem 2.3   The boiling points of the five isomers of formula C_6H_{14} are 50°, 58°, 60°, 63°, and 69°. The compound of b.p. 60° is 2-methylpentane. Write the structures of the five isomers, give their names, and match the boiling points to the structures.

2.3   Conformation

It was emphasized in Chapter 1 that ethane is a single compound, with no isomers, because two sp^3 hybridized atoms can be bonded in only one way to form C_2H_6. However, we also know that a σ bond has cylindrical symmetry, and thus rotation of one end of the C—C bond will result in no decrease in bonding (*i.e.*, orbital overlap). Now if we look at a three-dimensional model of ethane, it is apparent that many structures can be obtained by rotation about the C—C bond owing to the differences in the relative positions of hydrogen atoms on the two carbons. Two of these structures, called **conformations,** are shown in Figure 2.1.

Conformations of a molecule are isomers that can be interconverted through rotation about a single (σ) bond. The two conformations of ethane

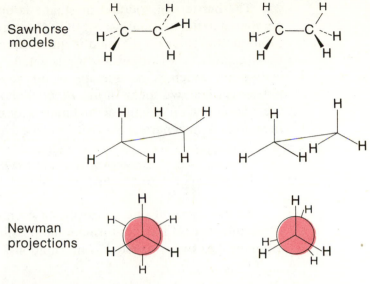

Sawhorse models

Newman projections

FIGURE 2.1

Sawhorse and Newman projections of the staggered and eclipsed conformations of ethane.

shown in Figure 2.1 are called **staggered** and **eclipsed.** The meanings of these terms can be seen in the Newman projection of ethane in which the molecule is viewed end-on, looking directly along the axis of the C—C bond. The carbon atoms are indicated by a circle; the C—H bonds of the front carbon are drawn to the center, and the C—H bonds of the rear carbon are drawn only to the edge of the circle.

The staggered conformation is slightly lower in energy and therefore more stable than the eclipsed conformation. This is largely due to electronic repulsion between the C—H bonds on one carbon and those on the other carbon, which is maximized in the eclipsed conformation and minimized in the staggered conformation. If we hold one CH_3 group fixed and rotate the other methyl group about the σ bond, the energy of the molecule rises and falls as it passes through alternating eclipsed and staggered conformations (Figure 2.2).

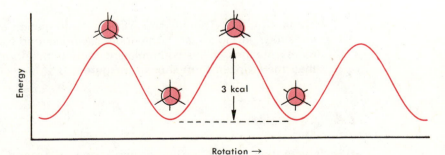

Energy

3 kcal

Rotation →

FIGURE 2.2

Rotational barrier in ethane.

The barrier for rotation in ethane is only 3 kcal/mole, which is small compared to the thermal energy of the molecule. Thus, at room temperature, rotation about the C—C bond is quite rapid and the individual conformations cannot be distinguished or isolated, so ethane behaves as a single compound which is an average of all possible conformations. Since the barrier is slight, we speak of *free rotation* about a single bond. As we replace the hydrogens on ethane with larger groups, the barrier to rotation will increase, but in general, rotation about a carbon-carbon σ bond occurs freely.

Problem 2.4   Draw sawhorse and Newman projections for both the highest and lowest energy conformations of CH_3CH_2Cl.

Replacement of one hydrogen on each carbon of ethane with another group introduces more conformational possibilities. In Figure 2.3 we see that *n*-butane has two staggered conformations. The staggered-*anti* form is of

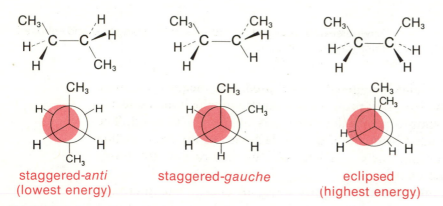

staggered-*anti*   staggered-*gauche*   eclipsed
(lowest energy)   (highest energy)

FIGURE 2.3

Some conformations of *n*-butane.

lower energy than the staggered-*gauche* form, since repulsive forces between the two methyl groups are minimized in the *anti* form. This is our first example of **steric hindrance**—interference caused by the spatial requirements of groups, a concept we will use many times.

Problem 2.5   There is one eclipsed conformation of *n*-butane that is not shown in Figure 2.3. Draw both sawhorse and Newman projection representations of this conformation. Why is this form of lower energy than the eclipsed form shown in Figure 2.3?

2.4   Cycloalkanes

Cycloalkanes are *n*-alkanes with one hydrogen removed from each terminal carbon and a ring formed with a new C—C σ bond. Thus, these alkanes possess the general formula C_nH_{2n}. Cycloalkanes are named simply by adding the prefix *cyclo-* to the corresponding alkane name. Since all positions in the ring are equivalent, a monosubstituted cycloalkane requires no numbering. With two or more substituents, the carbon bearing one group is given the number 1, and the other carbon atoms are numbered consecutively around the ring so as to provide the lowest numbers in the name.

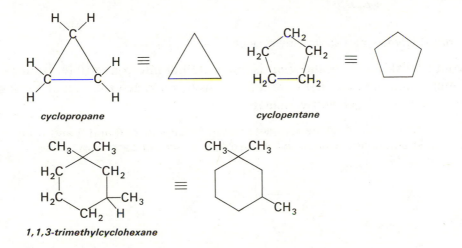

cyclopropane   cyclopentane

1,1,3-trimethylcyclohexane

Only the three-membered ring, cyclopropane, has all carbons in the same plane—it has no choice. Since this planar arrangement requires C—C—C bond angles of 60°, far from the desired tetrahedral bond angles of 109.5°, it is not surprising that this **strained ring** readily undergoes reactions in which the ring is opened. Larger rings **pucker** to try to achieve the normal bond angle, but the major reason for ring puckering is to avoid the eclipsed interactions of hydrogens or other substituents on adjacent carbons. This is illustrated for cyclobutane in Figure 2.4. However, it is often convenient to draw ring structures as if they were planar.

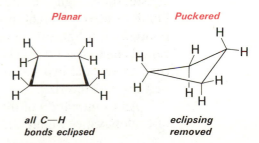

Planar   *Puckered*

all C—H   *eclipsing*
bonds eclipsed   *removed*

FIGURE 2.4

Removal of eclipsed H—H interactions by puckering the cyclobutane ring.

2.5   Geometric Isomerism

An important type of isomerism is encountered in substituted cycloalkanes. As we can see in Figure 2.5, the methyl groups of 1,2-dimethylcyclopentane can be located on the same (*cis*) or opposite (*trans*) sides of the ring. Since there cannot be free rotation about the ring C—C bonds (the ring

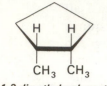
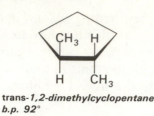

cis-1,2-dimethylcyclopentane
b.p. 99°

trans-1,2-dimethylcyclopentane
b.p. 92°

FIGURE 2.5

Geometric (*cis, trans*) isomers of 1,2-dimethylcyclopentane.

would break!), the *cis* and *trans* isomers are not interconvertible. Isomers in which atoms are held in different positions by restricted rotation about some bond are called **geometric isomers.**

Problem 2.6   Write complete names for structures **A** and **B** and write a structure and a name for a geometric isomer of each:

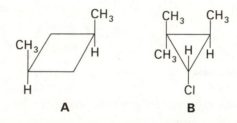

By far the most important cycloalkane ring system is the six-membered ring, found in a great many of the hydrocarbons made by plants and animals. Cyclohexane is more stable than the smaller cycloalkanes because it can exist with no angle strain and no eclipsed interactions. By twisting about the C—C ring bonds, cyclohexane can adopt several different conformations with normal tetrahedral bond angles. The lowest energy conformation is called the **chair** form. The conformation referred to as the **boat** form (Figure 2.6) is of considerably higher energy. In the chair form, each carbon has one C—H bond directed straight up or straight down, with the other C—H bond pointed off to the side. The former are called **axial** bonds and the latter are called **equatorial** bonds. Therefore, we have three axial hydrogens above the ring, three axial hydrogens below the ring, and six equatorial hydrogens in a belt around the ring.

An examination of the Newman projections of the chair and boat forms of cyclohexane (Figure 2.7) clearly shows that the chair form eliminates all

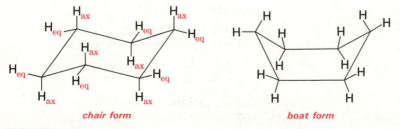

chair form   *boat form*

FIGURE 2.6

Chair and boat conformations of cyclohexane.

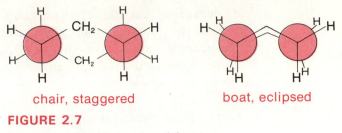

chair, staggered   boat, eclipsed

FIGURE 2.7

Newman projections of cyclohexane.

C—H bond eclipsing (refer to Figure 2.1) to present a perfectly staggered conformation.

By twisting the chair form into a boat form and then back again, we can interconvert one chair form to another. This can be visualized as simply pulling one end down while pulling the other end up. The consequence of this action is to convert all equatorial bonds into axial bonds and vice versa (Figure 2.8). Thus, "flipping" the ring of 1-methylcyclohexane converts the

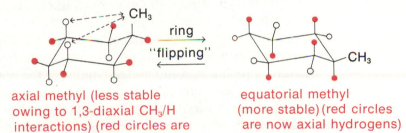

ring
"flipping"

axial methyl (less stable owing to 1,3-diaxial CH_3/H interactions) (red circles are equatorial hydrogens)

equatorial methyl (more stable) (red circles are now axial hydrogens)

FIGURE 2.8

Interconversion of the two chair conformations of 1-methylcyclohexane.

methyl group from equatorial to axial. The equatorial position is less crowded because axial—H/axial—CH_3 interactions are avoided, and the conformation with the large group in the equatorial position is of lower energy. At room temperature, 95 per cent of the 1-methylcyclohexane molecules will be in the equatorial methyl form, while only about 5 per cent will be in the axial methyl form at any instant.

Problem 2.7   Draw the two possible chair conformations of *trans*-1,4-dichlorocyclohexane. Which one would be more stable?

2.6 The Nature of Reactions

The major business of chemistry is the study of reactions. The drugs, fibers, and resins provided by the chemical industry are products of chemical reactions in which compounds are broken down, built up, or tailored for specific needs. The biochemistry of living organisms is an enormously complex network of reactions of organic compounds. It is necessary to understand the basic nature of reactions, with relatively simple models, before we can hope to understand the chemistry of actual compounds.

To understand a reaction and use it most efficiently, answers are needed to three basic questions:

1) *What* chemical changes occur in this reaction? What bonds are broken or formed?
2) *Why* does the reaction occur? What is the driving force?
3) *How* does the reaction take place? What is the mechanism by which the reaction proceeds?

We will not try to examine every reaction for answers to all these questions. Depending on the reaction and the context in which we are viewing it, one question may be of more consequence than the others, or it may be unnecessary to deal with any of them explicitly. However, it is useful at the outset to consider each of these points in a general way.

The first question is the most obvious one. It usually means determining what products are formed in the reaction—what bonds are broken in the starting compound and what new bonds are formed to give the product. The "reaction" may turn out to be several competing reactions that lead to a mixture of products. A full understanding of the process may permit selection of reaction conditions that give the desired product in highest yield.

The second question has to do with the energy changes that occur in the reaction. Any process, whether it is water flowing downhill or a chemical reaction, will take place if the change is in the direction of lower energy (*i.e.*, greater stability). An **exothermic** reaction, in which heat is given off, may occur spontaneously. It is also possible to carry out an **endothermic** reaction, in which heat must be supplied, in the same way that water can be pumped to a higher level by the expenditure of energy.

The overall change in energy that accompanies a reaction can be expressed in terms of the overall bond energies of reactants and products. In the reaction $AB + C \longrightarrow A + BC$, the energy required to break the A—B bond must be supplied, and the bond energy of the B—C bond is liberated (Figure 2.9). If the latter value is larger, the reaction is **exothermic** (the change in heat energy, ΔH, is negative). The formation of the stronger B—C bond provides a **driving force** for the reaction.

The final question—*how* the reaction takes place—involves the actual bond changes and the *rate* at which they occur, *i.e.*, the **mechanism.** A reaction mechanism is a detailed, step-by-step picture of exactly how a reaction proceeds. In one type of reaction, we start with a molecule of AB and one of C and obtain as products A + BC. For this to happen, AB and C must collide with sufficient energy to break the A—B bond. The highest energy which must be attained by the molecules during this collision is represented by the **transition state,** where the A—B bond is partially broken and the new B—C bond is partially formed. As the temperature increases, more collisions will be of sufficient energy to form the transition state complex, A---B---C, and "cross over the hill" to products.

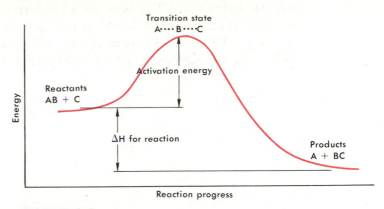

FIGURE 2.9

Reaction diagram for $AB + C \rightarrow A + BC$, an exothermic reaction.

This situation can be visualized by a schematic diagram as in Figure 2.9. The horizontal axis represents the "progress" of the reaction on the way from starting reactants to products. The vertical axis represents the conversion of kinetic energy to potential energy as the reaction occurs. The difference in potential energy between the starting reactants and the transition state, the height of the hill, affects the **rate** of the reaction and is called the **activation energy.** If the activation energy is very high, the reaction will be slow even though it may be highly exothermic.

To overcome the problem of a high activation barrier, a **catalyst** can sometimes be used. A catalyst is a substance which lowers the activation barrier of a reaction by becoming involved in the reaction, but is itself unchanged at the end of the reaction. Since one catalyst molecule can catalyze a large number of reactions, only a small amount of catalyst is required. An idealized scheme for our reaction of AB and C both with and without catalyst is shown in Figure 2.10.

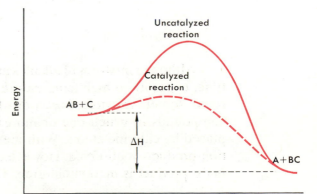

FIGURE 2.10

Catalyzed and uncatalyzed reactions.

Usually, a catalyst lowers the energy barrier of a reaction by first combining with one reactant to form an **intermediate** which is more reactive than the original reactant. For example, in Figure 2.11, AB reacts with the catalyst to form an intermediate A—B—cat which reacts with C much more easily than does AB. Although there are two hills to climb, neither is as high as the hill in the uncatalyzed reaction in Figure 2.10.

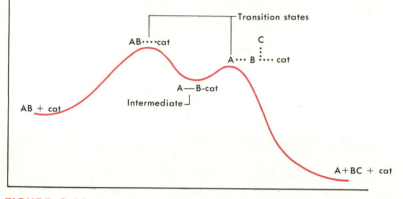

FIGURE 2.11

Reaction proceeding through an intermediate containing a catalyst.

REACTIONS OF ALKANES

An alkane molecule consists of saturated (sp^3) carbon atoms encased in a sheath of C—H bonds. The possibilities for reactions are quite limited, and alkanes are often used as solvents for the reactions of more reactive organic molecules which contain π bonds or other functional groups. The only reactions available to alkanes involve breaking a C—H or C—C σ bond. Since the bonded atoms are of similar electronegativity, such bond breaking is normally homolytic (forming free radicals).

2.7   Chlorination

Although mixtures of alkanes and chlorine can be kept together for some time, exposure to high temperature or ultraviolet (UV) light brings about a rapid, sometimes violent reaction. The products are hydrogen chloride and compounds in which one or more of the alkane hydrogens have been replaced by chlorine atoms. With methane, a mixture of all possible chlorination products is observed. However, if a large excess of methane is used, the main product is methyl chloride, CH_3Cl.

$$CH_4 + Cl_2 \xrightarrow[\text{UV light}]{\text{heat or}} CH_3Cl + HCl$$

methyl
chloride

We know that this reaction does not involve a direct interaction of CH_4 and Cl_2 molecules, since a mixture of these compounds in the dark remains unchanged. In the presence of UV light, however, chlorine molecules dissociate into chlorine atoms by homolytic cleavage of the Cl—Cl bond (step 1). It is these atoms, the Cl· radicals, that attack the methane molecule to remove a hydrogen atom and form the methyl radical, ·CH_3 (step 2). The methyl radical can now react with a Cl_2 molecule to abstract a chlorine and form another chlorine radical (step 3).

(1) Cl_2 $\xrightarrow[\text{UV light}]{\text{heat or}}$ 2 Cl·   *initiation*

(2)   Cl· + CH_4 $\longrightarrow$ HCl + ·CH_3 ⎫ *propagation of*
(3)   ·CH_3 + Cl_2 $\longrightarrow$ CH_3Cl + Cl· ⎭ *chain reaction*

Thus, every time a molecule of methyl chloride is formed, a chlorine atom is also generated and the sequence (steps 2 and 3) can continue until one of the starting reagents is completely consumed. This type of process, which is called a **chain reaction,** produces many molecules from a single initiation and is typical of free radical reactions. Eventually the chain is terminated when any two free radicals combine (step 4).

(4)   ·CH_3 + Cl· $\longrightarrow$ CH_3Cl ⎫ *possible chain*
         ·CH_3 + ·CH_3 $\longrightarrow$ CH_3CH_3 ⎬ *termination*
         Cl· + Cl· $\longrightarrow$ Cl_2 ⎭ *steps*

A reaction diagram for chlorination of methane is shown in Figure 2.12. There are three distinct steps in the overall process, and each has an activation barrier. The free radicals Cl· and ·CH_3, which are intermediates in the reaction, appear in the diagram as valleys. Although they are highly reactive species, each has a finite lifetime and can be detected by suitable experiments.

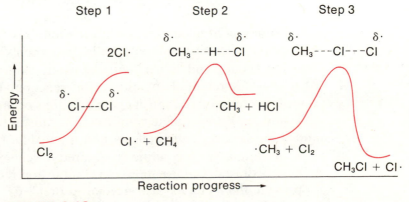

FIGURE 2.12

Reaction diagram for chlorination of methane.

Chlorine atoms can also abstract a hydrogen from methyl chloride to give dichloromethane, and the reaction can then proceed to the tri- and tetrachloro compounds.

$$CH_3Cl \xrightarrow{Cl_2} \underset{\textit{dichloromethane}}{CH_2Cl_2} \xrightarrow{Cl_2} \underset{\textit{chloroform}}{CHCl_3} \xrightarrow{Cl_2} \underset{\substack{\textit{carbon} \\ \textit{tetrachloride}}}{CCl_4}$$

CCl_4 is an end product in any process involving organic compounds and chlorine, including municipal water treatment. For this reason, carbon tetrachloride can be detected in drinking water if the source is a river that has been polluted with organic material. This is a matter of concern, since a number of compounds with a high chlorine content, including $CHCl_3$ and CCl_4, have toxic effects and can lead to kidney and liver damage.

Chlorination of larger alkanes can be controlled to give monochloro derivatives, but all possible isomers are obtained, and the reaction is not a useful method for preparing individual alkyl chlorides.

Problem 2.8   In methane and ethane, all hydrogens are equivalent, and there is only one monochloro alkane. Write structures for all the monochloro isomers that could be formed in the chlorination of (a) 2-methylbutane and (b) methylcyclohexane (including geometric isomers).

2.8   Oxidation

The reaction of alkanes with oxygen is without question their most important reaction. When ignited in the presence of an excess of oxygen, alkanes are oxidized to carbon dioxide and water. The important product of this reaction is the great quantity of heat that is liberated. The combustion of alkanes can be summarized by the equation

$$-(CH_2)- + \tfrac{3}{2}O_2 \longrightarrow CO_2 + H_2O + 156 \text{ kcal}$$

Oxidation of alkanes, like chlorination, is a free radical process with a high activation energy. Mixtures of hydrocarbon vapors and air, between the limits of about 5 and 80 per cent alkane, are explosive; but heat in the form of a spark is required to initiate the reaction. In the resulting flame, a complex series of free radical chain reactions occurs, leading to the products CO_2 and H_2O. Incomplete combustion results in formation of carbon (soot) and the extremely dangerous carbon monoxide, CO.

In summary, the conditions required for chemical attack on alkanes are relatively drastic. Chlorination and combustion require a large energy input to provide the 80 to 100 kcal needed to break C—H and C—C bonds. It is significant, therefore, that alkanes (and more reactive compounds as well) are readily oxidized under mild conditions by bacteria. This is one of many reactions we will encounter that are brought about in living organisms by enzymes, which are efficient catalysts for chemical reactions.

Petroleum hydrocarbons (Section 2.9) are decomposed only slowly in sewage treatment plants because of the presence of soluble, more readily oxidized organic matter. Under suitable conditions, however, microbiological oxidation of alkanes is efficient and is of considerable importance in petroleum waste disposal, and potentially, also as a source of useful products. Strains of soil bacteria multiply rapidly with n-alkanes as their only source of carbon. The reaction can be carried out to give oxygenated compounds with the same number of carbons as the reactants, or it can proceed completely to CO_2. It is even possible to produce nutritionally useful protein from hydrocarbon fermentation.

2.9 Petroleum

The ultimate source of all organic carbon is **photosynthesis** from CO_2 by green algae and higher plants. Each year about 50 billion tons of carbon pass through the cycle of photosynthesis and subsequent decay and combustion back to CO_2. Photosynthesis is catalyzed by the green pigment **chlorophyll** and various enzymes, with the energy supplied by the sun.

$$n\ CO_2 + n\ H_2O \xrightarrow[\text{chlorophyll}]{\text{sunlight}} (CH_2O)_n + n\ O_2$$

During ancient geological ages, an extremely small fraction of this organic matter accumulated in the sediments of oceans and inland seas. The combined action of anaerobic bacteria and increasing compaction led to the gradual removal of oxygen and the formation of hydrocarbons. These compounds accumulated in porous rocks and finally, when geological formations permitted, in pools of petroleum, which is a complex mixture of alkanes and cycloalkanes. Accompanying reservoirs of petroleum, and also occurring separately, is natural gas, consisting largely of methane with lesser amounts of ethane and propane.

Petroleum is a very convenient fuel, and its exploitation for this use has increased at an extremely rapid rate. The contribution of petroleum and natural gas to the total energy use in the U.S. rose from about 10% to 70% from 1920 to 1970, and the total energy produced per year increased about eight-fold during this period.

Today, everyone is aware that the supply of petroleum is limited and that the reserves of this unique resource are dwindling. Petroleum is not only a vital fuel; **petrochemicals** (compounds derived from petroleum) supply over 90 per cent of the starting materials that are used to produce synthetic fibers, synthetic rubber, detergents, plastics, drugs, and other products of the chemical industry. Burning of petroleum for the generation of electricity represents a deplorable waste of this valuable material, and we must not wait until petroleum is gone before developing other practical sources of power.

2.10 Gasoline and Petroleum Refining

Crude petroleum contains a very wide range of compounds, and must be separated into fractions according to boiling points before use. The fractions obtained by distillation of a typical crude oil are listed in Table 2.5.

TABLE 2.5   DISTILLATION FRACTIONS OF A TYPICAL CRUDE OIL

Distillation Fraction	Boiling Range, °C
Natural gas (C_1 to C_4)	less than 20
Petroleum ether (C_5 and C_6)	30–60
Ligroin (C_7)	60–90
Gasoline (C_6 to C_{12})	85–200
Kerosene (C_{12} to C_{15})	200–300
Heating fuel oils (C_{15} to C_{18})	300–400
Lubricating oil, greases, paraffin wax, asphalt (C_{16} to C_{40})	above 400

One of the most important components in crude oil is gasoline. The history of gasoline production for use in the internal combustion engine is an interesting illustration of the interplay of technology, organic research, and environmental problems.

Octane rating is a term widely used in connection with gasoline, but the origin and precise meaning of octane rating are not as well known. Gasoline fractions from various petroleum sources vary widely in engine performance, and the difference lies in the structure of the hydrocarbons. Straight-chain alkanes such as *n*-heptane are poor fuels, and cause engine "knock" or "ping." On the other hand, the branched alkane 2,2,4-trimethylpentane, known incorrectly as **"isooctane,"** is one of the best fuels for internal combustion engines. Isooctane is given a rating of 100 on an arbitrary performance scale, and *n*-heptane is rated at zero. Gasoline mixtures are compared to mixtures of isooctane and heptane to obtain their octane rating.

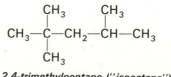

2,2,4-trimethylpentane ("isooctane")

In order to keep pace with the production of higher and higher compression engines, the antiknock properties of gasolines have since the 1930's been improved through the addition of *tetraethyllead,* $(CH_3CH_2)_4Pb$. To prevent the accumulation of lead deposits in the engine, fuel producers add 1,2-dibromoethane to convert the lead into volatile lead bromide.

$$(CH_3CH_2)_4Pb + BrCH_2CH_2Br + 16\ O_2 \xrightarrow{heat} PbBr_2 + 10\ CO_2 + 12\ H_2O$$

Unfortunately, this process both cleans up engines and floods the environment with toxic lead. This is one reason for the interest in "low-lead" or "no-lead" gasolines which we now see in the United States. (Another reason is that lead compounds "poison" the catalyst used in the emission control devices called catalytic converters.) However, the use of unleaded gasolines requires additional refining of petroleum to obtain the higher octane hydrocarbons. The presence of branched-chain alkanes and alkenes (compounds with carbon–carbon double bonds, see Chapter 3) in gasoline dramatically improves the octane number. To meet the demand for high-octane gasoline without lead additives, the "natural gasoline" is augmented by **cracking** and **isomerization** of the higher alkanes.

On heating with a metal catalyst at about 800°, alkanes undergo degradation to smaller molecules, most of which are (π bonded) alkenes. This process is called "cracking" of ethane to ethylene (an alkene) and hydrogen; it is the key step in the preparation of a number of major chemical products.

$$CH_3-CH_3 \xrightarrow[\text{heat}]{\text{catalyst}} CH_2{=}CH_2 + H_2$$

ethane   *ethylene*

With higher alkanes, breaking of the carbon chain becomes important. For example, catalytic pyrolysis* of *n*-hexane gives a mixture of hydrogen, methane, and C_2, C_3, and C_4 compounds containing double bonds. Pyrolysis conditions can be adjusted so that ethylene and propylene are the major products. These reactions are very complex, stepwise processes and lead to rearrangement of the carbon chain as well as chain-breaking.

$$CH_3CH_2CH_2CH_2CH_2CH_3 \xrightarrow[\text{heat}]{\text{cat.}} H_2,\ CH_4,\ CH_2{=}CH_2,\ CH_3CH{=}CH_2,$$

ethylene   *propylene*

$$CH_3CH{=}CHCH_3 + CH_3\underset{\underset{\displaystyle CH_3}{|}}{C}{=}CH_2 + CH_2{=}CHCH_2CH_3$$

butenes

Isomerization processes, known as **reforming** in the petroleum industry, have been developed to convert straight-chain alkanes to their cleaner burning branched-chain isomers. The catalysts used are strong Lewis acids such as aluminum chloride. Unlike most other reactions of alkanes, isomerization proceeds by carbocation intermediates rather than radicals.

---

*The ending *-lysis* means breaking apart, and will be encountered in several words—pyrolysis, breaking by heat; hydrolysis, breaking by water; photolysis, breaking by light.

FIGURE 2.13

Petroleum refining. An important part of a modern refinery is the extraction unit, in which high octane gasoline, produced by catalytic reforming, is separated from other components. (Courtesy Getty Refining and Marketing Company.)

While the detailed pathways are complex, the driving force for these carbon skeleton rearrangements is the greater stability of the branched-chain carbocations, as discussed in Section 3.5.

$$CH_3CH_2CH_2CH_3 + AlCl_3 \xrightarrow{\text{heat}} CH_3-\underset{\displaystyle \overset{\displaystyle CH_3}{|}}{CH}-CH_3$$

n-butane   *2-methylpropane*
                                       (isobutane)

Alkenes from petroleum cracking and branched-chain alkanes from reforming can be added to gasoline; but, of course, the cost goes up along with the octane rating.

SPECIAL TOPIC: Nonpetroleum Carbon Sources of Energy

Coal

Under somewhat different geological conditions than those which produced petroleum, masses of plant material were buried and compressed to form coal. Seams of coal are much more widely distributed than are oil deposits, and the amount of carbon in coal over the earth's surface is probably at least 25 times that in gas and oil.

Coal varies much more widely in composition than does petroleum. Low grade peat has a $C:H:O$ atomic ratio of about $1:2:0.8$, whereas the $C:H$ ratio in anthracite coal is about $2.5:1$ with only a trace of oxygen. Very little is known about the detailed chemical structure of coal, as it is extremely complex and varies greatly with the type of coal. Even so, organic raw materials have long been obtained by heating coal in the absence of oxygen to yield the volatile *coal tar* and *coke*. Coke is almost pure carbon and is used for the reduction of iron oxide in steel manufacture; coal tar is distilled to produce a wide variety of organic compounds, mostly aromatic hydrocarbons (Chapter 4).

For the past 20 years the energy sources in the U.S. have been largely oil (44%) and gas (31%), with coal accounting for 21% and all other sources adding up to only 4%. In contrast, coal accounts for 75% of our nation's fossil-fuel resources (50% of the world's coal reserves are located in North America!). Unfortunately, coal is not an ideal fuel because it is not liquid and it burns less cleanly than oil or gas. A particularly serious problem in burning coal for energy is the fact that some coals contain rather large amounts of sulfur, which is discharged as SO_2 upon burning. Thus, the coal must be ''cleaned'' in some way to avoid liberating large amounts of harmful SO_2 into the atmosphere. Overall, the amount of SO_2 produced by fuel combustion is estimated to be only $\frac{1}{20}$ of the total amount produced by organic decay. However, local levels of SO_2 above the tolerable limit of 0.03 ppm (parts per million) frequently result from fuel combustion because the SO_2 concentration in stack gases is many thousands of times higher than this value.

To reach the U.S. markets efficiently, coal must be converted into oil and gas. The conversion of coal to oil was accomplished during World War II in Germany, and today in South Africa this is being done on a moderate scale. Unfortunately, neither of these processes can be scaled up to meet U.S. needs. At peak production the Germans processed only about 600 tons of coal per day, while plants envisioned for current U.S. requirements would process at least 25,000 tons per day. The conversion of coal to oil or gas requires adding hydrogen to the coal. The ratio of hydrogen atoms to carbon atoms in average coal is 0.8 to 1, while in oil it is 1.75 to 1. The hydrogen can be produced by the ''water gas'' reaction in which water is reduced by carbon at high temperatures. For example, carbon (from coal) can be converted to methane by the following sequence of reactions:

$$C + H_2O \rightleftharpoons CO + H_2 \text{ (''water gas'' reaction)}$$
$$CO + H_2O \rightleftharpoons CO_2 + H_2$$
$$CO + 3 H_2 \longrightarrow CH_4 + H_2O$$
$$C + 2 H_2 \longrightarrow CH_4$$

The chemistry of this process has been developed, but major engineering and economic factors remain to be solved.

Biomass

Additional hydrocarbon fuels and raw materials for organic synthesis can be obtained directly from vegetable matter without waiting eons for their conversion into petroleum or coal. Green plants are the most efficient converters of solar to chemical energy, and many methods of extracting synthetic fuels from **biomass** (plants, animal wastes, garbage, etc.) are now being investigated. Methane is produced by anaerobic bacterial decomposition of vegetable matter. Water-based plants offer extremely high growth rates. For example, water hyacinths (Color Plate 1) produce 60 dry tons per acre per year and are easily harvested. An industrial system has been designed to produce 200 million cubic feet of methane per day. In this design, the harvested hyacinths would be converted by anaerobic digestion to methane, which has a heat of combustion of about 600 Btu per cubic foot of gas. It has been estimated that, with intensive cultivation of about 6 per cent of the continental U.S., green plants could provide the equivalent of the oil and gas requirements for our country with existing technology.

Animal wastes are also being investigated as petroleum sources. Heating manure with carbon monoxide at high pressure yields a crude oil from which hydrocarbons for organic synthesis can be obtained. One ton of manure produces three barrels of this crude oil; in 1971 alone, over two billion tons of chicken, pig, and cattle manure were produced in the U.S.

More than 50 biomass conversion processes are now in pilot plant, demonstration, and commercial stages. Most of them are related to waste disposal and are operated as anti-pollution measures. Hopefully, biomass processes will reduce our dependence on fossil fuels for organic compounds.

SUMMARY AND HIGHLIGHTS

1. **Alkanes** are a homologous series of hydrocarbons containing only C—C and C—H bonds, with the general formula C_nH_{2n+2}. The chains are straight (*n*-alkanes) or branched; names are based on the longest straight chain.

2. Physical properties depend on chain length and degree of branching.

3. **Conformations** are structures that differ by rotation around a single bond; individual conformational isomers cannot be distinguished or isolated.

4. **Cycloalkanes** contain a *ring* or closed chain; **geometrical isomers** can be formed by two substituents on the same side (*cis*) or opposite sides (*trans*) of the ring.

5. Chemical reactions in which the products are more stable (larger total bond energy) than reactants are **exothermic;** if products are less stable, energy must be supplied and reaction is **endothermic.** In most reactions, an **activation energy** must be supplied; the lower the activation energy, the *faster* the reaction.

6. Alkanes are very stable, unreactive compounds. Among the few reactions they undergo are: **cracking** to smaller fragments at high temperatures, **oxidation,** and **chlorination.**

7. Chlorination of alkanes occurs by a **free-radical chain process:**

$$RH + Cl\cdot \longrightarrow R\cdot + HCl; \ R\cdot + Cl_2 \longrightarrow RCl + Cl\cdot$$

8. Coal and petroleum are major sources of carbon for chemical manufacturing and energy.

ADDITIONAL PROBLEMS

2.9 Write structural formulas for the following:

a) 2,7-dimethyloctane

b) ethyl radical

c) *trans*-1-chloro-4-methylcyclohexane

d) 2,3,4-trimethylpentane

e) *cis*-1-chloro-2-ethylcyclopropane

f) 3,3-dimethyl-4-ethylnonane

g) 2-butyl carbocation

h) isopropylcyclohexane

2.10 In each of the following cases the names are incorrect. Draw the indicated structures and rename the compounds correctly.

a) 2-ethyloctane

b) 3,3-dimethylbutane

c) isopropane

d) *cis*-1,5-dimethylcyclohexane

e) tetraethylmethane

f) 3,4-dibromopentane

2.11 Give complete names for the following compounds:

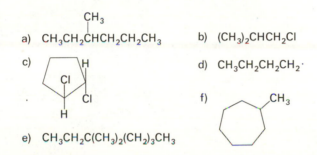

a) $CH_3CH_2\overset{\overset{\displaystyle CH_3}{|}}{C}HCH_2CH_2CH_3$

b) $(CH_3)_2CHCH_2Cl$

c)

d) $CH_3CH_2CH_2CH_2\cdot$

f)

e) $CH_3CH_2C(CH_3)_2(CH_2)_3CH_3$

2.12 Using Newman projections, illustrate both the most and the least stable conformations of 1,4-dibromobutane.

2.13 *Trans*-1,2-dimethylcyclohexane is shown below in a planar structure and in the chair conformation, with both CH_3 groups equatorial. Draw conformational structures for *cis*-1,2-dimethylcyclohexane and for the *cis* and *trans* isomers of 1,3- and 1,4-dimethylcyclohexane, with the maximum number of equatorial CH_3 groups in each case. The *trans* isomer is the more stable one for 1,2- and 1,4-dimethylcyclohexane, but for 1,3-dimethylcyclohexane the *cis* isomer is the more stable. Suggest an explanation.

2.14 Combustion of cyclopropane liberates 500 kcal per mole. Compare this value with that calculated for three —CH_2— groups of a straight-chain alkane, which liberates 156 kcal per —CH_2— group (Section 2.8), and account for the difference.

2.15 As discussed in the Appendix, the molecular formula C_4H_9, with an odd number of hydrogen atoms, corresponds to a free radical. Write structures for all possible radicals with this formula.

2.16 In the chlorination products of methane, trace amounts of chlorinated ethanes can be detected. Suggest how the carbon–carbon bond could be formed during the chlorination.

2.17 How many possible monochloro compounds ($C_6H_{13}Cl$) could be obtained from each of the five isomers of C_6H_{14}? (It is not necessary to write structures of all chloro compounds; write the carbon chain for each hydrocarbon and count the number of different positions.)

2.18 Chlorination of 2-methylpropane gives a mixture of 64 per cent 1-chloro-2-methylpropane and 36 per cent 2-chloro-2-methylpropane. Compare this ratio with the composition that would be obtained if all C—H bonds in the hydrocarbon were substituted at the same rate. The difference observed is due to the lower activation energy for formation of the more stable radical. Which is the more stable radical?

3

UNSATURATED HYDROCARBONS

Alkenes (also called **olefins**) are hydrocarbons containing a double bond and have the general formula C_nH_{2n}; **alkynes** contain a triple bond and have the general formula C_nH_{2n-2}. Since alkenes and alkynes have fewer hydrogens than the corresponding alkanes (C_nH_{2n+2}), they are often referred to as **unsaturated** hydrocarbons. These multiple bonds are the functional groups responsible for the characteristic properties of the compounds. The alkenes include some of the most important compounds in industrial organic chemistry, and their reactions are of great practical importance.

Alkenes are obtained on a large scale in the refining and cracking of petroleum (Section 2.9), although they are not present in significant amounts in native crude oil or natural gas. The simplest alkene, ethylene, is produced in small amounts by fruits and blossoms during ripening (Section 13.4). A number of dienes and polyenes (many double bonds) are found in nature as terpenes (Special Topic, page 74).

Double bonds are a key feature in a number of kinds of biologically important compounds. Fats and vegetable oils, for example, have structures with long alkyl chains (Section 10.16). The presence of double bonds in these chains affects both the physical properties and ease of metabolism of these compounds. The substitution in the diet of highly unsaturated vegetable oils in place of the more saturated animal fats is recommended as a measure to reduce the chance of atherosclerosis.

3.1 Names

To form the systematic name of a specific unsaturated hydrocarbon, the **-ane** ending of the parent alkane name is replaced by **-ene** for alkenes and **-yne** for alkynes. The parent name is determined by finding the longest continuous carbon chain which contains the double or triple bond. The chain is then numbered so as to give the multiple-bonded carbons the smallest numbers, and the *lower* of these numbers is used in the name to indicate the position of the double bond. If two double bonds are present in the chain, the ending becomes **-adiene.**

$$\overset{1}{C}H_3\overset{2}{C}H=\overset{3}{C}H\overset{4}{C}H_2\overset{5}{C}H_3$$

2-pentene
(not 3-pentene)

$$\overset{4}{C}H_3\overset{3}{C}H\overset{2}{C}H=\overset{1}{C}H_2$$
$$|$$
$$CH_3$$

3-methyl-1-butene
(not 2-methyl-3-butene)

$$CH_3CH_2CH_2\overset{3}{C}H\overset{4}{C}H_2\overset{5}{C}H_2\overset{6}{C}H_3$$
$$|$$
$$\overset{}{C}H=CH_2$$
$$\quad 2 \quad 1$$

3-n-propyl-1-hexene
(note that the 7-carbon chain does not contain the double bond)

$$CH_2=\overset{Cl}{\underset{|}{C}}-CH_3$$

2-chloropropene

$$CH_3C\equiv CH$$

propyne

4-methylcyclohexene

$$\overset{1}{C}H_3\overset{2}{C}H=\overset{3}{C}H-\overset{4}{C}H=\overset{5}{C}H\overset{6}{C}H_2\overset{7}{C}H_3$$

2,4-heptadiene

$$CH_3C\equiv C\overset{Br}{\underset{|}{C}}HCH_3$$

4-bromo-2-pentyne

As frequently happens, for some of the simplest compounds the trivial names are preferred and are generally used instead of the systematic ones. The names ethylene, propylene, and acetylene are almost always used for these three compounds.

	$CH_2=CH_2$	$CH_3CH=CH_2$	$HC\equiv CH$
preferred (systematic)	*ethylene (ethene)*	*propylene (propene)*	*acetylene (ethyne)*

There are also special names for two important groups derived from ethylene and propylene. The $CH_2=CH-$ group is called the **vinyl** group, and $CH_2=CH-CH_2-$ is called the **allyl** group. The names describe the structural type of the systems $-\overset{|}{C}=\overset{|}{C}-X$ and $-\overset{|}{C}=\overset{|}{C}-\overset{|}{C}-X$, and are also used in names of specific compounds.

$$CH_2=CHCl \qquad CH_2=CH-CH_2Cl \qquad CH_3-CH=CH-CH_2Cl$$

vinyl chloride — *allyl chloride* — *1-chloro-2-butene (an allylic chloride)*

Problem 3.1 Write structures for the following compounds: a) 3-ethyl-3-hexene, b) 3-chlorocyclopentene, and c) vinylacetylene. d) Give a systematic name for vinylacetylene.

3.2 Alkene Geometry

As described in Chapter 1 (Figure 1.8), the double bond is a combination of a σ bond, symmetrically located between two sp^2 carbons, and a π bond, formed by the overlap of p orbitals. Because the p orbitals can overlap only when they are lined up side-by-side, we cannot rotate the carbons with respect to each other without breaking the π bond. For example, if we tried to convert A to B by rotation around the axis of the σ bond, we would have to

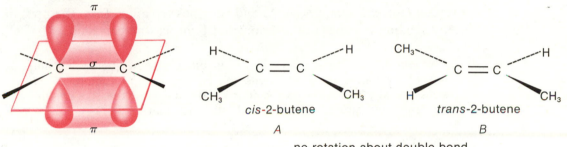

cis-2-butene   trans-2-butene

A   B

no rotation about double bond

overcome the energy of the π bond to arrive at the mid-point of the rotation. The π bond thus serves as a lock to prevent free rotation around a double bond, and the structures represent *different* compounds. A and B are **geometric isomers,** with the same relationship as that between the *cis*- and *trans*-isomers of 1,2-dimethylcyclopentane (Figure 2.5, page 34).

Any alkene with different groups attached to *both* of the sp^2 carbons exists as *cis*- and *trans*-isomers. Illustrations of *cis*- and *trans*-alkenes are the isomers of 1-bromopropene or 3-heptene. The *cis*-isomer has the two like

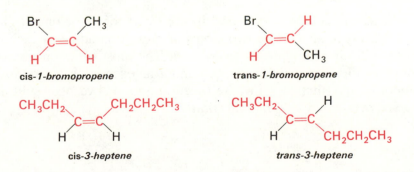

cis-*1-bromopropene*   trans-*1-bromopropene*

cis-*3-heptene*   trans-*3-heptene*

groups on the *same* side of the double bond; in the *trans*-isomer the two like groups are *across* from each other.

Problem 3.2 Write structures for a) *cis*-2-butene, b) *trans*-1-chloro-2-pentene, c) *trans*-3-methyl-3-hexene.

Problem 3.3 Which of the following compounds can exist as *cis*- and *trans*-isomers and which cannot? a) 1,1-dichloroethylene, b) 2-methyl-2-pentene, c) 3-chloro-3-hexene, d) cyclohexene, e) 1,3-pentadiene.

The shapes of *cis*- and *trans*-isomers differ considerably when the double bond is in the center of a chain, since the two isomers are bent differently at the double bond. The geometry in the vicinity of a π bond can have important effects on the physical and biological properties of unsaturated compounds. An interesting example is found in the chemistry of vision.

Receptor cells in the eye contain a complex pigment called rhodopsin, which consists of the polyene 11-*cis*-retinal bound tightly to a protein. When light is absorbed by rhodopsin, the energy supplied is sufficient to overcome the activation barrier for rotation around a π bond, and the *cis*-retinal is converted to the *all trans*-isomer. As the double-bond isomerization occurs, the retinal molecule changes shape and becomes separated from the protein, thereby sending a signal to the brain that light was perceived.

Problem 3.4 The structure of 11-*cis*-retinal is shown below. Locate the 11,12-double bond and write the corresponding structure of *all trans*-retinal ("*all*" refers to the four double bonds in the side-chain). Compare your structure with that of vitamin A (page 75).

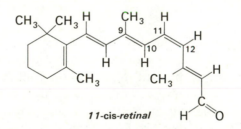

11-cis-retinal

3.3 Synthesis of Alkenes

Alkenes are generally prepared in the laboratory by **elimination reactions.** In this type of reaction two atoms or groups on adjacent carbons are split out to form a carbon-carbon π bond. The most common elimination reactions are those in which hydrogen and a halogen atom are removed by treatment of alkyl halide with base (Section 6.8), or hydrogen and an OH group are eliminated as water (Section 7.12).

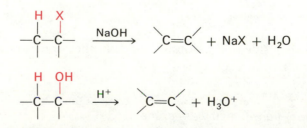

We will come back to these reactions and examine them in more detail in the chapters dealing with alkyl halides and alcohols.

ADDITION REACTIONS OF ALKENES

Since a carbon-carbon π bond is weaker than the σ bond, a double bond is a reactive center in an alkene. The characteristic reaction of alkenes is **addition** of two groups across the double bond, the reverse of the elimination reaction just mentioned. The driving force for this reaction is the energy released by the formation of two new σ bonds.

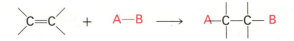

3.4   Hydrogenation

One of the simplest and most important reactions of alkenes is addition of hydrogen to form an alkane.

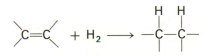

Although this reaction is quite exothermic, the activation energy is high and a catalyst is needed for the reaction to proceed at a reasonable rate; a finely divided metal catalyst, usually nickel, platinum, or palladium, is used. These metals adsorb hydrogen, and the interaction between H_2 and the metal both weakens the H—H bond and orients the hydrogen atoms for addition. Both hydrogen atoms are added to the same face of the double bond. In the hydrogenation of 1,2-dimethylcyclopentene, for example, the product is *cis*-1,2-dimethylcyclopentane.

FIGURE 3.1

Catalytic hydrogenation of 1,2-dimethyl-cyclopentene.

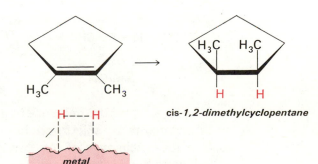

The amount of hydrogen absorbed by a weighed sample of a compound can easily be measured, and this is a useful analytical procedure for determining the number of double bonds in a compound. Thus, if a compound has the molecular formula C_6H_{10}, with four hydrogens less than the six-carbon saturated alkane ($H_{2n+2} = 14$), it could contain two double bonds, or one double bond plus one ring. If on hydrogenation only one mole of hydrogen is consumed per mole of compound, we know that the compound is not a diene, and presumably contains one double bond and one ring, as in cyclohexene, for example.

Problem 3.5   A compound has the molecular formula $C_{10}H_{16}$. On hydrogenation, one mole of hydrogen is consumed. How many rings are present?

3.5   Addition of Acids

In most addition reactions of alkenes, an **electrophilic reagent** (Section 1.10) "attacks" the easily accessible electron pair of the π bond. In a subsequent step, a nucleophile may then donate an electron pair to carbon to complete the reaction. One of the simplest and most important types of electrophilic addition is seen in the reaction with a Brönsted-Lowry acid HA, as in the addition of HCl to ethylene.

$$CH_2{=}CH_2 + HCl \longrightarrow CH_3CH_2Cl$$

When the alkene is not symmetrical, the proton (hydrogen ion) will bond to the *less* carbon-substituted (branched) sp^2 carbon atom, and the conjugate base, A^-, ends up attached to the more highly substituted carbon of the double bond.

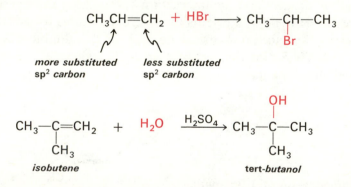

To understand the influence of alkene structure on the direction of these additions and to be able to predict the outcome of related reactions, let us look at the mechanism of the third reaction shown above. In this process, called **hydration**, water adds to the double bond of isobutene in the presence of 30 per cent aqueous sulfuric acid to produce an alcohol. In the first step of the reaction, the alkene functions as a base by donating an electron pair to a proton. The π-bonding electrons then form a σ bond between carbon and hydrogen. Since both π electrons are used in forming this new σ bond, but only one carbon is involved, the other sp^2 carbon becomes a carbocation. The hydration reaction is completed by donation of an electron pair from water, which serves as a nucleophile, to form a σ C—O bond. Loss of a proton gives the alcohol product.

Path actually followed:

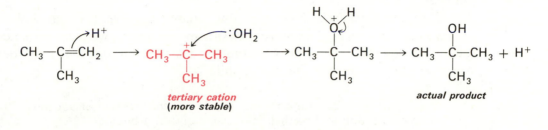

Path *not* followed:

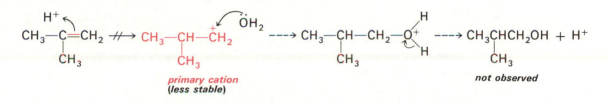

In this process two different sp^2 carbons are available in the alkene, but only the product that arises from bonding of a proton to the CH_2 carbon is obtained. Why? The direction of addition is determined by *which carbocation is formed* in the first step. From the structure of the alcohol product it can be seen that the more highly branched tertiary cation is favored. Carbocations can be classified, according to the degree of substitution on the electron-deficient carbon, as **primary, secondary,** or **tertiary.** The same terms are also used to designate alcohols, alkyl chlorides, and other compounds with different numbers of alkyl groups attached to the carbon which bears the functional group (Table 3.1). In each case, primary denotes a compound with one carbon attached to the carbon bearing the charge or group, secondary means two carbons attached, and tertiary means three carbons attached.

TABLE 3.1   CLASSIFICATION OF CATIONS, ALCOHOLS, AND ALKYL CHLORIDES

	Primary (1°)	Secondary (2°)	Tertiary (3°)
Cation	RCH_2^+	R_2CH^+	R_3C^+
Alcohol	RCH_2OH	R_2CHOH	R_3COH
Chloride	RCH_2Cl	R_2CHCl	R_3CCl

The **stability** of carbocations decreases in the order tertiary > secondary > primary, the most highly substituted being the most stable. When the addition of a proton (or any Lewis acid) to a double bond presents a choice between formation of two carbocations, the reaction proceeds by way of the more stable one since there is a lower activation energy for its formation. In the hydration of 2-methylpropene, the choice was between a tertiary and a primary cation. The more stable tertiary ion was formed and the tertiary alcohol was the final product, because this is the *lower energy pathway*.

With this background we can now predict the product that will be obtained in the addition of any acid HA to an alkene. All we need do is add the proton to the π bond to give the most stable carbocation, and then combine that cation with the nucleophile A^-.

Problem 3.6   What are the products from the reaction of aqueous sulfuric acid with a) 1-methylcyclohexene and b) 1-butene?

Problem 3.7   Using curved arrows to indicate electron pair movement, write a complete, stepwise mechanism for the addition of HCl to propylene.

Problem 3.8   Write the structures of the carbocations that would be formed in the reaction of sulfuric acid with a) 4-methyl-1-pentene and b) 2-methyl-2-pentene.

When more concentrated sulfuric acid and a slightly different procedure are used in the addition reaction with 2-methylpropene, the product, instead of the alcohol, is a mixture of two alkenes with the formula C_8H_{16} plus small amounts of compounds with the formula $C_{12}H_{24}$. The C_8H_{16} compounds are **dimers** (made up of two units) of the alkene, and are formed by reaction of the carbocation as the electrophile with a second molecule of alkene. The initial addition follows the same path as seen before; a new tertiary carbocation is formed, and under the reaction conditions, this cation *loses a proton* to give either of two C_8H_{16} isomers. This type of dimerization is an important method for obtaining highly branched hydrocarbons for high octane gasoline (Section 2.9).

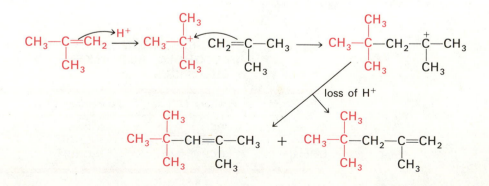

Problem 3.9   In Chapter 2 (Section 2.9) we learned that gasoline octane rating is based on a value of 100 for "iso-octane," 2,2,4-trimethylpentane. How could you prepare this alkane starting from 2-methylpropene?

3.6   Addition of Halogens

Bromine and chlorine act as electrophiles toward alkenes, and addition of Br_2 or Cl_2 occurs at room temperature without a catalyst to give 1,2-dihaloalkanes. The red color of bromine rapidly disappears when a solution of bromine in CCl_4 reacts with an alkene, providing a simple test for double bonds. This reaction has been used to analyze for the extent of unsaturation in fats and oils.

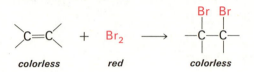

colorless   red   colorless

The addition of halogen appears similar to hydrogenation, but the mechanism must be different because the halogen atoms add exclusively to *opposite* faces of the double bond. This ***trans*-addition** is illustrated in the reaction of cyclopentene with bromine to give only *trans*-1,2-dibromocyclopentane.

The **stereochemistry** of the addition (*i.e.*, the geometric relationship of the reactants and products) shows that the two C—Br bonds are not formed simultaneously. The overall addition of Br_2 is a stepwise process in which both sp^2 carbon atoms become attached to one bromine atom as the relatively weak Br—Br bond is broken. The resulting three-membered ring, a **bromonium ion,** is then attacked by Br^- *from the side opposite to the first bromine* to produce the *trans*-1,2-dibromide.

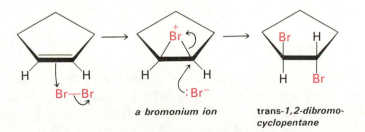

a bromonium ion   *trans*-1,2-dibromo-
                          cyclopentane

The second step of this process, as in the addition of acids, is attack by the nucleophile Br^- on the carbocation. The two-step mechanism suggests the possibility that the addition might be completed by some nucleophile other than Br^-. This does, in fact, occur. For example, if the addition of bromine to an alkene is carried out in aqueous solution, Br^- cannot compete with the much more abundant nucleophile H_2O, and the product is an alcohol with the halogen atom on the adjacent carbon, commonly called a **halohydrin.** The direction of addition is again determined by the formation of the more stable cation. With propylene, for example, opening of the cyclic bromonium ion gives the secondary carbocation, and thus the 1-bromo-2-hydroxy compound.

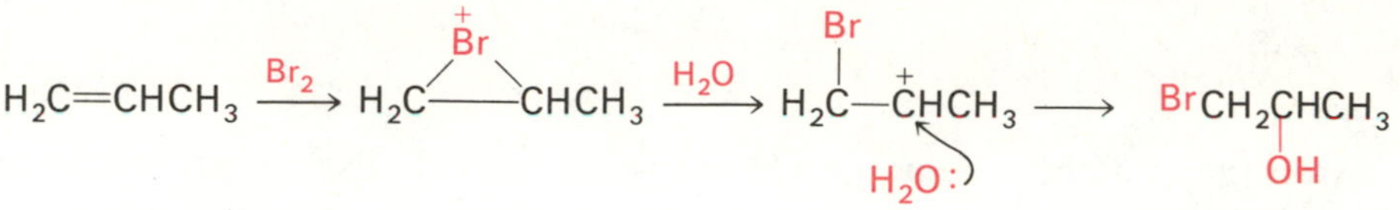

$$H_2C=CHCH_3 \xrightarrow{Br_2} H_2C\overset{\overset{+}{Br}}{\diagup\diagdown}CHCH_3 \xrightarrow{H_2O} H_2C-\overset{+}{C}HCH_3 \longrightarrow BrCH_2\overset{OH}{C}HCH_3$$

1-bromo-2-propanol
(propylene bromo-
hydrin)

Problem 3.10 Write the structures of the products that would be obtained in the following reactions: a) propylene plus bromine, b) 1-methylcyclopentene plus chlorine, c) 1-butene plus bromine in aqueous solution.

3.7 Epoxide Formation

A reaction very similar to the addition of bromine is the oxidation of alkenes to **epoxides.** The most useful reagent for carrying out the reaction on a laboratory scale is a peroxy acid, $R\overset{\overset{O}{\parallel}}{C}-O-OH$. The O—O bond, like that in Br—Br, is weak and readily breaks to give the stable anion RCO_2^- (see Chapter 10). The result is the formation of a three-membered oxide ring. The product is called an epoxide, or alkene oxide.

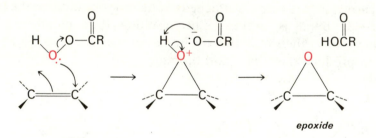

epoxide

The oxidation of ethylene is carried out on a very large scale industrially to give the important raw material **ethylene oxide.** As in most industrial reactions, the oxidation is a continuous catalytic process. In this case the olefin and oxygen are passed over a bed containing a silver oxide catalyst. The uses of ethylene oxide are discussed in Section 7.8.

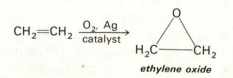

ethylene oxide

3.8   Addition of Ozone

Ozone, O_3, is a highly reactive electrophile and reacts with the carbon-carbon double bond to yield a five-membered ring. Since the —O—O—O— unit is quite unstable, this intermediate quickly rearranges with rupture of the C—C bond to give the more stable **ozonide**. This ozonide is cleaved with zinc in aqueous acid to produce aldehydes and/or ketones—compounds with a C=O functional group, discussed in Chapter 8.

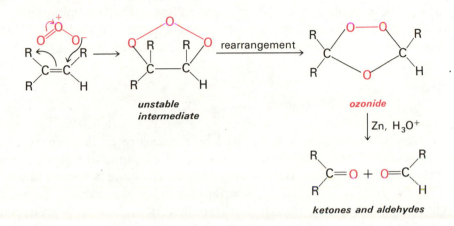

unstable intermediate

ozonide

Zn, H_3O^+

ketones and aldehydes

Ozonolysis provides a method for determining the position of the double bond in an unknown alkene. Since the overall reaction simply cleaves the molecule at the double bond, we need only to mentally remove the oxygen atoms from the products and reconnect the fragments with a double bond to reconstruct the alkene. For example, if an unknown alkene gives equal amounts of CH_3CH=O and $(CH_3CH_2)_2C$=O upon ozonolysis, the structure is immediately established as 3-ethyl-2-pentene.

$$CH_3-\underset{\underset{H}{|}}{C}=O \ + \ O=\underset{\underset{CH_2CH_3}{|}}{C}-CH_2CH_3 \ \xleftarrow[H_3O^+]{Zn} \ \xleftarrow{O_3} \ CH_3-\underset{\underset{H}{|}}{C}=\underset{\underset{CH_2CH_3}{|}}{C}-CH_2CH_3$$

aldehyde   *ketone*   *3-ethyl-2-pentene*

Problem 3.11   Write structures and give names for the alkenes that give the following products on ozonolysis:

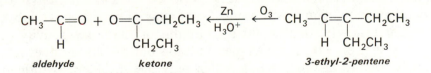

a) $CH_3-\overset{O}{\overset{||}{C}}-CH_3 \ + \ CH_3CH_2-\overset{O}{\overset{||}{C}}-H$ b) CH_2=O $+ \ CH_3CH_2\underset{\underset{CH_3}{|}}{CH}-\overset{O}{\overset{||}{C}}-CH_3$

c) $CH_3CH_2CH_2-\overset{O}{\overset{||}{C}}-H$ only   d) $H-\overset{O}{\overset{||}{C}}-CH_2CH_2CH_2-\overset{O}{\overset{||}{C}}-H$

The significance of ozone goes far beyond its use in this special laboratory technique. Ozone is generated from O_2 molecules in an electric discharge or by high-energy radiation, and it is formed in low concentration in the upper atmosphere. This ozone layer serves as a protective filter against the higher energy ultraviolet solar radiation. Any thinning of the earth's shield can be expected to cause an increase in the incidence of skin cancer due to radiation, and the decomposition of ozone by chemicals that diffuse into the upper atmosphere is a potentially serious health hazard.

Ozone is also produced in the reactions of automobile exhaust gases in sunlight. These gases contain as primary pollutants carbon monoxide, unburned hydrocarbons, and nitric oxide. Upon exposure to sunlight, ozone is produced as a secondary pollutant in this photochemical smog.

$$NO_2 \xrightarrow{\text{sunlight}} NO + O$$
$$O + O_2 \longrightarrow O_3$$

Ozone concentration can be used as a measure of the degree of air pollution, and it may reach five to ten times the normal atmospheric level under adverse conditions. At these concentrations, ozone is toxic, and causes respiratory problems including interference with the normal protective mechanism of the lung against bacterial infection. Ozone is particularly destructive to rubber and other elastomers (Section 3.16). In areas where the ozone pollution is particularly high, the lifetime of automobile tires, insulated wire, and other rubber products is significantly reduced.

3.9   Alkynes

The chemistry of alkynes is similar to that of alkenes but is less extensive. Alkynes may be hydrogenated to alkanes under the same conditions used for alkenes. It is possible to use a "poisoned" palladium catalyst to stop the hydrogenation of an alkyne at the alkene stage. The poison is a substance (sulfur, for example) that is added to inhibit or lower the activity of the catalyst so that the alkyne undergoes addition but the alkene does not. An important feature of this reaction is that only the *cis*-alkene is formed.

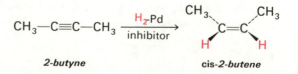

CH₃—C≡C—CH₃ (*2-butyne*) → H₂-Pd / inhibitor → *cis-2-butene*

Although the triple bond has a high electron density, alkynes are actually *less* reactive than alkenes toward electrophilic addition. This is because the vinyl carbocation formed in the first step is particularly unstable and thus difficult to form.

$$H^+ + RC \equiv CR \longrightarrow RCH = \overset{+}{C}R$$

vinyl
cation

The addition of hydrogen halides can be stopped after the addition of one molecule of HX, since the vinyl halides are considerably less reactive toward electrophilic addition than is a simple alkene.

$$R-C\equiv CH \xrightarrow{HCl} R-\overset{\overset{\displaystyle Cl}{|}}{C}=CH_2 \xrightarrow{HCl} R-\overset{\overset{\displaystyle Cl}{|}}{\underset{\underset{\displaystyle Cl}{|}}{C}}-CH_3$$

a vinyl chloride

The most important alkyne is the first compound in the series, **acetylene.** Acetylene has an unusually large heat of combustion and is used in welding to obtain a very hot flame. Acetylene is produced industrially by the pyrolysis of methane at very high temperatures, and is the starting material for a number of important processes.

$$2\ CH_4 \xrightarrow[\text{flow reactor}]{1500°} HC \equiv CH + 3\ H_2$$

Higher alkynes are usually prepared from acetylene by reactions discussed in Section 6.9. Triple bonds are not as commonly encountered as many other functional groups among naturally occurring compounds, but a number of hydrocarbons containing several triple bonds are present in certain plants and fungi. An example of these remarkable structures is the dienetetrayne, $CH_3CH=CH(C\equiv C)_4CH=CH_2$, found in dahlias.

An important property of 1-alkynes is the acidity of the hydrogen attached to the terminal *sp* carbon. The acidity of a proton in a C—H bond increases as the *s* component of the hybridized carbon orbital increases.* Acetylenic hydrogens are more acidic than those in ammonia, and can therefore be removed by amide anion, which is the conjugate base of ammonia.

$$2\ NH_3 + 2\ Na \longrightarrow 2\ Na^+\ \overset{..}{:}NH_2 + H_2$$

sodium amide

$$NaNH_2 + HC\equiv CH \longrightarrow HC\equiv C\overset{..}{:}\ Na^+ + NH_3$$

sodium acetylide

Acetylide anions are useful reagents, as will be seen in later chapters. Metal acetylides are also formed from acetylene and metal ions such as Ag^+ and Cu_2^{2+}. The formation of an insoluble precipitate of the silver acetylide

---

*An *sp* orbital is closer to the nucleus than an *sp²* or *sp³* orbital, and an electron pair in an *sp* orbital is therefore attracted more strongly to the positive carbon nucleus.

after treatment with silver hydroxide in ammonia can be used as a test for a 1-alkyne, but care must be taken not to isolate or dry the precipitate since these derivatives are dangerously explosive compounds.

$$RC{\equiv}CH + Ag(NH_3)_2^+ OH^- \longrightarrow RC{\equiv}C-Ag + 2NH_3 + H_2O$$

silver
acetylide

3.10 Dienes and 1,4-Addition

Dienes in which the two double bonds are separated by one or more saturated carbon atoms react in the same way as simple alkenes. For example, in 1,4-pentadiene, the double bonds undergo addition reactions independently of each other; reaction with chlorine gives initially 4,5-dichloro-1-pentene and then the tetrachloropentane.

$$CH_2{=}CH-CH_2-CH{=}CH_2 \xrightarrow{Cl_2} \underset{\substack{Cl \quad Cl}}{CH_2-CH_2-CH_2-CH{=}CH_2} \xrightarrow{Cl_2} \underset{\substack{Cl \quad Cl \qquad Cl \quad Cl}}{CH_2-CH-CH_2-CH-CH_2}$$

1,4-pentadiene   4,5-dichloro-1-pentene   1,2,4,5-tetrachloropentane
(a non-conjugated diene)

If the double bonds in a diene are separated by only a single bond, the double bonds are said to be **conjugated.** Conjugation of double bonds significantly changes the chemical properties of a 1,3-diene relative to those of a non-conjugated diene. For example, when 1,3-butadiene is treated with a limited amount of chlorine, a mixture of 3,4-dichloro-1-butene and 1,4-dichloro-2-butene is obtained. The somewhat surprising latter product is the result of addition to the ends of the four-carbon diene unit. This **1,4-addition** is often the major reaction of 1,3-dienes with electrophiles.

$$CH_2{=}CH-CH{=}CH_2 \xrightarrow[25°]{Cl_2} \begin{cases} CH_2{=}CH-CHCl-CH_2Cl \quad \text{1,2-Addition} \\ \qquad\qquad 25\% \\ ClCH_2-CH{=}CH-CH_2Cl \quad \text{1,4-Addition} \\ \qquad\qquad 75\% \end{cases}$$

In order to account for 1,4-addition, we must examine the mechanism. We know that the first step in addition of an electrophile to a double bond is formation of the most stable carbocation. Addition of chlorine would be expected to form the secondary carbocation (an allylic cation), but addition of Cl^- to this cationic carbon would produce only the minor product of the reaction.

$$Cl-Cl + CH_2{=}CH-CH{=}CH_2 \longrightarrow Cl^- + Cl-CH_2-\overset{+}{CH}-CH{=}CH_2$$

allylic cation

The mechanism that produces the major product is discussed in the next section.

3.11 Resonance

The explanation of 1,4-addition lies in the fact that our representation of this carbocation is not adequate. The allylic cation is actually a new type of species called a **resonance hybrid.** A resonance hybrid is a structure which combines the features of two or more structures that differ only in their electronic arrangements. In fact, these structures, called **contributing resonance forms,** are hypothetical, and only the resonance hybrid actually exists. If we look at a simplified orbital picture of this allylic cation (Figure 3.2), we

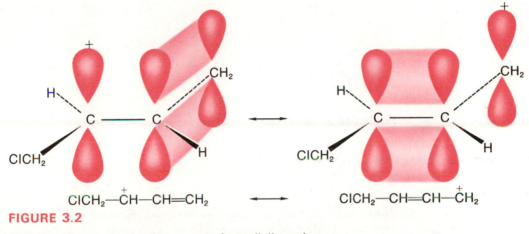

FIGURE 3.2

Contributing resonance structures of an allylic cation.

see that the π bond can be perfectly lined up to donate electron density to the vacant p orbital of the adjacent carbon. Complete donation of the π electrons would produce a new allylic cation with the charge on the terminal carbon. In fact, the actual structure of the allylic cation is a hybrid of these two structures, with the positive charge distributed between the 2- and 4-positions.

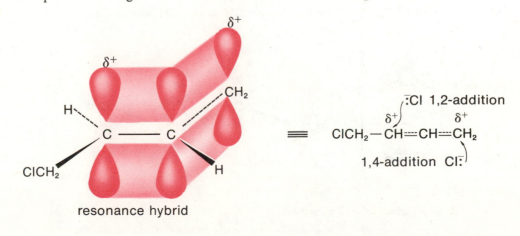

resonance hybrid

The resonance hybrid is more stable than either of the two contributing resonance forms would be, because the positive charge is **delocalized** or dispersed over two atoms, C-2 and C-4. Delocalization of charge always produces a more stable (lower energy) ion than if the charge were forced to reside on a single atom. With this more accurate picture of the carbocation, it is not at all surprising that the chloride anion attacks at both the 2- and 4-positions.

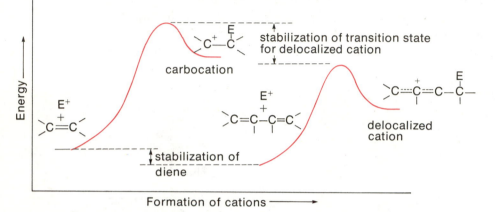

FIGURE 3.3
Energy relationships of diene and delocalized cation.

As shown in Figure 3.3, conjugation also slightly lowers the energy of the diene compared to that of a simple alkene, although the difference is less than that between the two carbocations. The greater stability of the diene is due to delocalization of the π bonds. The p orbitals of C-2 and C-3 can overlap in the same way as those at C-1—C-2 and C-3—C-4. As a result, the π bonding extends over all four atoms (Figure 3.4). The 1—2 and 3—4 bonds

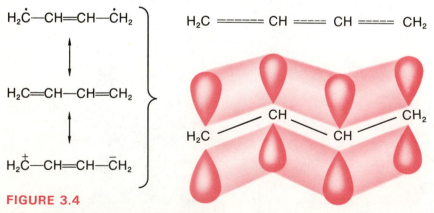

FIGURE 3.4
π Bond delocalization in 1,3-butadiene.

are slightly less than full double bonds, and the 2—3 bond now has some double bond character as shown by the 2—3 bond length, which is only 1.48 Å compared to the 1.54 Å of a normal C—C single bond.

It is extremely important to realize that a resonance hybrid is *not* an equilibrium between two or more structures. Note that a double-headed arrow, ⟷, denotes resonance structures while ⇌ describes an equilibrium process. Resonance structures differ only in the position of electron density—the relative positions of the atoms in the structures are identical.

Problem 3.12   Nitrate ion, NO_3^-, is stabilized by resonance since the negative charge can be delocalized over all three oxygen atoms. a) Write electronic formulas for the three contributing resonance structures. b) Write electron structures for two resonance forms of nitrite ion, NO_2^-.

Problem 3.13   Predict the products from addition of dry HCl to a) 1,3-hexadiene, b) 2,4-hexadiene.

3.12   Diels–Alder Reaction

Conjugated dienes undergo another type of 1,4-addition in which the double bond of an alkene, which is designated as the dienophile, adds to the 1,3-diene system to give a six-membered ring containing one double bond. This process is known as the Diels-Alder reaction. The net result is the formation of two new σ bonds and one new π bond at the expense of the three original π bonds.

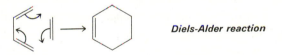

Diels-Alder reaction

When a cyclic 1,3-diene is used in this reaction, the product is a *bicyclic* structure, with a "bridge" across one ring.

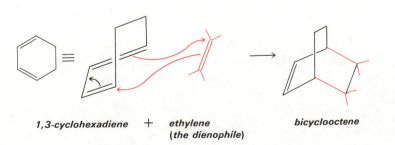

1,3-cyclohexadiene   +   *ethylene*   *bicyclooctene*
                            (the dienophile)

Several well known insecticides are prepared by the Diels-Alder reaction, using hexachlorocyclopentadiene and a cyclic alkene. These highly chlorinated compounds are effective in combating certain agricultural pests, but they are also toxic to higher organisms (Chapter 6).

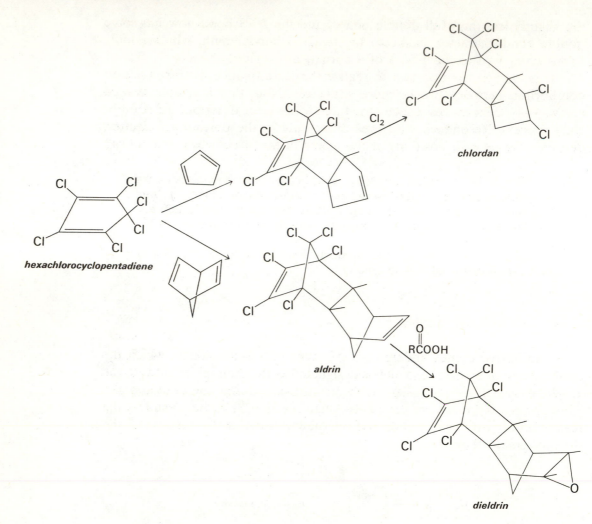

hexachlorocyclopentadiene

chlordan

aldrin

dieldrin

Problem 3.14 What products would be obtained from the Diels-Alder reaction of a) 1,3-pentadiene plus ethylene, b) 1,3-butadiene plus propene?

3.13 Free Radical Reactions of Alkenes

Free radicals can react with alkenes in two ways: (1) *addition* to the π bond to form a σ bond and generate a new radical center, and (2) *abstraction* of an allylic hydrogen atom to form an **allylic radical.**

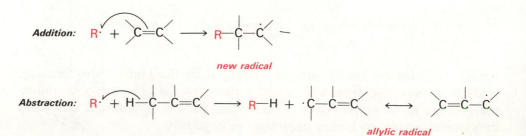

Addition: new radical

Abstraction: allylic radical

An allylic radical is stabilized in the same way as an allylic cation, through resonance delocalization, this time by delocalization of an unpaired electron. This special stability of an allylic radical makes it relatively easy to abstract an allylic hydrogen when free radical conditions (Section 2.7) are used. Under these conditions, free radical substitution at an allylic position is a high yield, selective reaction.

$$CH_2{=}CH{-}CH_3 \xrightarrow[-HCl]{Cl\cdot} CH_2{=}CH{-}\overset{\cdot}{C}H_2 \xrightarrow{Cl_2} CH_2{=}CH{-}CH_2Cl + Cl\cdot$$

propylene　　　　　*allyl radical*　　　　*allyl chloride*

Problem 3.15　　In addition to the insecticides shown in Section 3.12, *heptachlor* is also prepared from hexachlorocyclopentadiene. Suggest a two-step synthesis of heptachlor. (Hint: use the same intermediate involved in the synthesis of chlordan.)

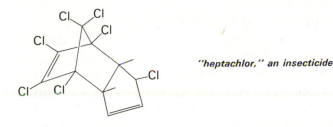

"heptachlor," an insecticide

Another reaction of alkenes that depends on the easy formation of allyl radicals is oxidation by air. Oxygen molecules are diradicals, *i.e.*, they possess *two* unpaired electrons. In the presence of ultraviolet light, O_2 reacts with alkenes to remove a hydrogen atom and produce an allylic radical, which is then trapped by more oxygen. The resulting **hydroperoxide** radical can then continue a chain reaction of allylic hydrogen abstraction and radical combination. This overall process is called **autoxidation**; it is responsible for the weathering and general deterioration of organic substances, such as oil-based paints, that are exposed to air and sunlight.

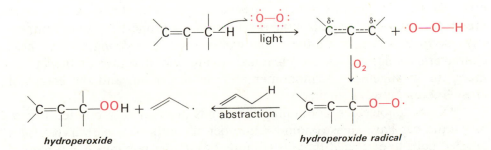

hydroperoxide　　　　　　　　　　　　　*hydroperoxide radical*

Free radicals such as $\cdot OH$ and $\cdot OOH$ are among the components of photochemical smog. The combination of these radicals with the alkenes, NO, and NO_2 that are present in automotive emissions leads to end-products such as peroxyacetylnitrate (PAN), $CH_3C-O-O-NO_2$, one of the princi-

$$\overset{\displaystyle \|}{O}$$

pal eye irritants in metropolitan areas where atmospheric inversion prevents air circulation.

ORGANIC POLYMERS

Polymers are very large molecules made up of repeating small units called **monomers.** Familiar naturally occurring polymers include proteins, rubber, cellulose, silk, wool, and starch. One of the major contributions of organic chemistry in the past 30 years is the wealth of new materials that have been developed from **synthetic polymers.** Polymers can be fabricated into fibers, molded plastics, rubber, protective coatings, thin films, or rigid, tough solids. The primary requirement for all these varied uses is a *high molecular weight* chain or network.

Polymers are obtained by one of two general processes: (1) repeated **addition** of units to a growing chain, with each step generating a reactive intermediate (cation, anion, or radical) for the next addition, or (2) **condensation** of compounds with two functional groups with loss of a small molecule, such as water. In this chapter we will look only at addition polymers; condensation polymers will be covered later when we study the functional groups that are involved in their formation. In the simplest type of addition polymer, alkene molecules combine according to the general reaction:

$$n\ CH_2{=}CHR \xrightarrow{\text{catalyst}} \left(\!\!CH_2-\overset{\overset{\displaystyle R}{\displaystyle |}}{CH}\!\!\right)_{\!\!n} \quad n = 100 \text{ to } 100{,}000$$

monomers   *polymer*

3.14   Structure and Properties

Polymeric substances can be classified according to their physical properties as flexible **plastics** which soften at an elevated temperature, rigid **thermosetting resins,** and **elastomers** with rubbery properties. A very important group of polymers are those that can be spun into strong fibers. These characteristics depend on the chemical nature and structure of the monomers, the arrangement of monomer units in the chain, and the extent of cross-linking between chains.

Polymer molecules, containing thousands of covalently bonded monomer units, can have an astronomical number of conformations, varying from tangled coils to extended threads (Figure 3.5). If the polymer has a regular repeating structure, attractive forces between chains favor an *ordered* array as opposed to completely random coils, and the solid polymer contains crystalline regions. A highly ordered, crystalline structure is the primary requirement for strong fibers. Examples of this type of polymer are polypropylene and condensation polymers, such as polyesters or nylon (Section 10.21).

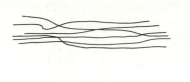

Oriented chain —
crystalline

Random coil —
amorphous

Rigid glass

FIGURE 3.5

Polymer conformations.

If the polymer framework is irregular or branched, an amorphous (noncrystalline) structure results. When the chain contains bulky groups that restrict flexibility, the resulting amorphous polymer is a stiff, tough glass, such as poly(methyl methacrylate) (Plexiglas) (Table 3.2). In thermosetting polymers which harden on heating, the chains are bonded together in a rigid three-dimensional network. The most important thermosetting materials are condensation polymers of formaldehyde (Section 8.15).

3.15 Vinyl Polymers

The major addition polymers are obtained by polymerization of **vinyl monomers,** with the general structure CH_2=CHR. As seen in Table 3.2, vinyl

TABLE 3.2   VINYL POLYMERS

Monomer	Polymer $\left(-CH_2-\underset{\underset{R}{|}}{CH}-\right)_n$	Use
CH_2=CH_2	polyethylene	molded plastics—containers, toys, bags
CH_2=$CHCH_3$	polypropylene	molded plastics and low-cost fibers for carpets
CH_2=$CHCl$	poly(vinyl chloride) ("PVC")	pipe, rigid panels, floor tile, electrical insulation
CH_2=$CHCN$	polyacrylonitrile (Orlon, Acrilan)	wool-like fiber
CH_2=$\underset{\underset{CH_3}{|}}{C}$—$CO_2CH_3$	poly(methyl methacrylate) (Plexiglas)	transparent, unbreakable sheets, ornamental trim
CH_2=CHO—$\overset{O}{\overset{\|}{C}}$—$CH_3$	poly(vinyl acetate)	water-soluble gums, adhesives, latex paints
CH_2=CH—N (pyrrolidone ring)	poly(vinyl pyrrolidone)	blood plasma extender
CF_2=CF_2	poly(tetrafluoroethylene) (Teflon)	greaseless bearings, liners for pots and pans

polymers include a wide variety of materials with properties ranging from sticky gums, as in poly(vinyl acetate), to hard, glassy solids such as poly-(methyl methacrylate). (Also see Color Plate 2.)

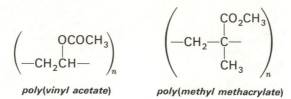

poly(vinyl acetate)   poly(methyl methacrylate)

Polymerization of vinyl monomers is accomplished by addition of an initiator to the double bond to give an intermediate, which then adds to a second monomer unit. Free radicals, generated by the thermal breakdown of a peroxide or other unstable compound, are one general type of initiator. Once begun, polymerization proceeds by a chain reaction, with each addition leading to a new radical. The reaction is exothermic since the energy released in forming a new strong C—C σ bond is greater than that required to break the relatively weak C—C π bond.

$$X_2 \xrightarrow[\text{light}]{\text{heat or}} 2\,X\cdot$$

$$X\cdot + CH_2=CHR \longrightarrow XCH_2-\dot{C}HR \xrightarrow{n\,CH_2=CHR}$$

$$X\!\left(CH_2CHR\right)_n\!CH_2\dot{C}HR \xrightarrow{\cdot X} X\!\left(CH_2CHR\right)_{n+1}\!X$$

vinyl polymer

To obtain a high molecular weight, the amount of initiator and the conditions must be carefully controlled. If too many chains are initiated, the monomer supply will be exhausted before the chains have grown to maximum length. If the polymerization is too slow, side reactions will limit growth of the chain. Termination may occur by capture of another initiator fragment X· or by coupling of two growing chains. The exact nature of the group X is usually not important, since the chain is many thousand units long, and the properties depend very little on the end groups. With a substituted ethylene ($CH_2=CHR$) the R group, regardless of its nature, stabilizes the radical —$\dot{C}HR$ relative to $\dot{C}H_2$—, and the polymerization occurs in a head-to-tail manner as shown previously.

The simplest and cheapest addition polymer is **polyethylene,** which is used as thin sheets or as molded articles for containers and protective coatings. A problem in the free radical polymerization of ethylene is the formation of **branches** in the polymer caused by the abstraction of a hydrogen atom from the chain by the terminal radical. When this occurs, the site of the radical propagation is transferred from the end to the interior of the chain, and a branch appears in the polymer chain (Figure 3.6). This same radical

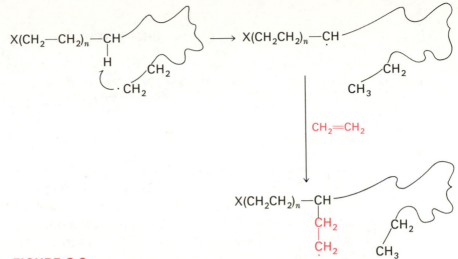

FIGURE 3.6

Generation of chain branching by internal hydrogen abstraction.

abstraction can also occur *intermolecularly,* with the radical of one molecule abstracting an atom from another molecule, thus terminating the chain prematurely. Special catalysts composed of trialkylaluminum and titanium tetrachloride have been developed to prevent branching. The polyethylene obtained with these catalysts has a higher melting point, higher density, and superior mechanical properties. An important use of high density polyethylene is in artificial joint implants (see Color Plate 2C).

> **Problem 3.16**   The free radical chain is only one way of achieving addition polymerization of alkenes. Write a stepwise mechanism for the acid-catalyzed polymerization of propylene. Is the polypropylene that is produced a head-to-tail or head-to-head polymer? Why?

Poly(vinyl chloride) is another low cost, large volume plastic. The polymer has a fairly high softening temperature, and can be molded into flame-resistant pipe and rigid panels. For uses in which flexibility is required, as in simulated leather, garden hose, or raincoats, a **plasticizer** is added. The plasticizer is a high-boiling organic liquid which is absorbed by the polymer, softening and essentially lubricating the chains to permit greater mobility. The fact that "vinyl" plastics gradually become brittle and crack is due to diffusion of the plasticizer out of the polymer.

Another means of modifying the properties of a polymer is **copolymerization** of two different monomers. Both units are incorporated in the chain, either in random sequence or in alternating blocks. Copolymers can be designed to have properties characteristic of both monomers, and different from those of a mixture of the two homopolymers prepared separately from the individual monomers. Thin film for food packaging (Saran Wrap) is prepared from a copolymer of vinyl chloride and vinylidene chloride (1,1-dichloroethylene). The dichloro units are distributed randomly along the chain and provide the flexibility and strength required for very thin films.

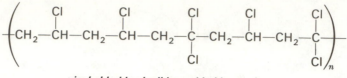

vinyl chloride-vinylidene chloride copolymer
(*Saran Wrap*)

3.16   Elastomers

Polymers with the property of long-range elasticity are called **elastomers;** they include various types of rubber and also specialized copolymers used in stretchable fibers. Elastomers can be stretched several-fold in length and then regain their original dimensions. This property requires a polymer structure in which the chains are *disordered* to permit movement, and yet are subject to a restoring force. Such a structure is achieved by selectively *cross-linking* the chains of an amorphous polymer. As the polymer is stretched, the chains become more ordered, and are pulled together (Figure 3.7). This motion requires energy, and is analogous to compression of a gas.* When the tension is removed, the chains resume the lower energy random arrangement.

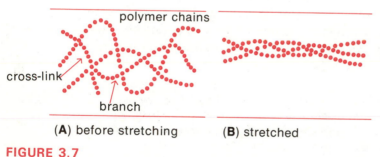

(**A**) before stretching   (**B**) stretched

FIGURE 3.7
Elastomer.

Natural rubber is a polymeric hydrocarbon obtained from the milky sap (latex) of certain trees in tropical South America and the East Indies. This latex is coagulated to give a spongy mass called crepe rubber. The structure of this material was determined by thermal decomposition and ozonolysis. On heating, rubber breaks down to give **isoprene** (2-methylbutadiene); on ozonolysis, *one* product, levulinaldehyde, is obtained in high yield. The latter result shows that the hydrocarbon chain consists of regular repeating units with a double bond and a methyl group together in each unit. The structure of natural rubber thus corresponds to a polymer in which isoprene monomer units are linked together by 1,4-addition. Thus, natural rubber is simply a large, regular terpene (see p. 74). The double bonds in natural rubber have the *cis* configuration with respect to the groups in the main polymer chain.

*Just as in compression of a gas, the temperature of rubber increases on stretching. To demonstrate, pull a rubber band quickly to the limit of its elasticity and touch the stretched band to the lips.

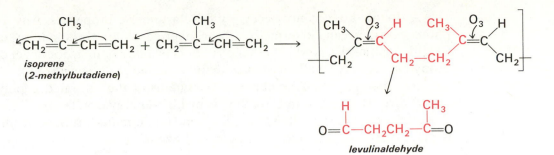

isoprene
(2-methylbutadiene)

levulinaldehyde

To obtain a useful elastomer, natural rubber is **vulcanized** by reaction with sulfur. This process, discovered accidentally by Goodyear long before the structure of rubber was known, introduces cross-links at positions activated by the double bonds. The number of bridges must be carefully controlled, since excessive cross-linking leads to a hard thermosetting material. The toughness and wear-resistance of rubber are greatly increased by addition of carbon black, which is strongly absorbed in the amorphous polymer network.

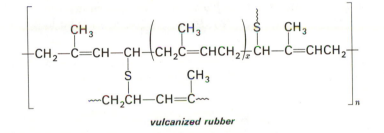

vulcanized rubber

Several types of synthetic rubber have been developed. A copolymer of 1,3-butadiene and styrene, called **SBR** (**S**tyrene-**B**utadiene-**R**ubber), has properties similar to those of natural rubber. Neoprene, a tough, oil-resistant rubber, is the *trans* isomer of poly(2-chloro-1,3-butadiene).

$$3n \ CH_2{=}CH{-}CH{=}CH_2 + n \ CH_2{=}CHC_6H_5 \longrightarrow$$

styrene

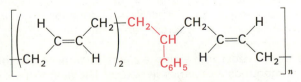

styrene-butadiene rubber **(SBR)**

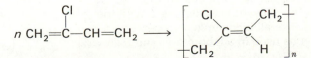

2-chloro-1,3-butadiene   *neoprene*

A double bond is not necessary for elastomeric properties. Any amorphous polymer can, in principle, be converted to an elastomer by cross-linking. One of the newer synthetic rubbers is a copolymer of ethylene and propylene. A very small amount of diene is added to provide double bonds for cross-linking sites. Alternatively, the chains of the saturated copolymer can be directly linked with C—C bonds by high-energy irradiation. Since this type of rubber has many fewer double bonds than natural rubber or SBR, it is more resistant to attack by oxygen and ozone.

SPECIAL TOPIC: Terpenes

The pleasant odors of pine needles, mint leaves, and other herbs are due to mixtures of compounds called **terpenes** that are present in the volatile oil of the plant. Terpenes may contain functional groups such as OH or C=O, but a number of these compounds are unsaturated hydrocarbons, often with cyclic structures. The characteristic feature of terpenes is a carbon chain that is made up of branched five-carbon segments called *isoprene units*. The units are arranged in a regular pattern with the branched "head" of one unit linked to the "tail" of another.

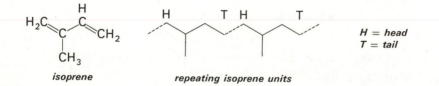

isoprene   *repeating isoprene units*   *H = head*
T = tail

Terpenes are classified according to the number of isoprene units they contain, which ranges from two to eight, as illustrated in Figure 3.8. A monoterpene contains two isoprene units, a sesquiterpene three units, a diterpene four, and so on. Regardless of functional groups and rings, the head-to-tail sequence of isoprene units can be seen by "dissecting" the structure (Problem 3.28).

One of the most interesting aspects of terpenes is the way that these complex compounds are formed in nature. The pathway provides an excellent illustration of some of the alkene reactions that we saw earlier in the chapter. As we would expect from their structural regularity, terpenes are built up in plant cells from five-carbon molecules. The biosynthesis occurs under the influence of catalysts called enzymes; but, like all biochemical processes, the reactions are the same as those that might be used in a laboratory, and follow the same principles.

Monoterpenes—two isoprene units

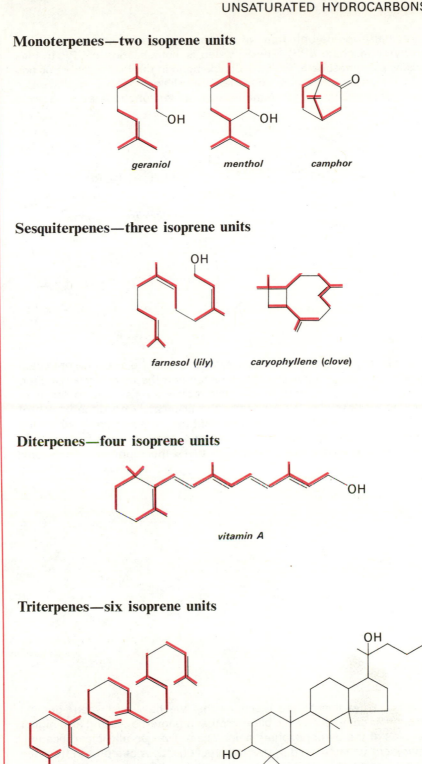

geraniol   *menthol*   *camphor*

Sesquiterpenes—three isoprene units

farnesol (lily)   *caryophyllene (clove)*

Diterpenes—four isoprene units

vitamin A

Triterpenes—six isoprene units

squalene   *dammarenediol*

FIGURE 3.8

Typical terpenes found in nature. (A short dash indicates a CH_3 group.)

The actual isoprene unit from which terpenes are formed is called iso-pentenyl pyrophosphate (IPP). The first step is isomerization of IPP to give dimethylallyl pyrophosphate, with the double bond in the more stable position (alkyl substitution on the double bond stabilizes alkenes). The pyrophosphate group (abbreviated OPP) in the dimethylallyl isomer functions as a "leaving

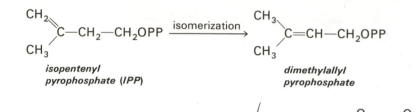

isopentenyl
pyrophosphate (*IPP*)

dimethylallyl
pyrophosphate

$$\left(-OPP = -O-\overset{\overset{\displaystyle O}{\|}}{P}-O-\overset{\overset{\displaystyle O}{\|}}{P}-O \right)$$

group"; that is, the C—O bond breaks heterolytically in the direction predicted by bond polarity to give a pyrophosphate anion and the resonance-stabilized allylic cation. This cation then acts as an electrophile for a second molecule of isopentenyl pyrophosphate, analogous to the acid-catalyzed dimerization of isobutene (Section 3.5), to give the most substituted carbocation (in which the newly formed carbon-carbon bond links the two isoprene units head-to-tail). Loss of a proton from the resulting cation gives the monoterpene *geranyl pyrophosphate*.

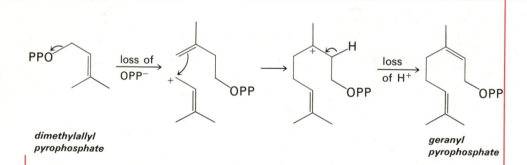

dimethylallyl
pyrophosphate

geranyl
pyrophosphate

By simple extension of the reactions that we have just seen, geranyl pyrophosphate is converted in the plant to other monoterpenes and also higher terpenes. Loss of the pyrophosphate anion again gives an allylic carbocation, and further steps can now lead in several directions. Loss of a proton from the cation introduces another double bond in the chain, giving the trienes *myrcene* and *ocimene* (see Problem 3.35). Another reaction of the carbocation is electrophilic addition to the double bond at the other end of the chain to give a cyclic structure (Figure 3.9). The cyclic carbocation can lose a proton, or it can undergo another intramolecular addition to form a bicyclic structure and then lose a proton.

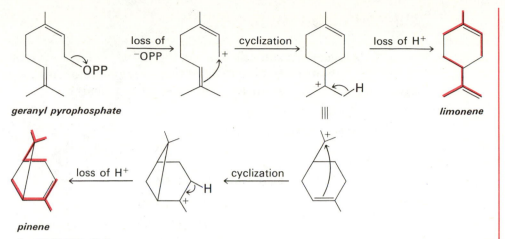

geranyl pyrophosphate

loss of
−OPP

cyclization

loss of H⁺

limonene

loss of H⁺

cyclization

pinene

FIGURE 3.9

Formation of cyclic monoterpenes.

Higher terpenes are formed by linking additional isopentenyl pyrophosphate units to the carbocation formed from geranyl pyrophosphate. After three units are linked head-to-tail, tail-to-tail coupling can then occur, leading to the triterpene **squalene.** Curiously, this hydrocarbon accumulates in large amounts in the liver of sharks.

Squalene is the precursor of the triterpenoids and also cholesterol (Chapter 7). These compounds have rather complex structures, with several rings fused together, but their formation from squalene follows the same basic process of electrophilic addition that we have seen for simpler terpenes. The carbocation is generated in this case by opening of an epoxide ring at one end of the squalene molecule. The chain is folded into the proper conformation, and the molecule is then "zipped up" by successive additions of the double bonds to a carbocation resulting from a previous step (Figure 3.10).

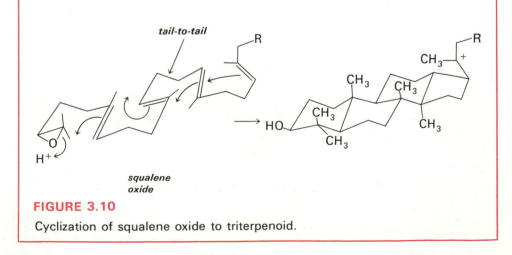

tail-to-tail

squalene oxide

FIGURE 3.10

Cyclization of squalene oxide to triterpenoid.

SUMMARY AND HIGHLIGHTS

1. **Alkenes** contain a *double bond*, C=C . The four groups attached to the sp^2 carbons in an alkene lie in the same plane; geometric isomers can exist when different groups are attached to both carbons.

2. **Hydrogenation** of alkenes: $\text{C=C} \xrightarrow{\text{H}_2 / \text{Pt}} \text{CH—CH}$

3. **Electrophilic addition** to alkenes:

$$\text{C=C} \xrightarrow{\text{H}^+} \text{CH—C}^+ \xrightarrow{\text{A}^-} \text{CH—C—A}.$$

Addition of HCl or H_3O^+ occurs so as to form the most stable carbocation: $R_3C^+ > R_2\overset{+}{C}H > R\overset{+}{C}H_2$.

4. Other electrophilic additions include:

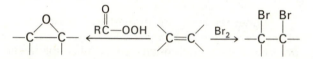

5. **Ozonolysis:** $\text{C=C} + O_3 \longrightarrow \text{C=O} + \text{O=C}$; locates position of double bond in chain.

6. **Alkynes** contain a *triple bond*, $-\text{C}\equiv\text{C}-$; $-\text{C}\equiv\text{C}-\text{H}$ is more acidic than other C—H bonds.

7. A compound is stabilized by **resonance delocalization** of electrons. The actual structure is a hybrid of hypothetical contributing structures.

8. **Conjugated dienes** undergo 1,2- and 1,4-addition because of **allylic resonance:**

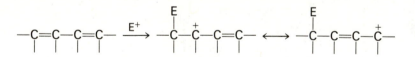

9. **Diels-Alder Reaction:**

diene   dienophile

10. **Free radicals** attack alkenes to give allylic radicals.

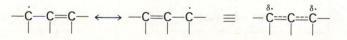

11. **Addition polymers:** $n \, CH_2{=}CHX \longrightarrow +CH_2CHR{+}_n$ **vinyl** polymerization. Properties depend on structure—crystalline or glassy. Elastomers such as rubber require *cross-links* between chains.

ADDITIONAL PROBLEMS

3.17 Write structures for the following compounds:

a)   3-methylcyclopentene   b) *cis*-3-octene

c)   2-hexyne   d) 2,3-dimethyl-2-butene

e)   allyl chloride   f)   tetrachloroethylene

g)   1,6-heptadiene   h) *trans*-1-chloro-1-propene

3.18 Write, or attempt to write, structures corresponding to the following names, and state why each name is incorrect.

a) *cis*-2-methyl-2-hexene   b)   3,3-dimethyl-2-pentene

c)   1-chloro-2-propene   d)   2-methyl-2-butyne

e) *trans*-1,2-dimethylcyclohexene   f)   1-pentene-2-yne

3.19 Give names for the following structures:

a)

b) $HC{\equiv}C{-}C{\equiv}C{-}CH_3$

c)

d) $CH_3{-}\overset{\overset{\displaystyle CH_3}{|}}{C}{=}\overset{\overset{\displaystyle Cl}{|}}{C}{-}CH_3$

3.20 Three of the following molecular formulas represent stable compounds that can be isolated; two of them correspond to molecules that are known only as unisolated reaction intermediates. Write a structure for each formula, showing all bonds and unshared electrons, and state whether it is a stable compound or not.

a) C_2H_2 b) CH_2

c) C_3H_5 d) C_3H_6

e) C_2H_3Cl

3.21 Write the structures of four isomers of formula C_4H_8 which on hydrogenation absorb one mole of H_2.

3.22 A compound C_9H_{16} absorbs one mole of hydrogen. Treatment of the original compound with ozone followed by reductive hydrolysis (Zn, H_3O^+) gives one product:

$$CH_3{-}\overset{\overset{\displaystyle O}{\|}}{C}{-}(CH_2)_5{-}\overset{\overset{\displaystyle O}{\|}}{C}{-}CH_3$$

Give the name and structure of the C_9H_{16} compound.

3.23 Write structures of the products that would be obtained in the following reactions:

a) propylene + Cl_2 $\longrightarrow$

b) 1-heptene + HCl $\longrightarrow$

c) 2-methyl-1-pentene + Br_2 and water $\longrightarrow$

d) cyclopentene + perbenzoic acid ($C_6H_5-\overset{\overset{O}{\|}}{C}-OOH$) $\longrightarrow$

e) 3-heptyne + H_2 (one mole); poisoned catalyst $\longrightarrow$

f) 1-butyne plus excess Cl_2 $\longrightarrow$

g) 3,3-dimethyl-1-butene plus ozone, then Zn, H_3O^+ $\longrightarrow$

h) cyclohexene plus water; H_2SO_4 $\longrightarrow$

3.24 *Cis*-2-butene and *trans*-2-butene give different epoxides on reaction with $C_6H_5CO_3H$. Write the structures of these oxides.

3.25 If the bromination of 2-methylpropene is carried out in the presence of lithium chloride, in addition to the 1,2-dibromide a significant amount of a compound with the formula C_4H_8BrCl is formed. Suggest a structure for this product and write a mechanism to show how it is formed.

3.26 The reaction of 4-methyl-1,3-pentadiene with HCl gives a mixture of two main products, both having the formula $C_6H_{11}Cl$, plus traces of other compounds. Write the structures of the two main products.

3.27 Show how the following conversions could be carried out; specify reagents and, where important, conditions.

a) $CH_3CH_2CHBrCH_2Br$ from 1-butene

b) $CH_3(CH_2)_3CH(CH_3)_2$ from 2-methyl-2-hexene

c) $CH_3\overset{\overset{OH}{|}}{C}HCH_2CH_3$ from 1-butene

d) *n*-heptane from 3-heptyne

e)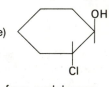

from cyclohexene

3.28 Show reactions by which the following compounds could be prepared from alkenes.

a) $CH_3\overset{\overset{OH}{|}}{C}HCH_3$

b) $CH_2{=}CHCH_2Cl$

c) $CH_3\overset{\overset{OH}{|}}{C}H-\overset{\overset{Br}{|}}{C}HCH_3$

d)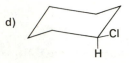

3.29 The compounds below are obtained by Diels-Alder reactions. Show the diene and dienophile for each compound. (Hint: in *C*, the dienophile is an alkyne.)

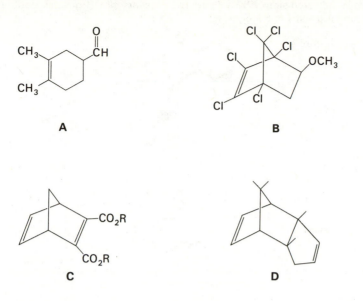

A   B

C   D

3.30 Indicate a simple "test tube" reaction that could be used to distinguish between the compounds in the following pairs; explain the basis for the test and state what observation would be made.

    a)   cyclohexane and cyclohexene

    b)   1-hexyne and 2-hexyne

3.31 Assuming that the rate of electrophilic addition depends on the first step (*i.e.*, formation of a carbocation intermediate), predict the relative rates of addition of HCl to the following compounds. (Hint: consider the relative stabilities of the carbocation formed in b and c *versus* that in a.)

    a)   1-butene

    b)   1-butyne

    c)   1,3-butadiene

3.32 Bicyclic alkenes with structures like *A* can be prepared when the number of carbons in the $(CH_2)_n$-bridge is greater than three, but the compounds are not stable when $n = 1$ or 2. Suggest a reason for this, based on the geometrical requirements for the formation of a π bond.

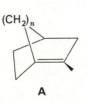

$(CH_2)_n$

A

3.33 The *isoprene rule* is a generalization that the carbon skeleton of naturally occurring terpenes (formed by cyclization of geraniol, farnesol, and so forth) can be dissected into isoprene units linked "head to tail." This rule has been widely used to decide between the likelihood of two alternative structures, both of which satisfy chemical data. Dissect structures

A to *E*, as shown in the example, marking off the isoprene units with heavy lines and the positions of junctions and ring closures with wavy lines. Remember that in terpene structures a CH_3 group is indicated simply by a dash. The presence of OH groups, double bonds, and similar functional groups is disregarded.

Example:

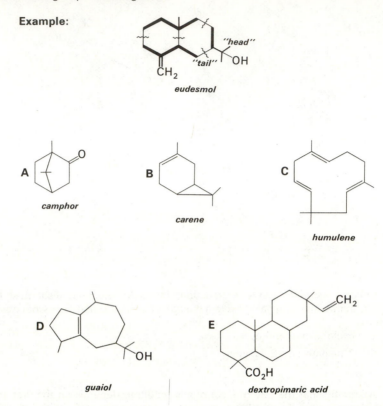

eudesmol

A   camphor

B   carene

C   humulene

D   guaiol

E   dextropimaric acid

3.34 Write the structure of the repeating unit in the polymer prepared from a) $CH_2\!=\!CHCN$ and b) $(CH_3)_2C\!=\!CH_2$.

3.35 Myrcene and ocimene are isomeric monoterpenes, with the molecular formula $C_{10}H_{16}$, derived by loss of a proton from geranyl carbocation (the carbocation formed from loss of ^-OPP in Figure 3.9). On azonolysis, myrcene gives 2 moles of $CH_2\!=\!O$ per mole (plus other products); ozonolysis of ocimene gives 1 mole of $CH_2\!=\!O$ per mole. Write the contributing resonance structures for geranyl carbocation and deduce the structures of the two monoterpenes.

3.36 Predict what products would be formed when each of the following dienes is treated with two moles of hydrogen chloride.

a)

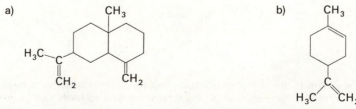

selinene
(found in celery oil)

b)

limonene

4
AROMATIC COMPOUNDS

The term **aromatic** was first used to designate a group of compounds derived from plant oils and gums, whose pleasant odors gave rise to the name. These compounds contained a variety of different groups, such as $-CH=O$, $-CO_2H$, and $-CH=CHCO_2H$, attached to a C_6H_5 unit which remained unchanged when the groups were chemically modified or interconverted. This same unit was found in the abundant hydroxy compound C_6H_5OH, called **phenol,** and in the parent hydrocarbon C_6H_6, called **benzene,** both of which were obtained by distillation of coal tar. Thus, "benzene derivative" or "benzenoid" and "aromatic" came to have much the same meaning, and to refer to the unusual chemical stability of these compounds. Organic compounds that are not aromatic are referred to as **aliphatic.**

4.1 Structure of Benzene

The nature of benzene and the C_6H_5 unit presented a major challenge to organic chemists for many decades. Early suggestions included prism structures with three- and four-membered rings, and a structure with six partial bonds extending to the center of the ring.

early suggestions for the structure of benzene

The idea of a ring containing six carbon atoms connected by alternating single and double bonds was put forth in 1865 by the German chemist August Kekulé. It was proposed that the double bonds oscillated between two different but equivalent structures:

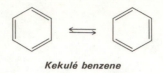

Kekulé benzene

However, benzene does not undergo the addition reactions typical of alkenes, and, unlike alkenes, it is quite resistant to oxidation. To account for these properties, despite the high degree of unsaturation, some modification of Kekulé's proposal was needed.

A more satisfactory representation of the benzene structure became available with the concept of resonance (Section 3.8). In resonance terminology, benzene is a **hybrid** of the two Kekulé structures—a species *between* these extreme representations as opposed to a *mixture* of the two. In this picture there are no double bonds or single bonds; the ring is *symmetrical,* and all carbon-carbon bonds are the same length, intermediate between single and double. Direct physical measurements (by electron diffraction) confirm that benzene is, indeed, a symmetrical molecule. As expected for a hybrid of the two Kekulé forms, the C—C bond lengths are all intermediate (1.40 Å) between normal double (1.33 Å) and single (1.54 Å) bonds. The resonance hybrid can be symbolized by the original Kekulé structures with the double-headed arrow, or as a hexagon with an inscribed circle. These representations will be used interchangeably throughout the text. A single Kekulé structure is often written for clarity in showing a reaction path, but it should always be understood to represent the resonance hybrid.

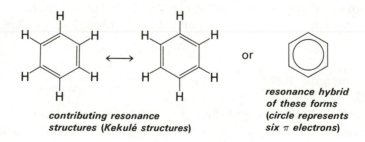

contributing resonance structures (Kekulé structures)

or

resonance hybrid of these forms (circle represents six π electrons)

All six carbon atoms in benzene are sp^2 hybridized and lie in the same plane. The remaining p orbitals of these carbons are perpendicular to this ring plane and overlap to form a circular π molecular orbital above and below the plane of the ring. This delocalized "π cloud" is equally distributed over the six atoms of the ring.

The structure of benzene can now be summarized in the following statements:

1) Benzene has the formula C_6H_6 with all 12 atoms in the same plane and the carbons forming a regular hexagon.
2) All carbons are sp^2 hybridized, and the remaining p orbitals overlap to form a delocalized, circular π molecular orbital above and below the plane of the ring.

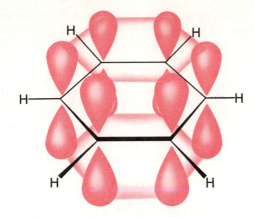

Orbital representation of benzene.

4.2 Aromatic Character

Theoretical calculations indicate that **aromatic properties** (unique stability) are associated with all sp^2 hybridized, planar rings which have $4n + 2\pi$ electrons. In this rule, n is an integer $(0, 1, 2, 3 \ldots)$; thus, for benzene $n = 1$. A triumph for this theory came when it was discovered that cyclopentadiene easily loses a proton to form an especially stable anion, which gains its stability through cyclic delocalization of the six π electrons. Likewise, cycloheptatriene gains aromatic character by loss of a hydride ion, H^-, to form a carbocation in which the vacant p orbital now allows complete cyclic delocalization of the six π electrons.

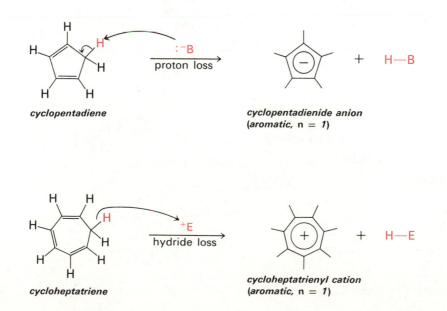

Problem 4.1 Predict which of the following structures would show cyclic delocalization of π electrons and aromatic character:

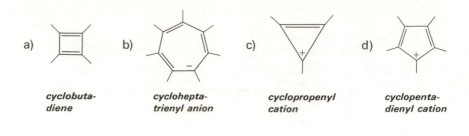

a) *cyclobuta-diene* b) *cyclohepta-trienyl anion* c) *cyclopropenyl cation* d) *cyclopenta-dienyl cation*

4.3 Resonance Energy of Benzene

As we saw in Chapter 3, the addition of H_2 to a double bond is an exothermic reaction. Hydrogenation of cyclohexene gives off 28.6 kcal per mole; thus, the product, cyclohexane, is 28.6 kcal more stable than the starting materials. Complete hydrogenation of 1,3-cyclohexadiene liberates 55 kcal/mole. This value is a bit less than twice the heat gained from two moles of cyclohexene, since a conjugated diene is slightly (2.2 kcal per mole) more stable than two isolated double bonds. From these numbers we would expect hydrogenation of cyclohexatriene to give off slightly less than 86 kcal/mole—three times the value for cyclohexene. The fact that hydrogenation of benzene to cyclohexane produces only 50 kcal/mole clearly shows that benzene is something quite different from the hypothetical cyclohexatriene. This difference of 36 kcal/mole (86 − 50) between the calculated energy of the contributing resonance structure and that of the actual resonance hybrid is called the **"resonance energy"** of benzene (Figure 4.1).

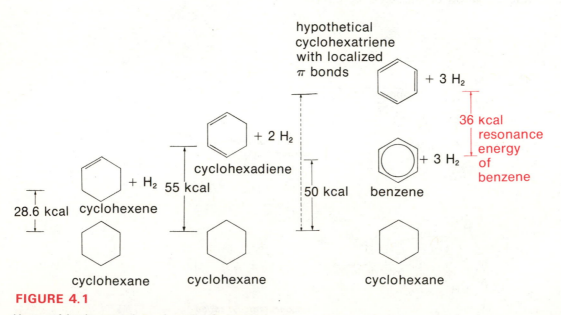

FIGURE 4.1

Heats of hydrogenation of cyclic C_6 species.

4.4 Names of Benzene Derivatives

The hydrocarbons benzene and toluene (methylbenzene) are the parent names for most of the compounds in this chapter. Since all carbons in benzene are equivalent, there can be *only one* toluene or *one* chlorobenzene. With two substituents on the benzene ring there are three possible arrangements, corresponding to 1,2-, 1,3-, and 1,4-disubstituted rings. These three arrangements are called *ortho* (*o*), *meta* (*m*), and *para* (*p*), respectively.

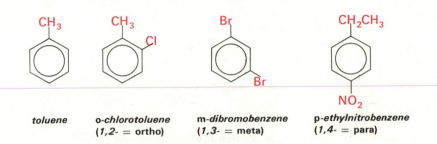

| toluene | o-chlorotoluene (1,2- = ortho) | m-dibromobenzene (1,3- = meta) | p-ethylnitrobenzene (1,4- = para) |

When three or more substituents are placed on the benzene ring, numbers must be used to designate their positions.

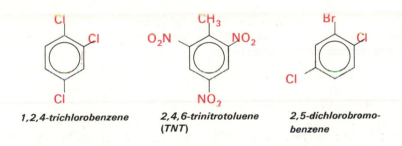

| 1,2,4-trichlorobenzene | 2,4,6-trinitrotoluene (TNT) | 2,5-dichlorobromo- benzene |

The benzene ring can also be named as a substituent on the rest of the molecule. When this is done, the C_6H_5— unit is called **phenyl.** Likewise, the group formed by removing a hydrogen from the CH_3 group of toluene, $C_6H_5CH_2$—, is called **benzyl.**

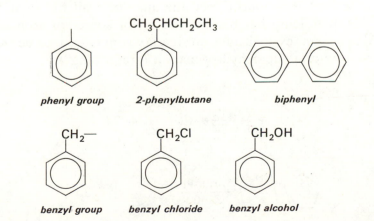

| phenyl group | 2-phenylbutane | biphenyl |
| benzyl group | benzyl chloride | benzyl alcohol |

As we shall see, the chemical properties of a substituent such as chlorine when attached to an aromatic ring are quite different from those of the same substituent on a corresponding alkyl compound (*e.g.*, chlorobenzene versus chlorohexane). For this reason, **Ar—** for **aryl** is used as a general symbol for an aromatic group to distinguish it from an alkyl group, **R—**. Thus, Ar—Br would be a general description for an aromatic bromide, while R—Br would indicate an aliphatic bromide.

Problem 4.2   Write correct names for the following structures.

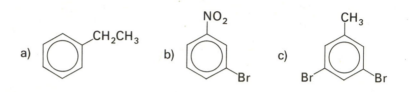

Problem 4.3   Provide structures for the following names.

a)   *m*-bromochlorobenzene
b)   *p*-nitrotoluene
c)   3,5-dibromonitrobenzene
d)   *o*-methylbenzyl chloride

Problem 4.4   a) Write structures and names for all benzene derivatives with the formula $C_6H_4Cl(CH_3)$. b) Write structures and names for all of the isomeric trichlorobenzenes.

4.5   Sources of Aromatic Compounds

Heating coal in the absence of air produces "coal gas" (largely H_2, H_2S, and small alkanes), coke (essentially elemental carbon, used in steel manufacture) and **coal tar.** Coal tar is a complex mixture of aromatic hydrocarbons, and until about 1940 it was the only important source of these compounds.

In the past 30 years the demand for aromatic compounds by the chemical industry has far outstripped the amount available from coal tar. Consequently, petroleum has become the major source of aromatic compounds. Alkanes and cycloalkanes are readily converted into benzene and alkylbenzenes by catalytic dehydrogenation ("reforming").

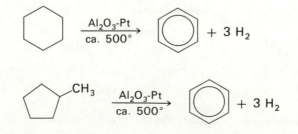

It is not necessary to have a cyclic precursor, since cyclization of an aliphatic chain occurs in the same process.

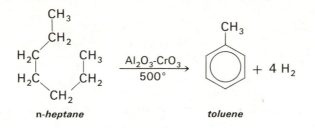

n-heptane   toluene

Benzene and toluene are the raw materials for the production of most benzene derivatives. Throughout this chapter we will look carefully at the methods for conversion of these starting materials into important substituted aromatic hydrocarbons.

4.6 Polycyclic Aromatic Hydrocarbons

The higher boiling fractions of coal tar distillate contain a number of polycyclic aromatic hydrocarbons. The most abundant single constituent of coal tar (11%) is **naphthalene**, $C_{10}H_8$, which consists of two fused benzene rings. Formerly used as a moth repellent and insecticide, its major use today is as a starting material in the synthesis of dyes, synthetic resins, and lubricants. The numbering of the naphthalene ring is shown on its structure, and you should note that there are only two different positions for substitution since positions 1, 4, 5, and 8 form one equivalent set and positions 2, 3, 6, and 7 form another. A third ring can be fused onto naphthalene in two ways to produce two important tricyclic coal tar constituents—**anthracene** and **phenanthrene.**

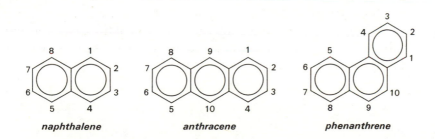

naphthalene   anthracene   phenanthrene

Problem 4.5   These polycyclic systems fit the $4n + 2$ rule for aromaticity. What is n for naphthalene and for anthracene?

Large polycyclic aromatic hydrocarbons contain networks of fused rings with structures that resemble a mosaic tile pattern. If extended indefinitely, these structures would become a flat sheet of sp^2 hybridized carbon atoms, and this is, in fact, the structure of **graphite.** Two typical polycyclic networks are **dibenzanthracene** and **benzpyrene.**

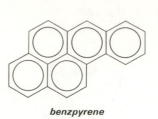

dibenzanthracene   *benzpyrene*

These hydrocarbons and related compounds are present in soot, smoke, and automotive emissions, and present a major health problem since they are potent **carcinogens**—cancer-producing substances. The atmospheric concentration of benzpyrene in an urban center, particularly where coal is burned, is 10-fold greater than in a rural area, and the incidence of lung cancer parallels these levels.

4.7   Electrophilic Aromatic Substitution

The most dramatic difference between aromatic hydrocarbons and alkenes is the tendency for the former to undergo **substitution reactions** with the same reagents which undergo addition reactions with simple alkenes. Although the aromatic ring is very stable, it is electron-rich and susceptible to attack by electron-seeking reagents, called **electrophiles.** A highly reactive electrophile is needed, however, since the resonance energy of the aromatic ring is lost when the cyclic π bonding is disrupted. Three typical electrophilic reagents are discussed in this section, but all their reactions with the aromatic ring occur by one general mechanism.

The general reaction of electrophilic aromatic substitution can be represented by using an electrophile E$^+$. The steps are:

1) Formation of the electrophile E$^+$.
2) Attack by E$^+$ on the π electron cloud of the aromatic ring to form a σ bond to a ring carbon and to generate a positive charge in the ring. This charge is stabilized by delocalization as in the case of the allylic cation (Section 3.11). Resonance stabilization of the charge partially compensates for the loss of aromaticity.

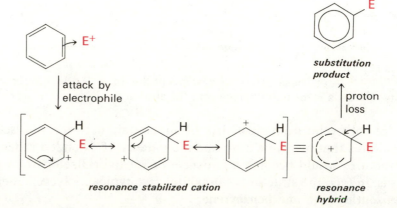

attack by electrophile

resonance stabilized cation

substitution product

proton loss

resonance hybrid

3) Formation of the final product by loss of the ring proton from the position of attack. The electrons from the C—H bonds are used in reforming the $4n + 2\,\pi$ system, and aromatic character is returned to the ring. The overall result is **substitution of the group E for hydrogen.**

Halogenation. Chlorine and bromine, which react readily with alkenes, react extremely slowly with benzene. In order to effect substitution, a Lewis acid (such as $AlCl_3$, $FeCl_3$, or $FeBr_3$) is added as a catalyst. The function of the catalyst is to make a more reactive electrophile. The ferric halide combines with halogen to give a complex salt containing the highly reactive X^+ ion. This powerful electrophile now attacks the π system of the ring, a proton is lost as HX, and C_6H_5X is produced.

Overall Reaction:

$$ArH \xrightarrow[\text{FeBr}_3]{\text{Br}_2} ArBr + HBr$$

Mechanism: $Br-Br \xrightarrow{FeBr_3} Br^+ + FeBr_4^-$

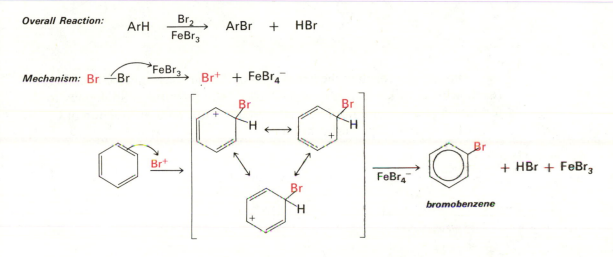

bromobenzene

Nitration. Nitration is the substitution of a nitro group, $-NO_2$, for a ring hydrogen. The attacking electrophile is the nitronium ion, $^+NO_2$, which is generated by the protonation of nitric acid by sulfuric acid.

Overall reaction:

$$ArH \xrightarrow[\text{H}_2\text{SO}_4]{\text{HNO}_3} ArNO_2 + H_2O$$

Mechanism: $H_2SO_4 + HO-N\overset{+}{\underset{O^-}{\overset{O}{\|}}} \rightleftharpoons HSO_4^- + H-\overset{+}{\underset{H}{O}}-\overset{+}{N}\overset{O}{\underset{O^-}{\|}} \rightleftharpoons H_2O +\,^+NO_2$

nitrobenzene

Alkylation.   Alkylation is the substitution of an alkyl group for a ring proton. We learned in Chapter One that carbocations are very strong electrophiles, and these are used in aromatic alkylation. One general method for generating a carbocation is the reaction of an alkene with a strong acid (Section 3.5). The most stable carbocation $(3° > 2° > 1°)$ is formed, and this cation can now attack the aromatic ring.

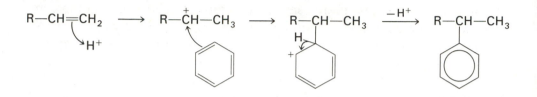

Another approach to the generation of the carbocation electrophile is the reaction of an alkyl chloride with $AlCl_3$. This reaction is analogous to the formation of electrophilic X^+ in halogenation, but here abstraction of Cl^- by $AlCl_3$ leaves behind a carbocation.

$$R{-}Cl \; + \; AlCl_3 \; \longrightarrow \; R^+AlCl_4^-$$
alkyl chloride

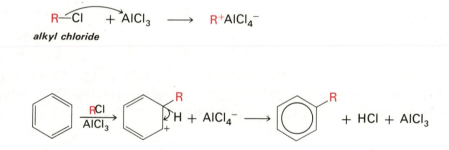

A complication in these reactions is the tendency for carbocations to rearrange to more stable carbocations. For example, if *n*-propyl chloride, $CH_3CH_2CH_2Cl$, is used as the carbocation source, the product from reaction with benzene is not the expected *n*-propylbenzene, which would require a 1° carbocation intermediate, but rather isopropylbenzene formed from the more stable 2° carbocation. As the chloride ion is removed by $AlCl_3$, a hydrogen atom from C-2, with the bonding electron pair, migrates to the electron-deficient primary carbon. This migration produces the secondary carbocation, which then attacks the benzene ring.

Overall reaction: $ArH + CH_3CH_2CH_2Cl \longrightarrow ArCH(CH_3)_2$

Mechanism:

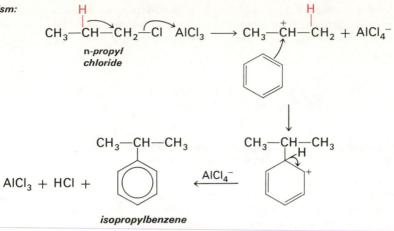

n-*propyl*
chloride

CH₃—CH—CH₃

isopropylbenzene

AlCl₃ + HCl +

Problem 4.6 Write structures for the alkylation products that would be formed in the AlCl₃-catalyzed reaction of benzene with a) chloroethane, b) 1-chloro-2-methylpropane, c) *n*-propyl bromide.

Sulfonation. This is another example of electrophilic aromatic substitution. Fuming sulfuric acid, which is a solution of SO_3 in H_2SO_4, provides the electrophile for attack of the aromatic ring. Although it is not a cation, sulfur trioxide is a powerful electrophile and reacts with the aromatic π system to form a neutral but dipolar intermediate. Transfer of a proton from the ring to the $-SO_3^-$ group re-aromatizes the ring and forms a benzenesulfonic acid.

Overall reaction:

$$ArH \xrightarrow[\text{H}_2\text{SO}_4]{\text{SO}_3} ArSO_3H$$

Mechanism:

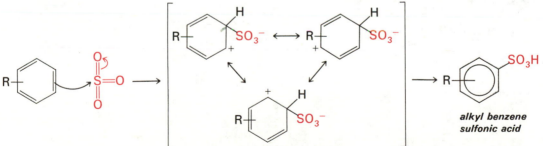

alkyl benzene
sulfonic acid

SPECIAL TOPIC: Alkylbenzenesulfonate Detergents

A surface active agent or **surfactant** reduces the repulsive forces that exist between water and a greasy surface or a second insoluble liquid. Surfactant compounds have structures containing a **hydrophobic** (water-repelling) group, such as a long alkyl chain, and a **hydrophilic** (water-attracting) group. The hydrophilic group may be ionic, *e.g.*, —$SO_3^-Na^+$, or non-ionic, such as a cluster of —OH groups. Thus, long-chain alkyl benzenesulfonates, prepared by the neutralization of alkyl benzenesulfonic acids, serve as surfactants, the key ingredient in a **detergent.**

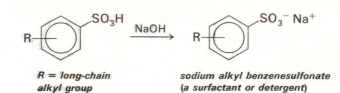

R = long-chain
alkyl group

sodium alkyl benzenesulfonate
(a surfactant or detergent)

Surfactants exert several distinct effects: wetting and penetration of water in a fibrous surface, dispersing or suspending insoluble particles, and causing (*or preventing*) foam. The most important property, which is to some extent a combination of the others, is *detersive* or cleansing action. **Detergents** actually are mixtures containing surfactants plus other substances called "builders" that enhance the cleansing action, but surfactants are often referred to simply as detergents.

The mechanism by which a detergent acts on a soft surface, such as cloth, is roughly as follows. Surfactant molecules become oriented at the interface between the detergent solution and the cloth, with the hydrophilic group directed toward the water and the hydrophobic end contacting the fiber or soil particle. Oily materials on the fiber "roll up" into aggregates with the detergent molecules, and water then dislodges these aggregates as well as solid particles. A specific property of surfactant compounds in water solution is the formation of **micelles,** which are spherical clusters of molecules with the hydrophobic chains in the center and the hydrophilic groups exposed (Figure 4.2). As soil

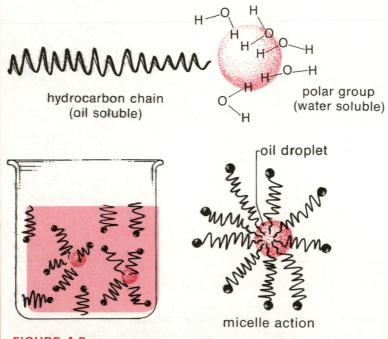

hydrocarbon chain
(oil soluble)

polar group
(water soluble)

oil droplet

micelle action

FIGURE 4.2

Behavior of detergent molecules.

and grease particles become dislodged, they are trapped in the micelles and kept in suspension. A complicating factor in the detergent action is the presence of multiply-charged metal ions, such as Mg^{+2}, Ca^{+2}, and Al^{+3}. These ions interfere with the removal of the soil and also promote redeposition of suspended materials.

Until the 1940's, the only surfactant compound available was soap, which is the sodium salt of an organic acid, $e.g.$, $CH_3(CH_2)_{16}CO_2^-Na^+$ (Section 10.17). Soap is a good detergent in "soft" water, but in "hard" water, with a high concentration of calcium ions, the calcium salt of the acid precipitates as a scummy solid. Household laundering with soap requires soaking, scrubbing, and repeated rinsing with *hot* water to remove the calcium soap deposits.

$$2\ RCO_2^-Na^+ + Ca^{+2} \longrightarrow (RCO_2^-)_2Ca^{+2} + 2\ Na^+$$

Synthetic detergents became available at about the same time as automatic washing machines, and they were essential to the development of automatic household laundry equipment. Alkylbenzenesulfonates do not form insoluble calcium salts, and significantly less hot water is needed. To obtain good laundry action, the troublesome calcium ions are "sequestered" or tied up by adding sodium polyphosphate, which forms a soluble complex ion.

The consumption of alkylbenzenesulfonates increased very rapidly during the 1950's. The compounds were cheaply produced by alkylation of benzene with the highly branched alkenes obtained by acid-catalyzed self-condensation

alkylbenzene sulfonates

of propylene. In the early 1960's, however, rivers and municipal water supplies began to show surfactant properties. It was found that bacterial degradation of synthetic detergents in sewage plants was slow and incomplete because of the branched alkyl groups. Normal oxidative breakdown of hydrocarbon chains occurs by oxidation of CH_2CH_2 units, and the methyl branches interfere with this process.

In response to the problem of surfactants in natural waters, chemical manufacturers developed more readily biodegradable detergents with *linear* alkyl chains. The switch to linear alkylbenzenesulfonates greatly reduced the level of surfactants in sewage effluents, but other problems remain.

linear alkyl sulfonates

Linear alkyl detergents still require substantial amounts of phosphate builders to tie up calcium ions, and this phosphate is eventually discharged in the environment. In localities where effluents enter lakes, this burden of phosphate, combined with surface run-off from fertilizers, stimulates excess algal growth and causes deterioration of water quality.

Ultimately, sulfonate detergents may be replaced entirely by non-ionic surfactants which do not require phosphates. This possibility is enhanced by the steadily increasing use of synthetic fibers which release soil more easily than does cotton, and the increasing use of disposable, non-woven materials. The problems that will arise with these developments cannot yet be clearly foreseen.

4.8 Effects of Ring Substituents in Electrophilic Substitution

The reactions discussed so far have been illustrated with benzene, in which all of the ring positions are equivalent. We should now ask the question: What is the effect on the substitution if the ring is already substituted?

A group G when attached to a benzene ring will affect (1) the **rate** at which C_6H_5—G undergoes substitution relative to benzene, and (2) the **position** at which further substitution occurs. If G can **donate electrons** to the ring, further substitution will occur *more rapidly* and predominantly at the *ortho* and *para* positions. If G is **electron withdrawing**, further substitution will be *slower* and occur largely at the *meta* positions.

G electron donating; activates the ring and directs ortho *and* para

G electron withdrawing; deactivates the ring and directs meta

A good example of an **electron donating** group is the methoxy group, —OCH_3. To understand the *ortho-para* directing effect of the methoxy group, let us compare the intermediate carbocations formed by attack of an electrophile such as Br^+ at the positions *para* to the OCH_3 group and *meta* to OCH_3. In the intermediate for *para* substitution, one of the resonance structures of the cation has a positive charge on the carbon adjacent to the methoxy group. The OCH_3 group can stabilize this charge by donating an unshared electron pair, and the charge is thus shared by the oxygen. The same is true for the intermediate for *ortho* substitution. In the intermediate for attack at the *meta* position, however, we cannot write a resonance structure with positive charge on oxygen.

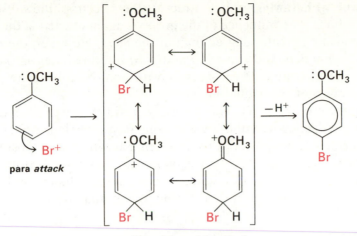

positive charge delocalized by oxygen

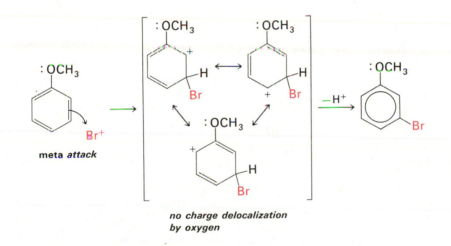

no charge delocalization by oxygen

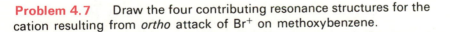

Problem 4.7   Draw the four contributing resonance structures for the cation resulting from *ortho* attack of Br⁺ on methoxybenzene.

Since the *ortho* and *para* intermediates have additional stabilization from the resonance contribution of the OCH₃ group, there is a lower activation energy for the formation of these intermediates than for the intermediate from *meta* attack, and substitution is directed to the *ortho-* or *para-*position (Figure 4.3). Moreover, since the *ortho-para* intermediates are stabilized more extensively than is the intermediate from benzene itself, *ortho-para* substitution in methoxybenzene has a lower activation barrier than substitution in benzene and occurs at a faster rate. The methoxy group therefore *activates* the ring.

Electron withdrawing groups direct further electrophilic substitution to the *meta* carbons. The influence of the group is again exerted at the *ortho* and *para* positions to a greater degree than at the *meta*. However, the effect is in the *opposite direction,* and the *ortho* and *para* intermediates are *destabilized* relative to the intermediate for *meta* substitution. Nitrobenzene provides a good illustration of this situation.

The nitro group, —NO_2, has a formal positive charge on nitrogen and has a strong tendency to withdraw electrons from the ring. Attack of Br^+ at either the *ortho* or *para* positions of nitrobenzene will place a positive charge on a carbon atom that already has a positively charged atom attached. Repulsion of like charges destabilizes these intermediates, and raises the activation energy required for their formation compared to the intermediate for *meta* attack.

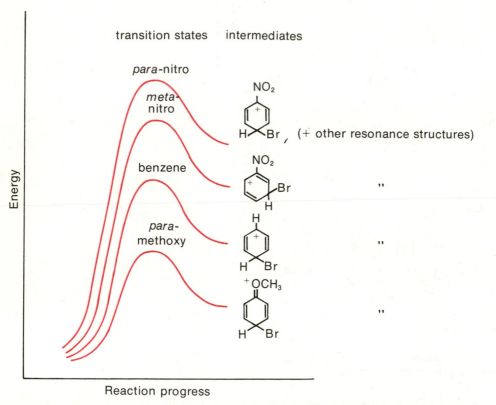

FIGURE 4.3
Activation barriers for electrophilic substitution.

Problem 4.8   Only one resonance structure is shown for each of the intermediates in Figure 4.3. How many other resonance structures can be written for each?

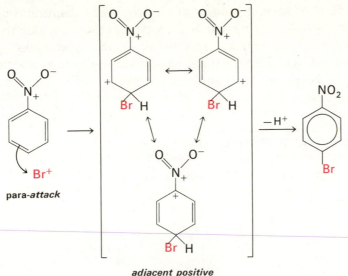

para-*attack*

adjacent positive
charges destabilize

The electron withdrawing influence of the nitro group also raises the activation barrier of the *meta* intermediate relative to that for benzene; nitrobenzene is therefore *deactivated,* and is much less reactive than benzene to electrophilic substitution.

To complete the picture, a bromo or chloro substituent must be considered. Like the —OCH$_3$ group, these substituents have unshared electron pairs capable of resonance stabilization of an adjacent positive charge; therefore, substitution of bromobenzene gives, as expected, the *ortho* and *para* products.

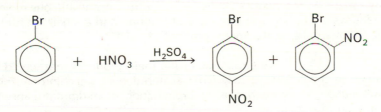

However, this reaction is slower than the nitration of benzene, and more vigorous conditions are required. The electronegative bromine atom deactivates the entire ring by electron withdrawal. When attack by an electrophile does occur, however, the intermediate for *ortho* or *para* substitution is stabilized by a resonance interaction with an unshared electron pair.

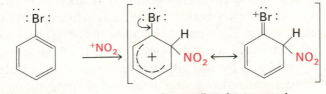

ring deactivated
by electron
withdrawal

intermediate from o-attack
stabilized by resonance
delocalization of charge

The effects of several groups on electrophilic substitution are summarized in Table 4.1. Activating and o-, p-directing groups include those with an unshared electron pair on the atom attached to the ring and alkyl groups such as —CH_3 or —C_2H_5. Deactivating, *meta*-directing groups are those with a C—O, C—N, N—O or S—O multiple bond.

TABLE 4.1   EFFECT OF SUBSTITUENTS ON ELECTROPHILIC SUBSTITUTION

Activating and o,p-directing: —ÖH, —N̈H₂, —ÖCH₃, —N̈HCOCH₃, —CH₃

Deactivating and o,p-directing: —Br̈:, —C̈l:

Deactivating and *meta*-directing: —NO₂, —C̈R, —C≡N, —SO₃H, —CCl₃, —CF₃

$$-N\overset{+}{\underset{}{}}\overset{O}{\underset{O^-}{}} \longleftrightarrow -N\overset{+}{\underset{}{}}\overset{O^-}{\underset{O}{}} \quad -C≡N: \longleftrightarrow -\overset{+}{C}=\overset{-}{N}: \quad -\overset{Cl}{\underset{Cl}{\overset{|}{\underset{|}{C}}}}+\!\!\!\rightarrow Cl$$

$$-\overset{O}{\overset{||}{C}}R \longleftrightarrow -\overset{O^-}{\underset{+}{C}}R \quad -\overset{O}{\underset{O}{\overset{||}{S}}}-OH \longleftrightarrow -\overset{O^-}{\underset{O}{\overset{|}{\underset{||}{S^+}}}}-OH$$

Problem 4.9   If a mixture of one mole of nitrobenzene and one mole of benzene were treated with one mole of chlorine and a small amount of AlCl₃, what would be the major product?

Problem 4.10   Write structures of the main products that would be obtained in the following reactions: a) nitration of p-dichlorobenzene, b) sulfonation of nitrobenzene, c) bromination of m-dibromobenzene, d) nitration of p-toluene.

4.9   Effect of Aromatic Ring on Substituents

The chemical behavior of any group attached to an aromatic ring is strongly influenced by the ring. Halogenation of alkyl benzenes under free radical conditions occurs very readily at the carbon bonded to the ring. These reactions have a low activation energy because the abstraction of a hydrogen atom leads to formation of a stable **benzyl radical.** In the same way that an allyl radical is stabilized by the adjacent π bond, the odd electron of the benzyl radical is delocalized over the *ortho* and *para* positions by resonance interaction with the π system of the aromatic ring. For example, chlorination of an alkyl side-chain in the presence of light (*hv*) takes place mainly at the carbon attached to the ring.

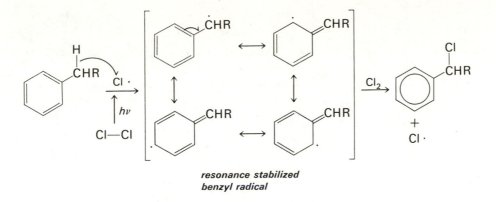

resonance stabilized
benzyl radical

Problem 4.11 Styrene, $C_6H_5CH=CH_2$, polymerizes on standing unless inhibitors are added to prevent formation of radical chains. Show the radical that would be formed from reaction of styrene with an initiator $Z\cdot$, and the structure of the resulting polystyrene.

A **benzyl cation** is similarly stabilized by resonance donation of the aromatic π electrons to the vacant p orbital, thus delocalizing the positive charge over the ring. Benzyl chloride undergoes reactions involving heterolytic cleavage (*i.e.*, ionization) of the C—Cl bond very easily because of the great stability of the resulting carbocation. For example, catalytic alkylation of benzene by benzyl chloride occurs very readily to give diphenylmethane.

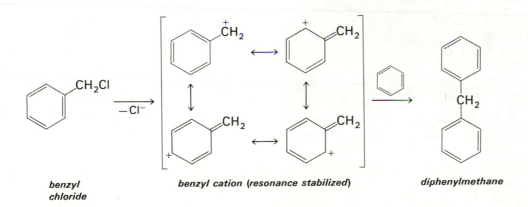

benzyl
chloride

benzyl cation (resonance stabilized)

diphenylmethane

4.10 DDT and the Insecticide Problem

Of the thousands of benzene compounds that have been prepared, none has received as much notoriety as **DDT** (*d*ichloro*d*iphenyl*t*richloroethane). DDT is prepared from two very cheap chemicals—chlorobenzene and trichloroacetaldehyde. The C=O group of the aldehyde is first protonated to serve as an electrophile for the aryl chloride.

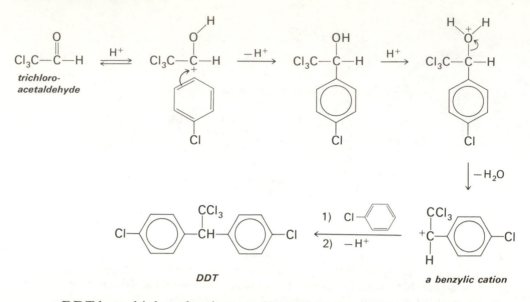

DDT has a high and rather specific toxicity for various insects, including ones that transmit disease, such as the house fly, mosquito, and tsetse fly. Although DDT was first prepared in 1874, its ability to act as an insecticide was not recognized until 1930, and its wide scale use did not begin until 1946. Production and distribution of DDT since then are estimated to have been about five billion pounds.

Even in the late 1940's it was evident that some insects were developing

FIGURE 4.4
The boll weevil, which causes severe damage to cotton crops, has been effectively controlled by DDT. However, this agricultural pest is rapidly becoming resistant to DDT.

a resistance to DDT, since larger and larger amounts were required for adequate control. This was most dramatically revealed by studies which showed that in 1965, cotton boll weevils were 30,000 times more resistant to DDT than in 1960. It is now believed that these resistant insects have "learned" to enzymatically dehydrochlorinate (remove HCl from) DDT to form the alkene **DDE,** which has no insecticidal activity.

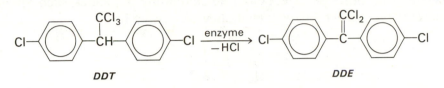

DDT is a very stable compound, and this has contributed to its notoriety, since its persistence has led to very wide distribution. In the 1950's, DDT could be detected in soils at a depth of many feet, and subsequently it was found in plankton and in the fatty tissues of fish, where it tends to concentrate. The compound is not readily metabolized and destroyed, and thus

FIGURE 4.5

Effects of DDT. The concentration of DDT in birds of prey, such as hawks, results in poor egg formation along with other abnormalities in the reproductive cycle. (From Jones, M. M., Netterville, J. T., Johnston, D. O., and Wood, J. L.: *Chemistry, Man and Society.* Philadelphia, W. B. Saunders Company, 1972, p. 255.)

accumulates at various points in the ecological food chain. The effect on birds can be devastating, since DDT interferes with calcium metabolism so as to produce eggs which are too thin to survive incubation (see Figure 4.5).

In response to these severe ecological problems, the use of DDT has been substantially reduced since the mid-1960's and is now approved in the United States for use with only a few crops. Similar problems are potentially associated with other persistent chlorine-containing pesticides, such as those prepared from hexachlorocyclopentadiene (Section 3.9).

SPECIAL TOPIC: Some Alternatives to DDT

Alternative types of insecticides are available that are rapidly broken down, but these compounds pose other problems. Thiophosphate esters such as parathion are effective, but they are toxic to animals also. Moreover, this type of insecticide is less selective, and eliminates beneficial insects as well as harmful ones.

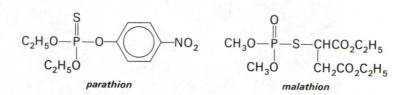

parathion

malathion

A sounder but more costly possibility for insect control is the use of compounds that are vital to the physiology of insects. Development of insect larvae depends on hormones which are complex aliphatic compounds. Synthetic analogs (closely related structures) with very high hormone activity cause disruption of the normal development, and application of these substances at the proper stage of the life cycle is lethal. A **juvenile hormone** was first isolated in 1956. This hormone, secreted by glands near the insect's brain, prevents, until the proper time, the metamorphosis of larvae to pupae and, eventually, to adults. To date, the juvenile hormones of about 20 insects are known; they represent only three compounds, all very similar in structure.

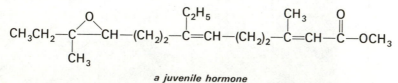

a juvenile hormone

It was reasoned that if a natural juvenile hormone or compounds with the same action were applied to the larvae at the time that they were supposed to proceed to the pupal stage—when their own secretion of juvenile hormone had ceased—the normal development would not take place and they should soon die. Indeed, it has been found that when larvae at the transition time are treated with juvenile hormone, or a synthetic mimic, they form grotesque organisms that are incapable of survival (Figure 4.6).

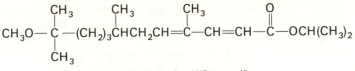

$$CH_3O-\underset{\underset{CH_3}{|}}{\overset{\overset{CH_3}{|}}{C}}-(CH_2)_3CHCH_2CH=\overset{\overset{CH_3}{|}}{C}-CH=CH-\overset{\overset{O}{\|}}{C}-OCH(CH_3)_2$$

synthetic analog ("Entocon")
of an insect hormone

For example, when juvenile hormone chemicals are applied to aquatic habitats, mosquitoes are prevented from completing their life-cycle. Thus, the presence of these chemicals during the larval-pupal or the pupal-adult molts results in abnormal mosquitoes that cannot develop normally and so eventually die.

FIGURE 4.6

Treated with a growth regulator, this yellow mealworm has an adult head and thorax, but an immature abdomen. (Courtesy of U. S. Department of Agriculture.)

Insect Pheromones. The term "pheromone" (from the Greek *pherein,* to carry, and *horman,* to stimulate) was coined in 1959 to describe compounds that animals secrete to affect the behavior of other animals of the same species. In the past decade it has become clear that the major means of insect communication is by chemical signals. Not surprisingly, it has been found that two of the things which insects seem compelled to communicate are the location of food and the availability of sexual partners. Early on, it was recognized that if the pheromones responsible for these messages could be identified and synthesized, they could be used to control harmful insects. Thus, the search is now on and represents one of the most exciting areas of organic chemistry today. In this brief survey we will look at a few examples which illustrate the potential of this method of insect control.

The first complete characterization of an insect pheromone involved extraction of the sex pheromone from the abdominal sections of 500,000 virgin female **silkworm moths** to obtain 0.02 gram of the attractant *trans-*10-*cis*-12-hexadecadiene-1-ol, known as "bombykol."

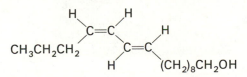

bombykol, female sex attractant of the silkworm moth

The attracting power of bombykol is amazing. From laboratory studies it can be calculated that 0.01 microgram is adequate to excite more than a billion male moths if the compound could be distributed with maximum efficiency. When male moths sense the pheromone, they simply fly upwind until they locate the female.

In the mid-1800's the **gypsy moth** was accidentally brought to Massachusetts. This moth is destroying forests in the northeastern U.S. and spreading to other areas. Even around the turn of present century it was recognized that the female gypsy moth could advertise her presence and attract the male moth. Only recently was it established that this was accomplished by the nineteen-carbon pheromone, 2-methyl-7-epoxyoctadecane—called "disparlure."

$$CH_3(CH_2)_9CH\overset{\displaystyle O}{\overbrace{\qquad}}CH(CH_2)_4CH(CH_3)_2$$

"disparlure," female sex attractant of the gypsy moth

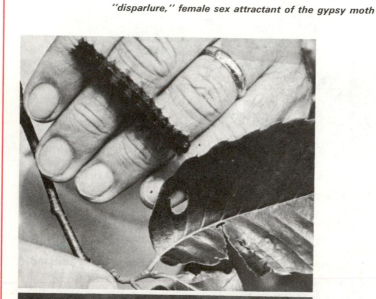

FIGURE 4.7

A, The gypsy moth caterpillar and a leaf that it has been eating. *B,* In late July and early August, a gypsy moth female may lay as many as 600 eggs. (Courtesy of U.S. Department of Agriculture.)

To date, the major commercial use of synthetic sex attractants is in baiting traps to monitor populations. For this purpose they are ideal, as they are highly specific and active at very low concentrations. Information from the traps can be used in timing the spraying of conventional insecticides. For example, in late summer female gypsy moths lay up to 600 eggs each (Figure 4.7). If the male population can be eliminated just before the time of mating, a tremendous reduction will be realized not only for that year but also the next.

The "confusion method" is another way to use sex attractants to control insect population. If the air is saturated with the pheromone at mating time, the males find it almost impossible to find females during this period. Like large scale trapping, this method suffers from its expense.

Some other known insect pheromones are shown below. An unknown, but presumably large, number remain to be identified.

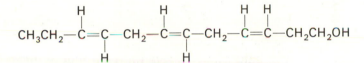

propylure, sex pheromone of the pink bollworm moth

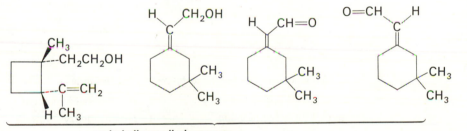

trail-marking pheromone of termites

male boll weevil pheromone
(no single component, only the mixture, is active!)

SUMMARY AND HIGHLIGHTS

1. **Aromatic** compounds contain a ring of all sp^2 carbon atoms and $4n+2$ π electrons in a continuous conjugated system. The ring does not have alternating single and double bonds but rather a **delocalized** π system.

2. **Benzene** and other aromatic hydrocarbons have large **resonance stabilization** energies. These compounds can be obtained from coal tar or catalytic reforming of alkanes.

3. A disubstituted benzene can be one of three isomers—*ortho* (1,2), *meta* (1,3), or *para* (1,4).

4. **Electrophilic aromatic substitution:**

$$Ar\!-\!H \xrightarrow{\ E^+\ } {}^+Ar\!\!\!\bigg\langle {\overset{\displaystyle E}{\underset{\displaystyle H}{}}} \xrightarrow{\ -H^+\ } Ar\!-\!E$$

can be accomplished with the electrophiles $^+NO_2$ (nitration), Br^+ or Cl^+ (halogenation), R^+ (alkylation), and SO_3 (sulfonation).

5. Electron releasing groups, such as —OR and —CH_3, make the benzene ring more reactive to further electrophilic substitution and direct that substitution preferentially to the *ortho* and *para* positions.

6. Electron withdrawing groups deactivate the ring, particularly the *ortho-para* positions, and direct *meta*. Cl and Br are unique in that they deactivate the ring but direct *ortho-para*.

7. **Alkylbenzenesulfonates, R—C_6H_4—SO_3^- Na^+,** are synthetic *detergents*.

8. The benzene ring stabilizes a positive charge or a radical on a carbon adjacent to the ring, as in $Ar\overset{+}{C}HR$.

9. Polyhalogenated aromatic compounds have been widely used as insecticides and in other commercial applications. Their stability makes these compounds persist for very long periods.

ADDITIONAL PROBLEMS

4.12 Write the structures of the following compounds:

a) *m*-chlorotoluene

b) *p*-toluenesulfonic acid

c) triphenylmethane

d) *p*-chlorobenzyl bromide

e) 1-bromonaphthalene

f) *o*-dinitrobenzene

g) 4,4'-dibromobiphenyl

h) 1,2,3,4-tetrahydronaphthalene

4.13 Give names for the following compounds:

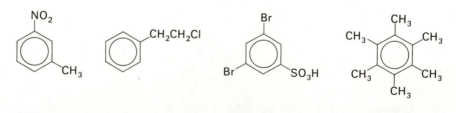

4.14 State how many isomers are possible for:

a) tetrachlorobenzene

b) dinitronaphthalene

4.15 a) Write structures and give names for four benzene derivatives with the formula C_7H_7Cl. b) There is another aromatic compound with the formula C_7H_7Cl which does not contain a six-membered ring. What is its structure? (Hint: it is a salt.)

4.16 Three isomers of diethylbenzene, **A, B,** and **C,** were obtained from the reaction of benzene with ethylene and sulfuric acid. On nitration, isomer **A** gave *three* mononitro derivatives, isomer **B** gave *one* nitro compound, and isomer **C** gave *two* nitro compounds. What are the structures of **A, B,** and **C?**

4.17 The reaction —CH$_2$—CH$_2$— $\xrightarrow{\text{cat}}$ —CH=CH— + H$_2$ is called **catalytic dehydrogenation.** It is the reverse of catalytic hydrogenation, and is normally an endothermic reaction, with an energy uptake of about 28 kcal/mole. From the data in Section 4.3 and Figure 4.1, calculate the energy change in the dehydrogenation of cyclohexadiene to give benzene. Is the reaction exothermic or endothermic?

4.18 Write the structures of the compounds that would be formed in the following reactions:

a) chlorobenzene + 2 HNO$_3$ $\xrightarrow{\text{H}_2\text{SO}_4}$ (dinitro product)

b) *m*-dimethylbenzene + SO$_3$ $\xrightarrow{\text{H}_2\text{SO}_4}$

c) chlorobenzene + CH$_3$CH$_2$Cl $\xrightarrow{\text{AlCl}_3}$

d) benzene propylene $\xrightarrow{\text{H}_2\text{SO}_4}$

4.19 To prepare a monoalkylbenzene by alkylation, it is necessary to use a fairly large excess of benzene (relative to alkyl chloride). Predict the result if equimolar amounts of benzene and halide were used, and explain why.

4.20 Write out in detail, with structural formulas, the steps in the conversion of chlorobenzene to *p*-chlorobenzenesulfonic acid, with SO$_3$ as the electrophile; show all contributing resonance forms in the reaction intermediates.

4.21 Bromination of 1,2,3,4-tetrahydronaphthalene under the two sets of conditions indicated gives different monobromo compounds (C$_{10}$H$_{11}$Br). Write the structures of the products in the two reactions and show the mechanism by which each is formed.

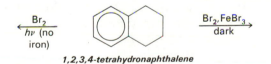

$\xleftarrow[\substack{h\nu\ (\text{no} \\ \text{iron})}]{\text{Br}_2}$ $\xrightarrow[\text{dark}]{\text{Br}_2, \text{FeBr}_3}$

1,2,3,4-tetrahydronaphthalene

4.22 Show how the following compounds could be prepared, starting from benzene or toluene and any aliphatic compounds. Specify the reagents and conditions. Assume that either an *ortho* or *para* isomer can be obtained from a mixture of the two.

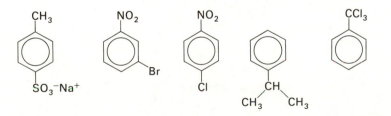

4.23 Two isomeric hydrocarbons, **A** and **B**, were isolated from the essential oil of a plant; both had the formula C$_9$H$_{10}$. Both compounds rapidly absorbed one mole of hydrogen to give the *same* product, C$_9$H$_{12}$. Compounds **A** and **B** were cleaved by ozone, and each gave two products. The lower boiling fragment was isolated in each case; **A** gave H$_2$C=O and **B** gave O=CHCH$_3$. Suggest the structures of **A** and **B**.

5
STEREOISOMERISM

Stereoisomers are compounds that have the same atom-to-atom bonding sequence, but differ in the arrangement of these atoms in space. One type of stereoisomerism has already been encountered in geometric isomers, in which atoms are held in different positions by a ring or double bond. In this chapter we will examine some other types of stereoisomers.

5.1 Symmetry and Enantiomers

Because of the tetrahedral structure of sp^3 carbon, a compound with at least two identical groups attached to a single sp^3 carbon must possess a **plane of symmetry** which divides the molecule into two mirror-image halves. Such a molecule is said to be *symmetric*. If a molecule, or any object, is symmetric, it will be *identical to,* and thus *superimposable on, its mirror image*. This is the case with 2-chloropropane, as seen in Figure 5.1. We can fit one structure on top of the other with complete correspondence of every atom.

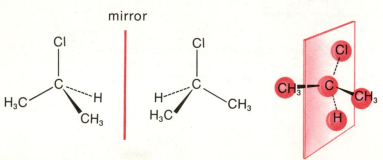

FIGURE 5.1

Superimposable mirror images of a symmetric molecule.

A different situation arises when *four different groups* are bonded to the same carbon atom, as in 2-chlorobutane (Figure 5.2). Regardless of how the structure is drawn, it cannot be divided into mirror-image halves. This molecule is asymmetric, and is *not* superimposable on its mirror image. Because of this lack of symmetry, 2-chlorobutane exists as two stereoisomers,

FIGURE 5.2

Nonsuperimposable mirror images, enantiomers of 2-chlorobutane.

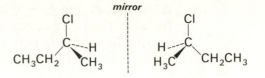

called **enantiomers,** which are related in the same way as are a right and a left hand. Any object that cannot be superimposed on its mirror image is said to be **chiral,** a term that comes from the Greek word *cheir,* meaning hand. Thus, 2-chlorobutane is a chiral molecule, while 2-chloropropane is not.

Many familiar objects have the property of chirality. The classic example is a hand. Hands look the same but differ in one respect—one is a left hand and one is a right hand; they are nonsuperimposable mirror images.

Problem 5.1 Which of the following objects are chiral and therefore have nonsuperimposable mirror images?

a) fork, b) shoe, c) ear, d) screw, e) cup, f) egg.

Problem 5.2 State which of the following compounds can exist in two enantiomeric forms:

a) 2-chloro-1-propanol   b) *m*-chlorotoluene

c) 2-methyl-1-butene   d) 1-phenyl-1-chloroethane

e) 3-methylhexane   f) 3-ethylheptane

Enantiomers are possible only for compounds whose molecules are chiral. However, it is not necessary to have a carbon substituted with four different groups (*asymmetric carbon*) to have a chiral molecule. Any compound that has a chiral structure can exist as enantiomers. Two examples of chiral molecules which do not contain asymmetric carbon are shown in Figure 5.3. Biphenyls in which bulky groups at the *ortho* positions prevent

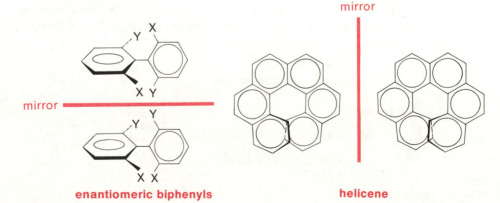

enantiomeric biphenyls   **helicene**

FIGURE 5.3

Enantiomers without asymmetric carbons.

rotation about the Ar—Ar bond are chiral. Another type of chiral molecule is found in the polynuclear aromatic hydrocarbons called *helicenes,* in which overlapping rings impart a helical or screw shape, as in a lock washer.

5.2   Optical Activity

Since enantiomers contain the same bonds and atoms in the same relationship (the same conformation will be most stable for each), they have identical energies. Properties such as boiling point, solubility, and density are identical. The only way that one enantiomer can be distinguished from the other is in a chiral medium, or in a reaction with a chiral reagent. One means of observing the difference between enantiomers is by their effect on a beam of plane polarized light.

Electromagnetic radiation consists of waves vibrating in all possible directions (Figure 5.4). If the light is passed through a special prism, the

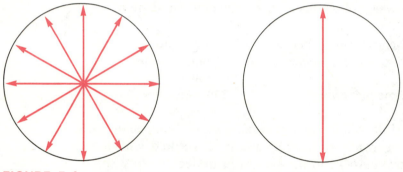

FIGURE 5.4
Oscillation of (a) ordinary light and (b) plane-polarized light.

intensity of all but one of these directions of vibration can be eliminated (the lenses of Polaroid sunglasses perform this function). The beam is said to be **polarized,** with all remaining waves parallel to one plane. When the polarized light beam is passed through a solution of a chiral compound in a polarimeter (Figure 5.5), the plane of the polarization is tilted or **rotated.** Two

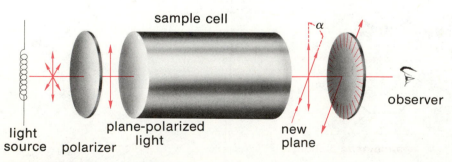

FIGURE 5.5
Diagram of a polarimeter for detecting and measuring optical activity of chiral compounds.

enantiomers cause exactly the same extent of rotation, but in opposite directions, *clockwise* (**dextrorotatory,** or +) in one case and *counterclockwise* (**levorotatory,** or −) in the other. Because of this property, enantiomers are sometimes called **optical isomers** and are said to exhibit optical activity.

The extent of rotation is proportional to the number of chiral molecules in the light path. Therefore, for a solution of an enantiomer, the magnitude of rotation depends on both the concentration and the length of the cell. The characteristic rotatory power of a given enantiomer with light of a given wavelength, λ, is expressed as the specific rotation [α].

$$[\alpha]_\lambda = \frac{\text{observed rotation (in degrees)}}{(\text{concentration in g per ml})(\text{length of cell in dm})}$$

A mixture of equal amounts of two enantiomers is called a **racemic mixture.** Such a mixture has zero rotation (does not rotate the plane polarized light) since the rotations of the two enantiomers exactly cancel. Consequently, if a substance shows no optical rotation, this does not necessarily mean that it is a symmetric compound, since it might be a racemic mixture.

5.3   Configuration of Enantiomers

Two enantiomers are designated as (+) or (−), according to the direction of optical rotation, but this does not indicate which enantiomer is which. The sign and magnitude of optical rotation are not related in any simple way to the structure of a chiral molecule. Therefore, knowing the direction in which one enantiomer rotates the plane of polarized light does not tell us how to draw it. All that can be said is that **two enantiomers have equal rotatory power in opposite directions.**

The arrangement of atoms around an asymmetric center is termed the **configuration.** Thus, enantiomers are said to have opposite configurations. For comparison of related compounds, it is important to be able to designate the configuration in a standard way, regardless of the optical rotation. The configuration of an asymmetric carbon atom in a given structure is specified by the Cahn-Ingold-Prelog (C.I.P.) rules:

1.   The four atoms attached to the asymmetric center are ranked *4-3-2-1, in order of decreasing atomic number*. If two or more of the attached atoms are the same, the adjacent atoms are ranked, and then the next adjacent, and so on. Thus, a $(CH_3)_2CH-$ group ranks above $CH_3CH_2CH_2-$ because the former has two adjacent carbons and one hydrogen, and the latter has only one adjacent carbon and two hydrogens. Double bonds count as two single bonds; thus, $CH_2=CH-$ ranks above CH_3CH_2-. *Hydrogen, if present, is always the lowest ranking atom.*
2.   The asymmetric center is then viewed with the *group of lowest rank held at the rear* of the tetrahedron, away from the eye, with the three

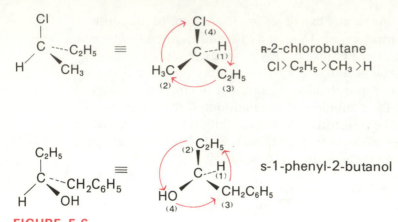

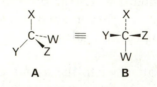

FIGURE 5.6

Determination of **R** and **S** configurations.

higher ranked groups radiating toward the viewer. The arrangement is analogous to the steering wheel of a car, with the lowest ranked atom being the one directed down the column (Figure 5.6)

3. If the three higher ranked atoms or groups are in descending order of rank in a clockwise direction, the configuration is called **R** (*rectus* or *right*). In the opposite configuration, with the rank decreasing in the counterclockwise direction, the configuration is **S** (*sinister* or *left*).

Three-dimensional perspective formulas such as those in Figure 5.6 can be oriented or viewed from any angle. Most of the perspective structures that we have seen so far have been written with the tetrahedron standing on three legs: one in front of the plane of the paper (wedge), one in the plane (solid line), and one behind (dashed line), with the fourth leg pointing up, as in **A**. However, we will also encounter structures oriented so that the tetrahedron is balanced on one leg, with two of the upper bonds in front and one behind the plane of the paper, giving a "bow tie" appearance as in **B**.

Another point to keep in mind in writing or looking at three-dimensional structures of asymmetric molecules is that the substituents for a given configuration can be written in quite a few ways. Thus, all of the following structures represent the **R** configuration of 1-bromo-1-chloroethane:

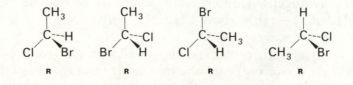

If we wish to redraw the structure, care must be taken to keep the relationship of the groups the same. If two groups are interchanged, the structure becomes that of the opposite enantiomer; when a second interchange of groups is made, the original enantiomer is obtained again. A simple model made from a gumdrop and toothpicks is helpful in visualizing the relationships.

Problem 5.3   Some of the following structures represent the **R**-enantiomer of 2-bromopentane and others represent the **s**-enantiomer. Label each structure **R** or **s**.

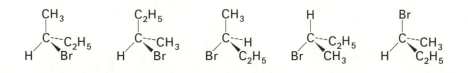

Problem 5.4   Name the following chiral structures. Be sure to specify the configuration. Draw the enantiomer of each. What are the configurations of these enantiomers?

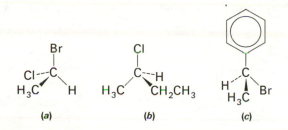

5.4   Absolute Configuration

If you were to find a glove on the street, you could immediately tell whether it was for the right or left hand, even though the two gloves in a pair differ only in being nonsuperimposable mirror images of one another. However, the examination of molecules is not so simple. The C.I.P. convention provides a means of designating configuration (that is, naming a particular arrangement), but it does not state whether a given enantiomer, say the (+) rotating one, is **R** or **s**. The answer to this question is termed the **absolute configuration,** and it is particularly important for naturally occurring compounds. For example, which structure, **R** or **s**, represents levorotatory (−)-malic acid, which occurs widely in fruits? This question of absolute configuration was answered in 1950 by an x-ray crystallographic technique with a related compound. When the absolute configuration of one compound is known, those of others can be determined by modifying groups in the original compound without breaking bonds attached to the asymmetric center. In this way, natural (−)-malic acid was shown to be the **s** isomer. Conversion of the CO_2H groups of natural malic acid to CH_3 groups gives

(−)-2-butanol.* Since the groups in the product differ in C.I.P. ranking, this isomer turns out to be **R**-2-butanol, but the relation of the hydroxyl group to the four-carbon chain is the same as that in **s**-malic acid.

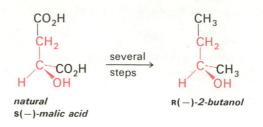

natural
s(−)-malic acid

R(−)-2-butanol

5.5 Natural Occurrence of Enantiomers

Practically all chiral compounds that occur in nature are found in only one enantiomeric form. Living organisms are chiral, and compounds that arise in metabolic processes are produced by enzymes (natural catalysts) which serve as chiral templates. This "lock-and-key" action of an enzyme surface allows only one enantiomer to interact properly, as illustrated in Figure 5.7.

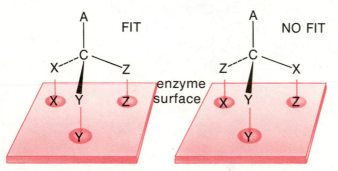

FIGURE 5.7

Different interactions of enantiomers with the chiral surface of an enzyme.

While you might not have thought of it in the terms we are now using, we often encounter the problem of the interaction of two chiral objects in everyday life. For example, try putting your left shoe on your right foot, screwing a bolt with a left-hand thread into a nut with a right-hand thread, or shaking hands using your left hand.

Our example from Section 5.4, malic acid from fruit, is formed biochemically by addition of water to an unsaturated diacid, *trans*-fumaric acid. The reaction is represented schematically in Figure 5.8, with the symmetric

---

*The fact that both **s**-malic acid and the butanol enantiomer that is obtained from it are levorotatory is a coincidence; two compounds with the same configuration at an asymmetric carbon do not necessarily have the same sign of rotation.

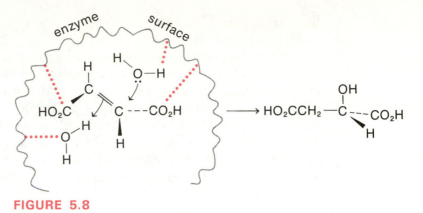

FIGURE 5.8

Enzymatic hydration of fumaric acid to form *only* **s**-malic acid.

fumaric acid held by the enzyme surface in a specific orientation. A water molecule is led in from a specific part of the surface to give only the s-isomer. It is interesting to note that the *cis* isomer of fumaric acid is unreactive under the same conditions.

> **Problem 5.5**   The reaction shown in Figure 5.8 is a *trans*-addition; that is, the —OH bonds to one face of the double bond and the —H to the other. Draw the product that would be formed if the enzymatic hydrolysis were done with D_2O in place of H_2O, and specify the configuration at *both* of the chiral centers.

The "lock and key" theory of enzyme selectivity of enantiomers has an interesting parallel in the hypothesis of odor sensitivity or smell. It is now generally believed that in the olfactory area of the nose there are receptor sites of different sizes and shapes, each of which corresponds to a type of primary odor. In some cases, only molecules fitting a particular site would exhibit that primary odor. An extension of this hypothesis is the possibility that enantiomers of a given compound might have different odors if the receptor sites themselves were chiral. This turns out to be the case for several optical isomers. A striking example is the odor difference between (+)-carvone from caraway seed oil and (−)-carvone from spearmint oil. These enantiomers are responsible for the characteristic odors of these oils, thus strongly indicating that our odor receptors are indeed chiral. For some reason, about 10 per cent of the population cannot tell the difference between the odors of these optical isomers.

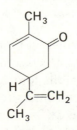

(+)-*carvone (caraway seed oil)*
(−)-*carvone (spearmint oil)*

5.6   Racemic Mixtures

When the hydration of fumaric acid is carried out in the laboratory, both R- and S-malic acids are obtained, unlike the enzyme catalyzed reaction. Here the fumaric acid is not bound to a catalyst surface and both faces of the molecule can be attacked, with equal probability, by water at either side of the double bond (Figure 5.9). Therefore, the racemic mixture, or **racemate,** containing exactly equal amounts of the two enantiomers is formed. This situation is encountered in *any* reaction in which a chiral molecule is generated from symmetric starting materials.

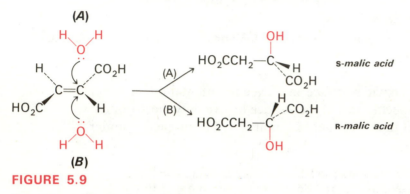

FIGURE 5.9

Formation of racemic malic acid.

When we discuss chiral compounds, it is usually in the context of one enantiomer; but it must be remembered that if a chiral compound is prepared by synthesis from a symmetric precursor, it will be obtained as a racemic mixture with zero optical rotation.

5.7   Two Asymmetric Centers—Diastereoisomers

When a molecule contains more than one asymmetric carbon, two enantiomeric configurations are possible at *each* center, and there are more than two stereoisomeric forms of the molecule. The *maximum* possible number of stereoisomers for a compound with n different asymmetric centers is 2^n. Thus, for a molecule containing two asymmetric carbon atoms there are four possible stereoisomers because there are two enantiomeric configurations possible for each center and thus four combinations (R_1,R_2; R_1,S_2; S_1,R_2; S_1,S_2). This is illustrated for 2-chloro-3-methylpentane in Figure 5.10.

Problem 5.6   Write structures showing configurations for all of the stereoisomers of 2,3-dichloropentane.

The four stereoisomers in Figure 5.10 consist of two pairs of enantiomers (two racemates), 2R,3S : 2S,3R and 2S,3S : 2R,3R. Stereoisomers which are not enantiomers are called **diastereoisomers.** Each of the four 2-chloro-3-methylpentanes has one enantiomer and two diastereoisomers (*e.g.,* the 2S,3S and 2R,3R isomers are both diastereoisomers of the 2R,3S compound).

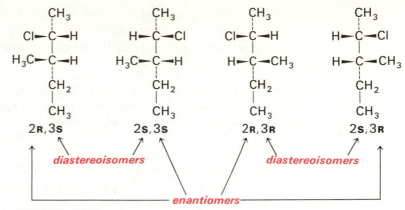

FIGURE 5.10

Stereoisomers of 2-chloro-3-methylpentane.

Since the atoms in diastereoisomers have different spatial relationships, the properties such as melting point, boiling point, solubility, and reactivity (all of which depend on molecular shape) are different. The two racemates in Figure 5.10 could be separated by fractional distillation, but the two enantiomers in each racemate could not be separated in this way.

The relationships among the stereoisomers of a compound with more than one asymmetric center are best understood in terms of their preparation from a molecule in which one of the centers is already present. Let's consider the hydrogenation of 1-methyl-3s-phenylcyclopentene to give 1-methyl-3-phenylcyclopentane. The product contains two asymmetric carbons, and four stereoisomers are thus possible. However, the two diastereoisomers with the **3R** configuration will not be formed because the starting material consists of the 3s phenyl enantiomer only. The hydrogenation product, therefore, is a mixture of the **1R**-methyl-3s-phenyl and **1s**-methyl-3s-phenyl diastereoisomers. These products could also be designated as *cis-* and *trans-*geometric isomers, respectively.

Now, what can be said about the relative amounts of these two isomers? In the hydrogenation, the cyclopentene must approach the catalyst surface where the hydrogen is adsorbed so that *cis*-addition can take place. The phenyl group on one side of the five-membered ring will make it much more difficult for addition to occur from this side, and the resulting **1s,3s** isomer will therefore be the *minor* product. The **1R,3s** isomer, formed by addition of hydrogen at the side opposite the phenyl group, is the major product of the

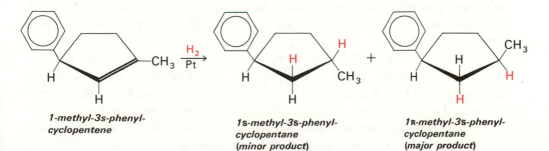

1-methyl-3s-phenyl-cyclopentene

1s-methyl-3s-phenyl-cyclopentane
(minor product)

1R-methyl-3s-phenyl-cyclopentane
(major product)

reaction. The starting material in this reaction was a chiral molecule, and introduction of the new asymmetric center is influenced by the chiral structure so that unequal amounts of the diastereoisomers are produced.

Problem 5.7 a) Write the structures and specify the configurations of the products that would be formed by hydrogenation of 1-methyl-3R-phenylcyclopentene. b) Predict the relative amounts of these products. c) State the stereochemical relationships of these products to each other and to the products obtained from 1-methyl-3S-phenylcyclopentene.

An important example of diastereoisomer formation is the polymerization of a vinyl monomer $CH_2{=}CHR$ (see Chapter 3). Every other carbon atom in the polymer chain is an asymmetric center, and the number of stereoisomers theoretically possible in a high molecular weight polymer is astronomical (at least 2^{5000}). The only meaningful way to consider the stereoisomerism in this case is the degree of regularity in configuration. Most vinyl polymers have a random or **atactic** structure in which the **R** and **S** configurations of the asymmetric carbons occur in an irregular way. With certain catalysts, highly stereoregular or **isotactic** chains can be obtained. These are more crystalline, higher melting, and stronger than those in the atactic polymer.

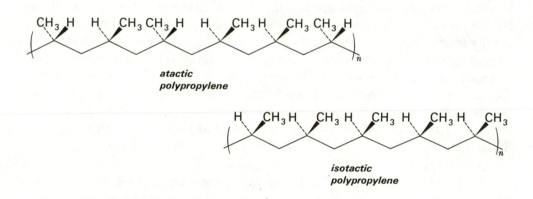

atactic
polypropylene

isotactic
polypropylene

5.8 Meso Isomers

A special situation comes up if a compound contains two asymmetric centers that have the *same* substituents. Let us look at the stereoisomers of 2,3-dichlorobutane. Both of the central carbons are asymmetric, and four structures can be written in two pairs, as in Figure 5.11. In this case, however, the 2S,3R and the 2R,3S structures are *identical*. This isomer is called the *meso* form.

The 2S,3R and 2R,3S isomers can be seen to be identical by rotating one structure 180° in the plane of the paper to give the other. A plane passed horizontally between C-2 and C-3 divides the molecule into two mirror-image halves. The *meso* isomer is therefore *symmetric,* and, of course, it has no optical activity. The total number of stereoisomers of 2,3-dichlorobutane is three—one pair of enantiomers and one *meso* isomer.

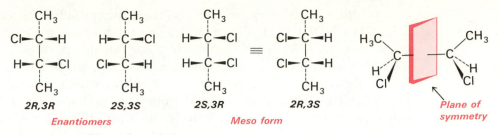

FIGURE 5.11

Stereoisomers of 2, 3-dichlorobutane.

5.9 Resolution of Racemic Mixtures

Separation of the enantiomers in a racemic mixture is called **optical resolution.** Since enantiomers differ only in chirality, resolution requires some method or reagent that can distinguish "right" and "left." The most general way to accomplish this is to convert the mixture of enantiomers to a mixture of diastereoisomers by reaction of the racemic mixture with a chiral compound, separate the diastereoisomers, and then regenerate the enantiomers.

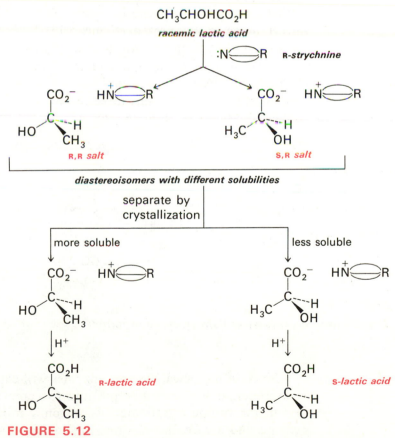

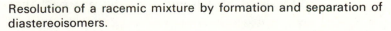

FIGURE 5.12

Resolution of a racemic mixture by formation and separation of diastereoisomers.

To illustrate this process, let us consider the resolution of racemic lactic acid, $CH_3CHOHCO_2H$ (Figure 5.12). The mixture of enantiomers is combined in solution with an asymmetric base. For this purpose, a commonly used base is the alkaloid strychnine (Section 13.6), which can be designated for simplicity as $:N\ominus R$. Strychnine has a bulky structure with several asymmetric centers. Since it is obtained from a natural source, strychnine is a single enantiomer, which we will designate arbitrarily as **R** base. Each of the **R** and **S** enantiomers of the racemic acid forms a salt with the base. These salts, [**R** base, **R** acid] and [**R** base, **S** acid], are diastereoisomers, with different shapes and different physical properties. The relationship between the two salts is analogous to that between two right hands clasping, and a right clasping a left. Because of differences in solubility, the diastereoisomeric salts can be separated by fractional crystallization. The individual salts are then treated with an inorganic acid to reverse the salt formation and give the separate enantiomers of the acid.

It is occasionally possible to achieve resolution of enantiomers by direct crystallization. Enantiomers in solution or in liquid form are distinguished only by molecular chirality; but with a solid, chirality is sometimes imparted to the crystal lattice. Louis Pasteur (1822–1895) was the first to separate enantiomeric crystals from a racemic mixture. He examined crystals of sodium ammonium tartrate by microscope and discovered that there were mirror image crystals present (Figure 5.13). By painstakingly hand-separating these crystalline forms, he was able to show that one form rotated the plane of polarized light to the left and the other to the right. This method of resolution is rarely possible, and it was Pasteur himself who developed the method of treating a racemic mixture with an optically pure chiral reagent to convert the two enantiomers into separable diastereoisomers.

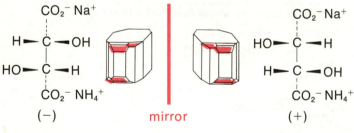

(−)   mirror   (+)

FIGURE 5.13

Enantiomeric crystals of sodium ammonium tartrate.

Since in biological systems enzymes will catalyze reactions of only one enantiomer, it is sometimes possible to achieve resolution by enzymatic destruction of one enantiomer of a racemic mixture. The complexes of an enzyme and a racemic substrate are in fact diastereomers. This, too, was discovered by Pasteur when he observed that when the mold *Pencillium glaucum* was allowed to grow in the presence of racemic tartaric acid, the (+)-isomer was utilized for growth and thus consumed, while the (−)-isomer could be recovered.

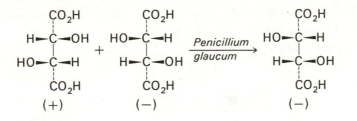

SUMMARY AND HIGHLIGHTS

1. **Stereoisomers** are isomers which have the same atom-to-atom bonding sequence but differ in the arrangement of these atoms in space.

2. **Enantiomers** are stereoisomers that are nonsuperimposable *mirror images* of each other; they are said to be **chiral.** Compounds with an **asymmetric** carbon atom (four different groups attached), and also compounds in which the overall structure is twisted, have enantiomeric forms.

3. Enantiomers differ only in reactions with chiral reagents and in the direction of rotation of polarized light.

4. The arrangement of atoms around a chiral center is termed the **configuration.** The configuration of an enantiomer is specified as **R** or **S,** depending on the sequence of the groups around the asymmetric center.

5. Chiral compounds usually occur in nature as one enantiomer, but mixtures of *equal* amounts of two enantiomers, called **racemic mixtures,** are obtained when a chiral compound is prepared from symmetric starting materials in a chemical reaction.

6. When a compound contains n asymmetric centers, the maximum number of stereoisomers is 2^n. The stereoisomers that are not enantiomers of each other are called **diastereoisomers.**

7. Diastereoisomers have different physical properties (boiling points, melting points, and so forth) and are usually formed in *unequal* amounts in chemical reactions.

8. If a molecule contains two asymmetric centers having the same substituents but opposite configurations, it contains a plane of symmetry and is called a *meso* isomer and is not optically active.

9. **Resolution** of a racemic mixture into the separate enantiomers can be accomplished by converting both enantiomers to diastereoisomers with a chiral reagent, separating the diastereoisomers, and regenerating the enantiomers.

ADDITIONAL PROBLEMS

5.8 Four of the following compounds are chiral. Identify the chiral compounds and write structures for both enantiomers, showing the mirror image relationship.

a) 2,3,3-trichloropentane

b) 2,4-dimethylpentane

c) 2-phenylbutane

d) *trans*-2-butene

e) 2-chloro-1-phenylpropane

f) *cis*-1,2-dimethylcyclopropane

g)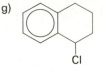

5.9 State which of the following structures have the **R** configuration and which have the **s** configuration.

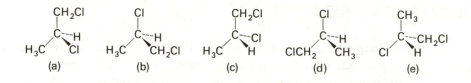

5.10 The structures written below represent the lowest energy conformation of 2-methylbutane. As drawn, these conformations are **enantiomeric**; *i.e.*, they are nonsuperimposable mirror images. Does this mean that 2-methylbutane is a chiral compound, capable of resolution into enantiomers? Explain.

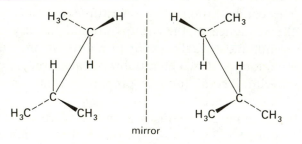

mirror

5.11 2-Phenylbutane was prepared by three reactions: (a) alkylation of benzene with *trans*-2-butene; (b) alkylation with 1-butene; and (c) hydrogenation of a sample of 3-phenyl-1-butene that had been obtained in several steps from a naturally occurring acid. The 2-phenylbutane from one of the three reactions was optically active; the product from the other two reactions had zero rotation. Which reaction, (a), (b), or (c), gave the optically active hydrocarbon?

5.12 Referring to the four stereoisomers of 2-chloro-3-methylpentane in Figure 5.10: (a) Which isomers are diastereoisomers of the 2**s**,3**R** isomer? (b) If the rotation of the 2**R**,3**s** isomer is +10°, and that of the 2**R**,3**R** is +5°, predict the rotations of the 2**s**,3**s** and 2**s**,3**R** compounds. (c) If 3-methylpentane were chlorinated (Cl$_2$, light) and the 2-chloro compound were separated from the other structural isomers, how many stereoisomers would be obtained? What can be predicted about the relative amounts of the isomers formed?

5.13 The stereochemical course of addition of HOCl to *cis*-2-butene can be represented by structures **1** to **3**. The groups can be rotated around the C_2-C_3 bond in structure **3** to give the eclipsed conformation **4**, and then rewritten in the projection formulas **5** or **6** to aid in visualizing the configuration.

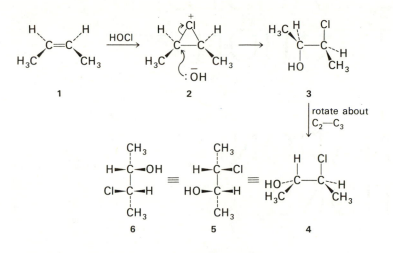

The addition of HOCl to *cis*-2-butene actually gives a racemic product, since it is a reaction of an achiral (*i.e.*, symmetric) compound with an achiral reagent.

    a) How many stereoisomers would be obtained in the reaction of *cis*-2-butene and HOCl?

    b) Write a series of structures analogous to **1** through **5** for the formation of other stereoisomers that would be obtained.

5.14 Depict the addition of HOCl to *trans*-2-butene using structures like those in problem 5.13, and compare the stereoisomeric products.

5.15 Depict the addition of Cl_2 to *cis*- and to *trans*-2-butene. (The reaction is the same as that in problem 5.13, with Cl^- instead of OH^- attacking structure **2**.) How many stereoisomers are obtained in each case?

6

ALKYL HALIDES; SUBSTITUTION AND ELIMINATION REACTIONS

Aliphatic halogen compounds range from simple alkyl halides with one halogen atom to compounds containing only carbon and halogen. We will look at a few of the uses of the latter compounds, but in our study of the reactions of halogen compounds we will concentrate on the simple mono-chloro and mono-bromo compounds.

6.1 Names

As we have seen in earlier chapters, simple halides can be named by combining the parent alkyl group name and the separate word chloride, bromide, and so forth. In systematic names the halogen atom is treated as a substituent on the alkane or alkene chain. Trihalomethane derivatives are named **haloforms.**

CH_3I

methyl iodide
or iodomethane

$CH_3CH_2CH_2Cl$

n-propyl chloride
or 1-chloropropane
(a primary halide)

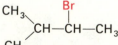

2-bromo-3-methylbutane
(a secondary halide)

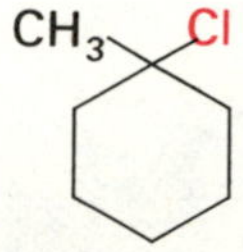

1-chloro-1-methyl-
cyclohexane
(a tertiary halide)

$CH_3CH{=}CHCH_2Cl$

1-chloro-2-butene
(an allylic halide)

CH_2Cl_2

methylene
chloride

$CHCl_3$

chloroform

CHI_3

iodoform

CCl_4

carbon
tetrachloride

6.2   Synthesis of Alkyl Halides

The main methods for preparing alkyl halides are summarized in the following chart. As we saw in Chapter 2, the free radical halogenation of an alkane is actually of limited use because a mixture of several products is usually obtained. The addition of HX to an alkene is satisfactory if the alkene is available and the addition can occur in only one direction, but these conditions are often not met. The most general and useful method for preparing alkyl halides is from the corresponding alcohol (method 3), which we will see in detail in the following chapter.

PREPARATION OF ALKYL HALIDES

1.   **Halogenation of alkanes**

$$RH + X_2 \xrightarrow[\Delta]{h\nu \text{ or}} RX + HX \qquad \text{(Sect. 2.7)}$$

2.   **Addition of HX to alkenes**

$$\Large{>}C{=}C{\Large<} + HX \longrightarrow -\overset{H}{\underset{|}{C}}-\overset{X}{\underset{|}{C}}- \qquad \text{(Sect. 3.5)}$$

3.   **Conversion of alcohols**

$$ROH + HX \text{ (or } PX_3 \text{ or } SOX_2) \longrightarrow RX + H_2O \qquad \text{(Sect. 7.11)}$$

6.3   Properties and Uses of Alkyl Chlorides

Alkyl chlorides have physical properties resembling those of alkanes. Compounds with several halogen atoms, such as CH_2Cl_2, $CHCl_3$, and CCl_4, are sweet-smelling liquids with densities greater than that of water. As the halogen content increases, the compounds become less susceptible to combustion in air. Because they are not very flammable and are good solvents for fats, oils, and greases, polychloro compounds—particularly trichloroethylene ($CHCl{=}CCl_2$) and tetrachloroethylene ($CCl_2{=}CCl_2$)—are widely used in cleaning and degreasing applications such as dry cleaning.

Chloroform and carbon tetrachloride ("carbon tet") are familiar compounds that have had many uses. At one time chloroform was used as an inhalation anesthetic and also as an expectorant and flavoring agent in cough medicine and toothpaste. However, both of these compounds, as well as several chlorinated ethanes, have been found to cause liver damage and cancer in experimental animals, and their applications have been sharply curtailed.

A special application of ethyl chloride is as a local anesthetic for minor local surgery such as stitching up a cut. The area is sprayed with ethyl chloride, which has a boiling point of 12°, and the rapid cooling due to evaporation causes temporary numbing of the tissue.

6.4   Organic Fluorine Compounds

Organic compounds containing fluorine have some very special properties and uses. The C—F bond is extremely strong (104 kcal/mole), and CF_2 and CF_3 groups are highly stable and unreactive. Moreover, compounds containing these groups generally have quite low toxicity.

Freons.   A series of chlorofluoro compounds called Freons are used in large amounts as refrigerant liquids and as aerosol propellants for dispensing such varied items as hair spray, topical medicines, air fresheners, furniture wax, insecticides and oven cleaners. An aerosol can contains a solution or emulsion of the material to be dispensed in a liquid Freon which boils below room temperature. The pressure in the can depends on the vapor pressure of the liquid, and remains constant as long as liquid is present. Rapid evaporation of the propellant liquid occurs when the contents are sprayed, providing a very fine mist. Freon propellants are odorless and nonflammable, and have very low toxicity. The most widely used propellants are CCl_2F_2 and $CFCl_3$.

A serious problem with Freons as propellants lies in the fact that when these very stable compounds are discharged from the spray can they diffuse away unchanged into the upper atmosphere. Photochemical breakdown from solar radiation can then occur to release chlorine atoms (radicals), which catalyze the decomposition of ozone (Section 3.8).

Fluorocarbons.   Completely fluorinated alkanes, with all hydrogens replaced by fluorine, are called **fluorocarbons.*** These substances represent a special class of organic compounds with exceptional stability and other properties. Because of the small size of fluorine (less than that of carbon) it is possible to obtain long chains of —CF_2— groups. In contrast, the accumulation of chlorine atoms on a carbon chain is limited by the much larger atomic radius of chlorine to three adjacent —CCl_2— groups; vigorous chlorination of longer chains leads to chlorinated alkenes, dienes and cyclic products.

Fluorocarbons have the unusual property of being incompletely miscible with other organic compounds, including hydrocarbons. Compounds with long completely fluorinated alkyl chains provide textile finishes that repel both oil and water, and thus remain "clean" under extremely severe conditions.

The polymer Teflon, —$(CF_2CF_2)_n$—, obtained by polymerization of

---

*The term "fluorocarbon" has unfortunately been widely misused to designate the compounds containing carbon, chlorine, and fluorine that react with radiation to liberate chlorine.

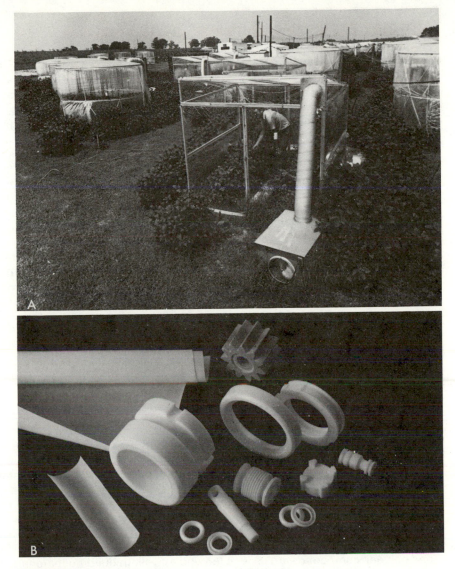

FIGURE 6.1

A, Growth chambers made of "Teflon" FEP fluorocarbon film are used to evaluate the effects of air pollution on soybean plants. "Teflon" is not affected by sunlight, weather, agricultural chemicals, fungus, or bacteria. *B,* Parts molded or machined from "Teflon" TFE fluorocarbon resin are used in many industries to provide chemical resistance, low friction, good electrical insulation, and service at extreme temperatures. (Photos courtesy Du Pont Company.)

tetrafluoroethylene, is valuable in applications such as lining of chemical process equipment and cooking utensils which require extreme inertness at high temperatures. Because of the peculiar surface properties of tightly-bonded fluorocarbon chains, molded Teflon surfaces have the lowest coefficient of friction of any material known. Bearing pads of Teflon are used in construction to permit buildings to undergo slippage during thermal expansion, and to slide off-shore oil rigs from a barge into drilling position.

SPECIAL TOPIC: Organic Halides in Fire Protection

Fire is one of the oldest and most deadly hazards of civilization, and improved methods for fighting fires, and particularly for prevention of fire, are major concerns of modern technology. Organic halogen compounds play an important role in both fire extinguishing and fire retarding agents.

The two traditional approaches to extinguishing fire are cooling and exclusion of oxygen. Flooding with water can act in both ways; this is, of course, still the major method of fire control, particularly with the addition of wetting and foaming agents. Other materials that are more effective for some fires are carbon dioxide and dry powders containing carbonates, phosphates, and other inorganic compounds. At one time carbon tetrachloride was used, but since a breakdown product of the hot vapor in air is the extremely toxic compound phosgene, $COCl_2$, fire extinguishers containing carbon tetrachloride are potentially quite hazardous.

A different approach to the control of fire is use of a chemical reaction to interfere with the actual combustion process. Burning occurs in a complex series of reactions involving species such as O_2^-, $\cdot OOH$, and $\cdot OH$, which propagate the spread of flames. In the presence of bromine compounds which release bromine atoms at the proper rate on heating, these chain reactions are interrupted. One possible mechanism is the sequence

$$RBr \longrightarrow R\cdot \ + \ Br\cdot$$
$$Br\cdot \ + \ RH \longrightarrow R\cdot \ + \ HBr$$
$$HBr \ + \ \cdot OH \longrightarrow H_2O \ + \ Br\cdot$$

A very effective extinguishing agent based on this approach is bromotrifluoromethane (Halon 1301®). With a concentration of 5 to 7 per cent of $CBrF_3$, the atmosphere of a room is inert and will not support combustion, but it is also nontoxic and can be breathed without danger. Thus, areas containing sensitive materials such as computer centers, libraries, hospital operating rooms, aircraft, and ships can be protected by systems that automatically flood the space with $CBrF_3$ in case of fire, without risk to occupants or damage to equipment.

Equally as important as extinguishing fire is its prevention or retardation, and halogen compounds are used to this end as well. In building materials, one of the main points at which fire retardant properties can be added is in the resins used to bond plywood, wallboard, and laminated panelling. These compounds are condensation polymers in which diols and diacids are combined to give polyesters (Section 10.21). In order to impart fire retarding properties, some of the usual components of the resin are replaced by highly halogenated compounds as in the examples shown. Bromine is more effective than chlorine; it is also much more costly.

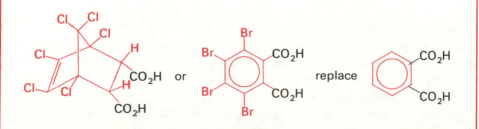

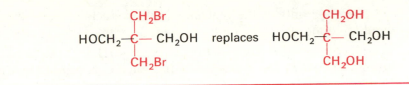

SUBSTITUTION AND ELIMINATION REACTIONS

The major reactions of alkyl halides can be classified into two very general processes: **(1) nucleophilic substitution** and **(2) elimination.** These reactions will be illustrated in this chapter with alkyl halides, but they occur with a variety of other reactants as well. Nucleophilic substitutions provide useful methods for preparing a number of types of compounds. It is important to recognize that these are all applications of *one* general reaction. If the general process and how it occurs are understood, the specific reactions do not have to be learned as individual facts.

In both substitution and elimination reactions, the carbon-halogen bond breaks to give the very stable halide anions Cl^-, Br^-, or I^-. In substitutions, halogen is replaced by a nucleophile which donates an electron pair to carbon. In eliminations, a base removes a proton on the β-carbon (adjacent to C—X), and a double bond is formed. Some reagents can act as either a nucleophile or a base, and the two reactions may thus compete.

Nucleophilic substitution: $Nu\!:^- + \ -\overset{|}{\underset{|}{C}}-X \longrightarrow Nu-\overset{|}{\underset{|}{C}}- + \ :X^-$

Elimination: $B\!:^- + H-\overset{|}{\underset{|}{C}}-\overset{|}{\underset{|}{C}}-X \longrightarrow BH + \ \overset{}{\underset{}{C}}{=}\overset{}{\underset{}{C}} + \ :X^-$

6.5   Nucleophilic Substitution

Alkyl halides can undergo nucleophilic substitution by two different mechanisms. Which mechanism is operative depends upon the structure of the alkyl halide, the nature of the solvent, and the nucleophilic strength of the attacking reagent. These two mechanisms can be illustrated by the conversion of *t*-butyl bromide and methyl bromide to the corresponding alcohols by hydroxide ion substitution.

$$CH_3-\overset{\overset{\displaystyle CH_3}{|}}{\underset{\underset{\displaystyle CH_3}{|}}{C}}-Br \ + \ :OH^- \longrightarrow CH_3-\overset{\overset{\displaystyle CH_3}{|}}{\underset{\underset{\displaystyle CH_3}{|}}{C}}-OH \ + \ :Br^-$$

t-butyl bromide   *t-butyl alcohol*

$$CH_3Br \ + \ :OH^- \longrightarrow CH_3OH \ + \ :Br^-$$

methyl bromide   *methyl alcohol*

Looking only at the reactants and products of these two reactions, we would assume that they both proceeded by the same general pathway. The overall course is the same—a nucleophile, $\bar{:}OH$, is substituted for the leaving group, $\bar{:}Br$. However, a dramatic difference in **reaction kinetics,** or dependence of reaction rate on concentration, shows that they proceed by quite different mechanisms.

Rate Data.   In a kinetic study, the rate of a reaction is measured by determining the disappearance of a reactant (RBr or OH⁻) or the formation of a product as a function of time. The rate depends on the concentration of molecules involved in the transition state which have sufficient energy to overcome the activation barrier (Section 2.6). If the concentration of a reactant does not have any effect on the reaction rate, we know that this species is not involved in the transition state.

The rate of the reaction of *t*-butyl bromide and hydroxide is found to be dependent *only* on the concentration of *t*-butyl bromide. Increasing the concentration of hydroxide ion has no effect upon the speed of this reaction.

$$\text{rate} \propto [(CH_3)_3CBr]$$

This means that the reaction whose rate we are actually measuring does not involve $\bar{:}OH$, although the overall process clearly does. When a reaction involves two steps, with one being slow and the other fast, the rate that is measured is simply that of the slower step. The slow reaction is referred to as the **rate determining step.** Since $\bar{:}OH$ does not affect the rate of the substitution reaction, only *t*-butyl bromide is involved in the rate determining step. A mechanism which fits these facts is one with two steps, (1) slow ionization of the alkyl halide to a carbocation, followed by (2) rapid combination of the carbocation and hydroxide ion to form the alcohol.

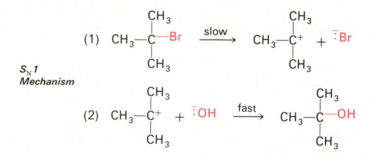

*$S_N 1$
Mechanism*

This type of mechanism, where the reaction rate is dependent only upon the concentration of the alkyl halide, is termed **substitution nucleophilic, unimolecular,** or **S_N1.**

In contrast, the rate of the methyl bromide reaction is found to be dependent upon the concentrations of *both* methyl bromide and hydroxide; if the concentration of either is doubled, the rate doubles.

$$\text{rate} \propto [CH_3Br][^-OH]$$

Thus, the reaction step being measured involves both of the starting materials. This can be explained by a one-step mechanism termed **substitution nucleophilic, bimolecular**, or **S_N2**. Here the carbon atom of methyl bromide is attacked by hydroxide ion to form a C—O bond and to break the C—Br bond.

S_N2
Mechanism   HO:⁻ + H₃C—Br ⟶ HO CH₃ + :Br⁻

6.6   Relative Reactivities in Substitution

Another sharp contrast between S_N2 and S_N1 reactions is found on comparing the reactivities of primary (1°), secondary (2°), and tertiary (3°) halides. As seen in Table 6.1, the relative rates of both processes are dependent upon the structure of the halide, but the trends are in *opposite directions*. As we shall see, this is exactly what we should expect.

TABLE 6.1   **ORDER OF HALIDE REACTIVITIES**

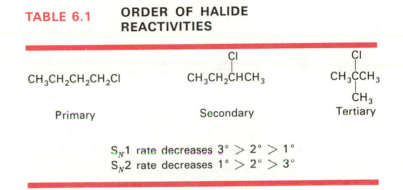

CH₃CH₂CH₂CH₂Cl   CH₃CH₂CHCH₃ (Cl)   CH₃CCH₃ (Cl, CH₃)

Primary   Secondary   Tertiary

S_N1 rate decreases 3° > 2° > 1°
S_N2 rate decreases 1° > 2° > 3°

S_N1 Reactivity.   The S_N1 order of reactivities is simply the order of carbocation stabilities, 3° > 2° > 1°. The slow reaction, which is the more difficult of the two steps, is the formation of a carbocation (Figure 6.2). The

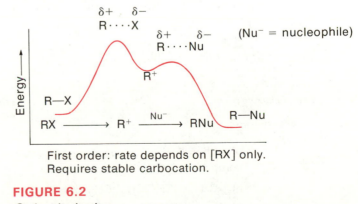

First order: rate depends on [RX] only.
Requires stable carbocation.

FIGURE 6.2
S_N1 substitution.

observed order of halide reactivities simply reflects the fact that the more stable the carbocation, the easier it is to form.

Reactivity in the S_N1 process provides a means of comparing the stability of carbocations. Allylic or benzylic halides are exceptionally reactive because the positive charge is stabilized by resonance delocalization. A carbocation bonded to an oxygen or nitrogen atom is stabilized by resonance donation of an unshared electron pair, as in the case of a chloromethyl ether (Table 6.2).

TABLE 6.2   RESONANCE-STABILIZED CARBOCATIONS

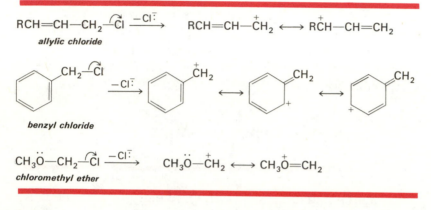

Compounds in which the halogen is attached *directly* to an sp^2 hybridized carbon, as in C=CHCl or ArBr, are very *unreactive* to both S_N1 and S_N2 displacements.

S_N1 substitutions occur most readily in polar solvents, such as water or alcohols, that can stabilize the resulting ions. These solvents can also react as the nucleophile to trap the carbocation. An alcohol is formed when the reaction is carried out in water; if an alcohol is used as the solvent, the product is an ether. A simple way to compare reactivities in S_N1 displacements is measurement of the rate of reaction of the alkyl halide in an ethanol solution of silver nitrate. The time required for appearance of a silver halide precipitate is an indication of the relative reactivity.

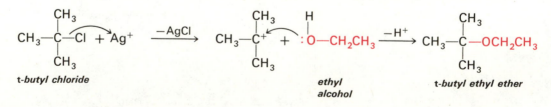

Problem 6.1   Predict the relative reactivities of the following series of compounds in S_N1 substitutions:

a) $CH_3CH_2CHClCH_3$, $CH_2=CHCH_2CH_2Cl$, $CH_3CHClCH=CH_2$

b) $CH_3OCHClCH_3$, $CH_3CH_2C(CH_3)_2Cl$, $CH_3CH_2CH_2Cl$

S_N 2 Reactivity. The S_N2 reactivity order is explained by the fact that this single-step reaction must involve a transition state which has a carbon atom attached to five groups, since it is partially bound to both the attacking nucleophile and the leaving group. As we proceed from 1° to 2° to 3°, the sizes of the other groups attached to this carbon increase as hydrogens are

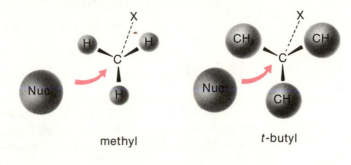

methyl   *t*-butyl

replaced by alkyl groups. Thus, the **steric crowding** about this carbon increases and the energy of the transition state (the height of the activation barrier) increases accordingly (Figure 6.3).

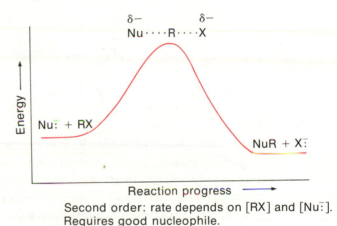

Second order: rate depends on [RX] and [Nu⁻].
Requires good nucleophile.

FIGURE 6.3

S_N2 substitution.

The relative reactivities of a series of alkyl chlorides or bromides in the S_N2 process can be compared by treatment of the halides with a solution of sodium iodide in acetone. Iodide ion is a good nucleophile, and causes displacement by the S_N2 mechanism to give the alkyl iodide. Sodium bromide and sodium chloride are insoluble in the reaction solvent, and appearance of a crystalline precipitate of these salts is evidence of reaction.

$$CH_3CH_2CH_2Br + Na^+I^- \longrightarrow CH_3CH_2CH_2I + Na^+Br^-$$

soluble in   *insoluble in*
acetone   *acetone*

In summary, alkyl halides which can ionize to form stable carbocations (3°, benzylic, allylic, etc.) undergo S_N1 processes most readily. Primary halides are most reactive to S_N2 processes, because they develop the least amount of steric crowding in the transition state for displacement.

Problem 6.2 a) Predict the relative order of reactivity of the following compounds in an S_N2 displacement by NaI:

$$(CH_3)_2CHBr, \quad C_6H_5CH_2Br, \quad CH_3Br.$$

b) Do the same for an S_N1 process with $^-OH/H_2O$.

6.7 Stereochemistry of Substitution.

Another important difference between S_N1 and S_N2 processes is observed if the reactions are carried out with a single enantiomer of an optically active alkyl halide with the halogen attached to the asymmetric carbon.

S_N1 Reaction. When an aqueous solution of s-1-phenyl-1-chloroethane is heated, only racemic alcohol is formed. Using the R-enantiomer likewise produces only racemic alcohol.

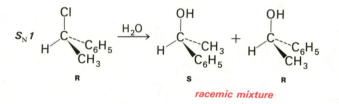

racemic mixture

The observation that S_N1 produces racemic mixtures from optically pure starting materials is consistent with a carbocation intermediate. In a carbocation the charged carbon is sp^2 hybridized and planar. Thus, the attacking nucleophile can bond from *either side* with equal ease to produce equal amounts of the enantiomeric products.

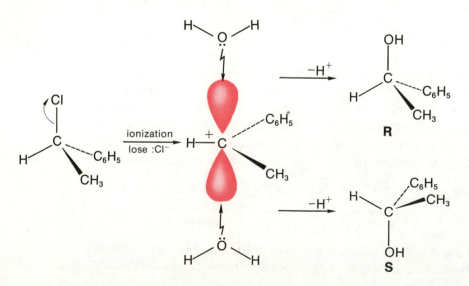

S_N 2 Reaction.   Quite a different stereochemical result is observed when substitution is carried out under S_N2 conditions. A good example is the reaction of a secondary alkyl chloride with cyanide ion as the nucleophile. When **R**-2-chlorobutane is warmed with sodium cyanide in alcohol solution, the product is the **s-cyano** compound. Similarly, with the **s**-chloride, only the **R**-cyano compound is obtained.

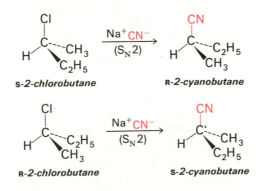

This **inversion of configuration** always occurs in an S_N2 displacement. It arises from "backside" attack by the nucleophile, $^-$CN in our example, to produce a transition state with the three remaining groups and the central carbon in one plane. As the chloride ion departs, these groups fold over in the other direction to complete the inversion of configuration. A good analogy is the inversion of an umbrella on a windy day.

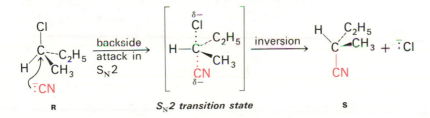

Problem 6.3   Write the structure of the product that would be obtained in the reaction of NaCN with *trans*-1-bromo-3-methylcyclopentane.

Problem 6.4   Predict, when possible, whether the following reactions would occur by S_N1 or S_N2 substitution, and write the structure and stereochemical configuration of the product.

a)   **R**-2-bromo-2-phenylbutane in boiling methanol gives $C_{11}H_{16}O$.
b)   1-bromo-3-methylbutane plus sodium iodide.
c)   **s**-*sec*-butyl chloride plus NaOH.

6.8   Elimination

Elimination of HX from alkyl halides to give an alkene can occur by two different mechanisms that are analogous to those of substitutions.

E2 Reaction   Like the S_N2 process, an E2 reaction (**elimination, bimo-lecular**) occurs in a single step. In the E2 transition state both the C—X bond and the C_β—H bond are partially broken, and the C=C π bond is partially formed (Figure 6.4). A strong base such as hydroxide ion is used to bring about the reaction by removing the β-hydrogen.

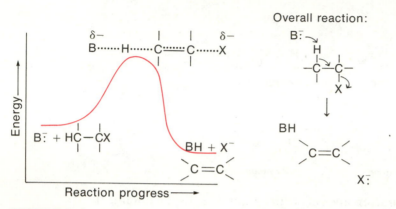

FIGURE 6.4

Mechanism of the E2 elimination.

Since an essential feature of the E2 reaction is removal of a proton from a β-carbon, the structural requirements for the alkyl halide are not the same in E2 elimination and S_N2 substitution. Secondary and tertiary halides are usually *more* reactive than primary halides toward elimination.

$$R_2C-CR_2 \ > \ R_2C-CHR \ > \ R_2C-CH_2X$$

tertiary   secondary   primary

$$R_2C=CR_2 \qquad R_2C=CHR \qquad R_2C=CH_2$$
$$+ \ H_2O + \ HX \quad + \ H_2O + \ HX \quad + \ H_2O + \ HX$$

Elimination with a primary halide leads to a single alkene; but with secondary or tertiary halides, a mixture of products may result since there can be more than one β-carbon. When different alkenes can be formed, the major product is the alkene with the largest number of alkyl groups attached to the sp^2 carbons.

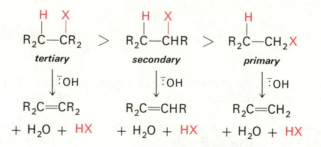

$$CH_3CH_2CH_2\overset{Cl}{\underset{|}{C}}HCH_3 + NaOH \longrightarrow CH_3CH_2CH=CHCH_3 + CH_3CH_2CH_2CH=CH_2$$

2-chloropentane   2-pentene, 75%   1-pentene, 25%

Problem 6.5 Write the structures of the alkenes that would be formed in the reactions of the following halides with base; if more than one alkene would be formed, indicate which would be the major product: a) 1-chloro-2-methylpropane, b) cyclopentyl bromide, c) 3-phenyl-2-chlorobutane, d) 2-bromo-3,3-diethylpentane.

In order to form the π bond in an E2 elimination, the p orbitals of the two carbons must be "lined up" so that they begin to overlap in the transition state. The C_β—H and C_α—X bonds strongly prefer to be **antiparallel,** *i.e.,* in the same plane and on opposite sides of the bond, as shown in Figure 6.5 for 1**R**,2**S**-1-bromo-1,2-diphenylpropane. Only one conformation of this halide has the antiparallel arrangement of C—H and C—Br bonds, and the elimination reaction occurs with this conformation, giving only the *trans*-alkene.

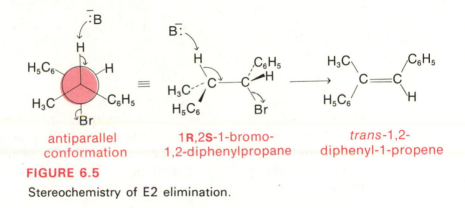

antiparallel   1**R**,2**S**-1-bromo-   *trans*-1,2-
conformation   1,2-diphenylpropane   diphenyl-1-propene

FIGURE 6.5

Stereochemistry of E2 elimination.

Problem 6.6 Write structures of the three other diastereoisomers of 1-bromo-1,2-diphenylpropane (1**R**,2**R**-, 1**S**,2**R**, and 1**S**,2**S**-), and write the structure of the elimination product in each case.

Elimination can be carried out with a vinyl halide under rather vigorous conditions to give an alkyne.

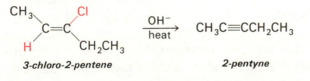

3-chloro-2-pentene   *2-pentyne*

E1 Reaction. The E1 process **(elimination, unimolecular)** occurs when a carbocation is formed and a proton from an adjacent carbon is then removed by a base.

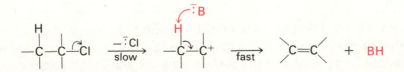

The slow, or rate-determining, step in the E1 reaction is the same as that for S_N1 substitution—dissociation of the alkyl halide to a carbocation plus halide ion. The same structural factors discussed for the reactivity of alkyl halides in the S_N1 reaction therefore apply also to the E1 reaction, and the reactivity order is $3° > 2° > 1°$. With tertiary halides, both elimination and substitution products are frequently obtained.

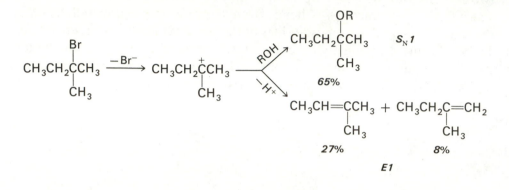

Elimination reactions are favored over substitution a) when a strong base is present, and b) by a higher reaction temperature. Under basic conditions, elimination from a secondary or tertiary halide may occur partly by the E1 mechanism and partly by the E2 process. The products are the same, and the distinction here between competing mechanisms is not important for our purposes.

Applications of Substitution Reactions

The most useful of the reactions that we have been examining in this chapter are S_N2 substitutions of alkyl halides. In these reactions, an alkyl group becomes bonded to a nucleophile, and the overall process is often called **alkylation.** The nucleophile may have an unshared electron pair on carbon (a carbanion) or on a more electronegative atom such as oxygen, nitrogen, or sulfur. Since the structure of the nucleophile can be widely varied, these substitution reactions provide a general method for the synthesis of many types of compounds. Since most nucleophiles, hydroxide ion for example, are also bases, elimination is likely to be an interfering side reaction with secondary halides, and is usually the major reaction with a tertiary halide. The most successful S_N2 reactions are therefore those with primary halides, but even with this restriction we can carry out a large number of useful preparations.

6.9   Formation of Carbon–Carbon Bonds

To obtain a C—C bond by nucleophilic substitution, we need a carbon with an unshared electron pair, *i.e.*, a carbanion (Section 1.11). The alkylation of carbanions can be used in a number of ways for building up carbon chains.

One simple case that we have already seen is the reaction of a primary halide with sodium or potassium cyanide to give an alkyl cyanide.

$$CH_3CH_2CH_2I + Na^+ \ ^-\!:CN \longrightarrow CH_3CH_2CH_2CN + NaI$$

n-propyl iodide   **n-propyl cyanide**

Another type of carbanion, very similar in structure to cyanide ion, is the acetylide anion, $HC\equiv C^-\!:$, obtained from acetylene by treatment with $NaNH_2$ (Section 3.9). The S_N2 reaction with acetylide ion leads to a 1-alkyne. A second alkylation step can be carried out by converting the resulting alkyne to the substituted acetylide anion and then allowing it to react with another alkyl halide. With this two-step process, we can build up a carbon chain with a triple bond at any position desired.

$$HC\equiv CH + NaNH_2 \longrightarrow HC\equiv C^-\!: Na^+ + NH_3$$

$$HC\equiv C^-\!: Na^+ + CH_3CH_2CH_2CH_2Br \longrightarrow CH_3CH_2CH_2CH_2C\equiv CH + NaBr$$

1-hexyne

$$CH_3(CH_2)_3C\equiv CH + NaNH_2 \longrightarrow CH_3(CH_2)_3C\equiv C^-\!: Na^+$$

$$CH_3(CH_2)_3C\equiv C^-\!: Na^+ + CH_3CH_2Br \longrightarrow CH_3(CH_2)_3C\equiv CCH_2CH_3 + NaBr$$

3-octyne

6.10   Bonds to Oxygen, Nitrogen, and Sulfur

Compounds with C—O, C—N, or C—S bonds are obtained by substitution reactions in which oxygen, nitrogen, or sulfur is the nucleophilic center. Earlier in the chapter we saw the formation of an alcohol by reaction of water with a carbocation in the S_N1 process. To obtain an alcohol by an S_N2 substitution, we need a more powerful nucleophile, and hydroxide ion is used for this purpose. From a preparative standpoint, this reaction is not often important since the *reverse*

$$RCH_2Cl + KOH \longrightarrow RCH_2OH + KCl$$

alkyl chloride   *alcohol*

process, going from alcohol to halide, is the most common method for obtaining alkyl chlorides and bromides (Chapter 7).

As will be discussed in Chapter 7, alcohols react with sodium metal to lose hydrogen and form sodium alkoxides. This is analogous to the reaction of water with alkali metals.

$$ROH + Na \longrightarrow RO^-\!: Na^+ + \tfrac{1}{2} H_2$$

sodium alkoxide

Like the hydroxide ion, alkoxide ions can react with alkyl halides to give substitution products; this is a general method for preparing **ethers,**

R—O—R. Since alkoxides are also strong bases, reactions with secondary or tertiary halides give significant amounts of elimination products. In summary, this substitution reaction is good for primary halides, fair for secondary halides, and useless for tertiary halides.

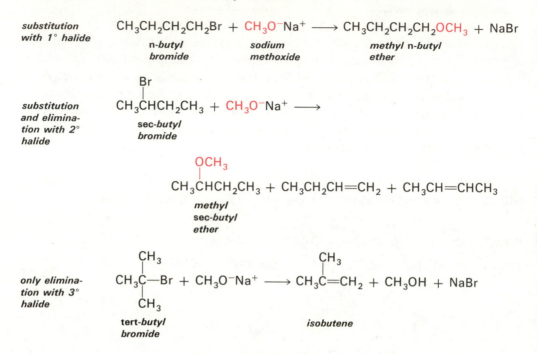

substitution with 1° halide

$CH_3CH_2CH_2CH_2Br + CH_3O^-Na^+ \longrightarrow CH_3CH_2CH_2CH_2OCH_3 + NaBr$

n-butyl bromide   sodium methoxide   methyl n-butyl ether

substitution and elimination with 2° halide

$\overset{\displaystyle Br}{\underset{\displaystyle |}{CH_3CHCH_2CH_3}} + CH_3O^-Na^+ \longrightarrow$

sec-butyl bromide

$\overset{\displaystyle OCH_3}{\underset{\displaystyle |}{CH_3CHCH_2CH_3}} + CH_3CH_2CH=CH_2 + CH_3CH=CHCH_3$

methyl sec-butyl ether

only elimination with 3° halide

$\overset{\displaystyle CH_3}{\underset{\displaystyle CH_3}{CH_3C-Br}} + CH_3O^-Na^+ \longrightarrow \overset{\displaystyle CH_3}{CH_3C=CH_2} + CH_3OH + NaBr$

tert-butyl bromide   isobutene

Problem 6.7   Write equations showing the formation of substitution and elimination products in the following reactions, and if possible predict which is the major product.

a) $CH_3CH_2Br + NaOCH(CH_3)_2 \longrightarrow$

b) $CH_3CH_2C(CH_3)_2Cl + NaOH \longrightarrow$

c) $CH_3CHBrCH_3 + NaOCH_2CH_3 \longrightarrow$

Sulfur nucleophiles that are useful in S_N2 reactions include $HS^{\bar{\cdot}}$, $RS^{\bar{\cdot}}$, $^{\bar{\cdot}}SCN$, and a number of others. These anions react very readily with alkyl halides to form C—S bonds; the products of these reactions are shown in Table 6.3. The simplest nucleophile leading to C—N bond formation is

TABLE 6.3   **REACTIONS WITH OXYGEN, NITROGEN, CARBON, AND SULFUR, NUCLEOPHILES**

Nucleophile	Reaction	Product
hydroxide ion	$HO^{\bar{\cdot}} + RCH_2X \longrightarrow {}^{\bar{\cdot}}X + RCH_2OH$	alcohol
alkoxide ion	$R'O^{\bar{\cdot}} + RCH_2X \longrightarrow {}^{\bar{\cdot}}X + RCH_2OR'$	ether
hydrosulfide	$HS^{\bar{\cdot}} + RCH_2X \longrightarrow {}^{\bar{\cdot}}X + RCH_2SH$	thiol
thiolate ion	$RS^{\bar{\cdot}} + RCH_2X \longrightarrow {}^{\bar{\cdot}}X + RCH_2SR$	sulfide
thiocyanate ion	$NCS^{\bar{\cdot}} + RCH_2X \longrightarrow {}^{\bar{\cdot}}X + RCH_2SCN$	thiocyanate
ammonia	$H_3N\colon + RCH_2X \longrightarrow {}^{\bar{\cdot}}X + RCH_2NH_3^+$	ammonium ion

ammonia. Reactions with ammonia provide us with a reminder about the changes in ionic charges that occur in an S_N2 substitution. When the unshared electron pair of the nucleophile displaces the halide anion, the charge on the nucleophilic atom becomes one unit more positive. Thus, the product of an S_N2 reaction with ammonia is the alkylammonium ion.

$$CH_3CH_2CH_2I + \overset{\cdot\cdot}{N}H_3 \longrightarrow CH_3CH_2CH_2\overset{+}{N}H_3I^-$$

n-propyl iodide   **n-propylammonium iodide**

6.11 Intramolecular Substitution; Cyclic Products

When a nucleophilic group and a halogen atom are present in the same molecule, an *internal* nucleophilic substitution (an *intramolecular* process) can occur to give a cyclic product. Cyclization to a five- or six-membered ring is particularly favorable for an S_N2 displacement.

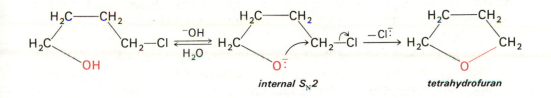

internal S_N2 *tetrahydrofuran*

Internal displacement is also possible when the groups are on adjacent carbons. The product in this case contains a three-membered ring. 2-Chloroethanol undergoes cyclization in base to form ethylene oxide (an epoxide, Section 3.7); the 2-chloroamine gives the carcinogen ethylene imine.

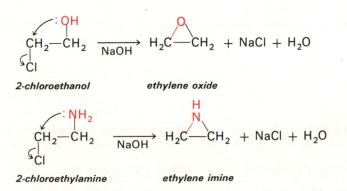

2-chloroethanol   *ethylene oxide*

2-chloroethylamine   *ethylene imine*

The reaction is carried out on a large scale industrially to prepare 3-chloro-propylene oxide, which is an intermediate in the preparation of epoxy resins, discussed in Section 7.8.

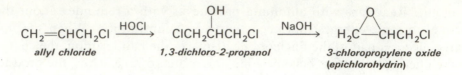

$$\text{CH}_2\!\!=\!\!\text{CHCH}_2\text{Cl} \xrightarrow{\text{HOCl}} \underset{\text{OH}}{\text{ClCH}_2\overset{|}{\text{CHCH}_2\text{Cl}}} \xrightarrow{\text{NaOH}} \underset{O}{\text{H}_2\text{C}\!\!-\!\!\text{CHCH}_2\text{Cl}}$$

allyl chloride   1,3-dichloro-2-propanol   3-chloropropylene oxide (epichlorohydrin)

Problem 6.8   Write the products that would be obtained from the reaction of base with a) 4-bromo-1-pentanol, b) 1-chloro-2-methyl-2-propanol.

SUMMARY AND HIGHLIGHTS

1. In **nucleophilic substitution** (S_N reaction), a *leaving group* such as halogen with its bonding electron pair is replaced by a group (the nucleophile) that can supply electrons to carbon.

2. Nucleophilic substitution can occur in two ways. In the S_N2 process the nucleophile "pushes off" the leaving group from the rear side; the configuration is inverted if the carbon bearing the leaving group is asymmetric. In the S_N1 process, the leaving group first dissociates, giving a **carbocation,** and the nucleophile then enters; racemization occurs if the carbon is asymmetric.

3. In the S_N2 reaction the halide reactivity sequence is $1° > 2° > 3°$. In the S_N1 process the sequence is the reverse $(3° > 2° > 1°)$ because carbocation stability is the major factor.

4. **Elimination** occurs by removal of a β-hydrogen with base and loss of the leaving group to give a *double bond*. In the **E2** process, the hydrogen is removed as the group departs; in **E1** reactions, the leaving group departs first and the base removes a proton from the carbocation.

5. Substitution and elimination may be competing reactions, since the nucleophile can also react as a base.

6. S_N2 reactions are useful for the preparation of *cyanides, alkynes, ethers, thiols,* and *amines*.

7. **Intramolecular** displacements can give cyclic products such as *epoxides*.

ADDITIONAL PROBLEMS

6.9 Nitrogen is vastly cheaper than Freon and is nontoxic and nonflammable. Why can't the liquid in aerosol cans be pressured by nitrogen instead of Freon vapor?

6.10 Write the structures of the following compounds:

     a)   vinyl chloride   b)   allyl bromide

     c)   diphenylmethyl chloride   d)   3-iodo-1-heptene

     e)   *p*-trifluoromethylnitrobenzene   f)   hexachloroethane

6.11 1-Bromo-2,2-dimethylpropane is a primary alkyl bromide, but it shows extremely low reactivity in S_N2 reactions. Suggest a reason.

6.12   Show how the following compounds could be obtained from halides:

a) $CH_3NH_3{}^+Br^-$

b) $C_6H_5CHOHCH_2C_6H_5$

c) $CH_3CH_2CH{=}CH_2$

d) $CH_3CH_2CH_2CH_2SH$

e) $CH_3CH_2C(CH_3)_2OCH_2CH_3$

6.13   Predict which of the compounds in the following pairs would be more reactive in the S_N1 reaction and explain why.

a)   1-chloro-2-butene or 2-chloro-1-butene

b) $(CH_3)_3CCl$ or $(C_6H_5)_3CCl$ (triphenylmethyl chloride)

c) $CH_3OCH_2CH_2Cl$ or $CH_3CHClOCH_3$

d)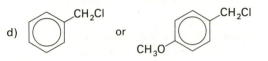

6.14   Compound **A**, C_3H_5Cl, is *less* reactive than either isomer of C_3H_7Cl toward alcoholic $AgNO_3$. Suggest the structure of **A**.

6.15   The *cis*- and *trans*-isomers of 1-bromo-2-methylcyclopentane both undergo dehydrohalogenation (loss of HBr) on treatment with KOH. From one isomer the major product is 3-methylcyclopentene, and from the other the main product is 1-methylcyclopentene. Write perspective structures of the two bromo compounds, showing the configurations, deduce which bromide gives which alkene, and state your reasoning.

6.16   Write structures for the products that would be formed in the following reactions:

a)   2-bromopropane plus $NaC{\equiv}N \longrightarrow$

b)   1-chloropentane plus $NaSCH_3 \longrightarrow$

c)   3-bromo-3-methylhexane plus $NaOCH_3 \longrightarrow$

d)   R-2-chloro-2-phenylbutane plus C_2H_5OH-$AgNO_3 \longrightarrow$

6.17   **3R**-Chloro-**2s**-butanol and **3s**-chloro-**2s**-butanol give different epoxides on reaction with base. Write the structure of the epoxide in each case. Would either epoxide be optically active?

6.18   Show how the following compounds could be prepared from the starting materials indicated plus any other compounds needed. In some cases, more than one step is required, and reactions covered in earlier chapters may be needed. In problems of this type, work backward from the final compound wanted, one step at a time.

a) $CH_3CH_2CH_2I$ from an alkyl chloride

b) $(CH_3)_2CHCH_2C{\equiv}CCH_3$ from acetylene and other compounds

c)   cyclohexene from cyclohexane

d) $\begin{array}{c} CH_2{-}CH_2 \\ |\qquad\ | \\ CH_2{-}O \end{array}$ from an open chain compound

e)   ⬡CH_2SH from toluene

f)   ⬡$C{\equiv}CH$ from styrene ($C_6H_5CH{=}CH_2$)

6.19   A compound $C_7H_{15}Br$ was treated with base, and a mixture of two alkenes with the formula C_7H_{14} was obtained. The alkenes could not be separated, so the mixture of the two was subjected to ozonolysis. Four compounds were obtained; in order of decreasing boiling point, these were:

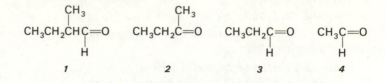

1   2   3   4

Write the structures of the two alkenes and suggest the structure of the original $C_7H_{15}Br$ compound.

6.20 The optical activity of a solution of **R**-2-iodopentane in acetone containing NaI decreases to zero on standing. Suggest an explanation for this observation.

7

ALCOHOLS, ETHERS, AMINES, AND THIOLS

The compounds discussed in this chapter contain alkyl groups attached to oxygen, nitrogen, or sulfur atoms. These compounds can be considered as organic derivatives of H_2O, NH_3, and H_2S, and although they represent several different homologous series, they have some important properties in common.

7.1 Structures and Names

Alcohols have the general formula **ROH,** in which R, as usual, denotes any alkyl group. The systematic name of an alcohol is formed by dropping the **-e** in the name of the parent alkane and adding the ending **-ol.** The chain is numbered to give the carbon atom bearing the —OH group the lowest number, and —OH takes precedence in the numbering even if a double bond is also present. Common names of simple alcohols can be formed from the alkyl group name and the separate word **alcohol.** As we have seen previously, the general structures RCH_2OH, R_2CHOH, and R_3COH are referred to as primary, secondary, and tertiary alcohols, respectively.

$CH_3CH_2CH_2OH$
1-propanol or
n-propyl alcohol

$ClCH_2CH_2CHOHCH_3$
4-chloro-2-butanol

$CH_2{=}CHCH_2OH$
2-propen-1-ol
or allyl alcohol

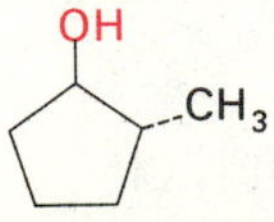

trans-2-methyl-
cyclopentanol

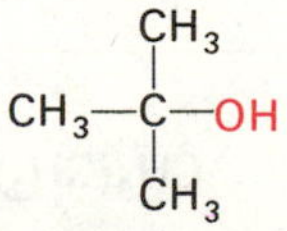

$$CH_3-\overset{\displaystyle CH_3}{\underset{\displaystyle CH_3}{C}}-OH$$
2-methyl-2-propanol
or tert-butyl alcohol

$HOCH_2CH_2CH_2CH_2OH$

1,4-butanediol

Ethers are compounds with two groups attached to oxygen, and have the general structure ROR. When the R groups are the same, or are very simple, the name is formed by indicating the two groups with **ether,** all as separate words. If the groups are different, we can consider the larger or more complex group as the parent compound, and name the other group as an **alkoxy** substituent.

$CH_3CH_2OCH_2CH_3$

diethyl ether

$CH_3OCH(CH_3)_2$

*methyl
isopropyl ether*

$CH_3CHCH_2CHCH_3$ with OCH_3 and CH_3 substituents

*2-methoxy-4-
methylpentane*

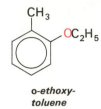

*o-ethoxy-
toluene*

Problem 7.1   a) Write structures and names for all alcohols with the molecular formula $C_5H_{12}O$. b) Classify each alcohol as primary, secondary, or tertiary. [Hint: work systematically, taking each of the isomeric pentanes in turn and placing an OH at each different position.]

Problem 7.2   How many *ethers* are there with formula $C_5H_{12}O$?

Amines are organic derivatives of NH_3 and are classified according to the number of hydrogens that are replaced by alkyl groups. **Primary** amines have the formula **RNH_2, secondary** amines **R_2NH,** and **tertiary** amines **R_3N.** (Note that primary, secondary, and tertiary when used with amines denote one, two, or three groups attached directly to *nitrogen,* in contrast to the use of these terms for alcohols, where they indicate the number of substituents on the carbon to which the —OH group is bonded.)

Formal names for primary amines can be formed by dropping the **-e** in alkane and adding **-amine** as an ending. In most cases, however, we will name amines by designating the alkyl groups together with the ending **-amine,** all as one word. Names for secondary or tertiary amines are often based on a parent compound with the prefix **alkylamino-** or **dialkylamino-.**

$CH_3CH_2CH_2CH_2NH_2$

*1-butanamine or
n-butylamine*

CH_3NH_2

methylamine

$(CH_3CH_2CH_2)_2NH$

di-n-propylamine

$(CH_3CH_2)_3N$

triethylamine

$(CH_3)_2NCH_2CH_2OH$

2-(dimethylamino)ethanol

cyclohexane with $NHCH_3$

methylaminocyclohexane

Quaternary ammonium salts, $R_4N^+X^-$, contain four groups attached to the nitrogen atom, and are named by specifying the four groups and the anion.

$(CH_3)_4\overset{+}{N}O\overline{H}$

*tetramethylammonium
hydroxide*

$CH_3(CH_2)_9\overset{+}{N}(CH_3)_3\ Cl^-$

*n-decyl-trimethylammonium
chloride*

Thiols, RSH, are the sulfur analogs of alcohols. The prefix **thio-** in any chemical name always means that an oxygen atom has been replaced by sulfur. Thus, we can name a compound of the formula RSH simply by changing the **-ol** in the alcohol name to **-thiol.** Another name that is occasionally used for thiol is **mercaptan.** The sulfur counterparts of ethers, RSR, are usually named by replacing the word ether by **sulfide.**

$$CH_3CH_2SH \qquad CH_3SCH_2CH_2CH_3$$

ethanethiol   *methyl n-propyl sulfide*
(ethyl mercaptan)

Problem 7.3 Write structures for: a) 2-ethyl-1-hexanol, b) 2-methoxyethanol, c) 2-butyne-1,4-diol, d) methyl benzyl ether, e) 1-methylcycloheptanol, f) *n*-octylamine, g) 2-dimethylaminobutane, h) 1-butanethiol.

7.2 Synthesis of Alcohols

For several reasons, alcohols are a particularly important class of compounds. As we will see in later sections, the OH group can react in several ways, and these reactions provide routes to other types of compounds. Moreover, alcohols and compounds derived from them have some important end uses, and a number of alcohols are produced industrially on a large scale. The major methods for preparing alcohols are summarized below; several of these involve reactions of other functional groups which will be discussed in later chapters.

PREPARATION OF ALCOHOLS

1. **Acid-catalyzed hydration of alkenes.**

$$RCH{=}CH_2 + H_2O \xrightarrow{H^+} \overset{OH}{\underset{}{R}CHCH_3}$$

(Sects. 3.5 and 7.2)

2. **Hydroboration-oxidation of alkenes**

$$RCH{=}CH_2 \xrightarrow{(BH_3)} (RCH_2CH_2)_3B \xrightarrow[OH^-]{H_2O_2} RCH_2CH_2OH$$

(Sect. 7.2)

3. **Hydrolysis of alkyl halides.**

$$RCl + NaOH \longrightarrow ROH + NaCl$$

(Sect. 6.10)

4. **1,2-Diols from epoxides.**

$$\overset{O}{\underset{}{-C-C-}} + H_2O \xrightarrow[OH^-]{H^+ \text{ or }} \overset{OH\ OH}{\underset{}{-C-C-}}$$

(Sect. 7.8)

---

5. Reduction of carbonyl compounds.

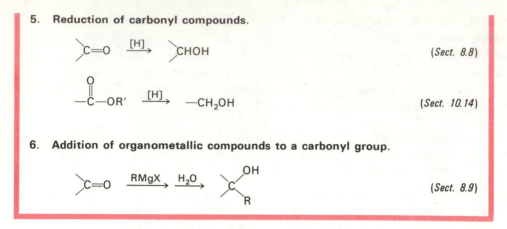

$$\text{>C=O} \xrightarrow{\text{[H]}} \text{>CHOH}$$ (Sect. 8.8)

$$\text{—C—OR'} \xrightarrow{\text{[H]}} \text{—CH}_2\text{OH}$$ (Sect. 10.14)

6. Addition of organometallic compounds to a carbonyl group.

$$\text{>C=O} \xrightarrow{\text{RMgX}} \xrightarrow{\text{H}_2\text{O}} \underset{\text{R}}{\overset{\text{OH}}{\text{>C<}}}$$ (Sect. 8.9)

Preparation of Alcohols from Alkenes. Alcohols can be obtained from alkenes in two ways. The first method, already discussed in Section 3.5, involves the direct acid-catalyzed addition of water to a double bond to give the alcohol with the OH group on the more highly substituted carbon atom. A major use of this method is the industrial preparation of isopropyl alcohol.

$$\text{CH}_3\text{CH=CH}_2 + \text{H}_2\text{O} \xrightarrow{\text{H}^+} \overset{\text{OH}}{\text{CH}_3\text{CHCH}_3}$$

The second method for preparing alcohols from alkenes is a two-stage sequence called **hydroboration-oxidation.** In the first reaction a B—H bond from borane, BH_3,* adds to the π bond of an alkene. Boron is less electronegative than hydrogen, and the BH bond is therefore polarized in the direction B+⟶H, opposite to that in acids such as HCl. For this reason as well as the fact that hydrogen is smaller than boron, the addition occurs with hydrogen becoming bonded to the more substituted carbon.

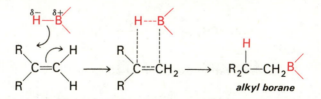

In the second step, the alkyl borane is treated with alkaline hydrogen peroxide, and the C—B bond breaks to give a C—OH group. The overall result of the two steps is addition of water to the double bond in the direction *opposite* that followed in the direct acid-catalyzed hydration.

Overall Hydroboration-Oxidation:

$$3\ \text{R}_2\text{C=CH}_2 \xrightarrow{\text{(BH}_3\text{)}} (\text{R}_2\text{CHCH}_2)_3\text{B} \xrightarrow[\text{OH}^-]{\text{H}_2\text{O}_2} 3\ \text{R}_2\text{CHCH}_2\text{OH} + \text{H}_3\text{BO}_3$$

---

*The actual reagent used is diborane, B_2H_6, which reacts as $2\,BH_3$.

Hydroboration-oxidation is a very useful process, particularly since it provides a means of synthesizing alcohols that cannot be obtained by direct addition of water to an alkene.

Problem 7.4   Write the structures of the alcohols that would be obtained by (1) acid-catalyzed addition of water and (2) hydroboration of a) styrene (phenylethylene), b) 2-methyl-2-butene, c) cycloheptene.

7.3   Special Sources of Alcohols

A few important alcohols are obtained by special methods. **Methanol** is manufactured by hydrogenation of carbon monoxide. Mixtures of CO and H_2 are obtained from coal and steam, and the mixture is passed over a catalyst at high temperature and pressure to produce methanol.

$$CO + 2\,H_2 \quad \xrightarrow[\text{350°, 4000 psi}]{\text{catalyst}} \quad CH_3OH$$

Methanol is a highly poisonous compound; amounts that are less than lethal can cause visual impairment and complete blindness.

The manufacture of **ethanol** by fermentation of grain is the oldest chemical process known, dating back at least 5500 years. In fermentation, the action of enzymes in yeasts causes the breakdown of starch to glucose, and then to ethanol plus carbon dioxide. The same reaction is responsible for the formation of alcohol in brewing or wine making, and of CO_2 in the leavening of dough.

$$C_6H_{12}O_6 \longrightarrow 2\,C_2H_5OH + 2\,CO_2$$
$$\textit{glucose}$$

Fermentation can proceed to a concentration of about 12 per cent alcohol, at which point the growth of the yeast is inhibited. This is the concentration of alcohol in most natural wines. More concentrated solutions are obtained by distillation of wine or grain mashes; beverages such as whiskey and brandy are 40 to 45 per cent alcohol (80 to 90 proof).* These distilled spirits derive their distinctive flavors from traces of other volatile compounds formed in the fermentation and by oxidation during the aging process.

A small fraction of ethanol for industrial use is produced by fermenta-

---

* "Proof spirit" refers to an eighteenth century method for determining the alcohol content of whiskey. A sample was poured over gunpowder, and if the alcohol content was high enough (about 50 per cent) to permit the moist powder to burn, the sample was designated "proof spirit." The term became standardized to mean a distillate containing 49.3 per cent by weight (57 per cent by volume) of ethanol.

tion of molasses, but most of it is prepared by acid-catalyzed addition of water to ethylene.

$$CH_2{=}CH_2 + H_2O \xrightarrow{H^+} CH_3CH_2OH$$

Glycerol or 1,2,3-propanetriol is obtained from fat, in which it is present in combined form. Treatment of the fat with NaOH (Section 10.16) liberates the glycerol and a soap.

$$\underset{\textit{a fat}}{RCOCH_2CHCH_2OCR} + 3\ NaOH \longrightarrow \underset{\textit{a soap}}{3\ RCONa} + \underset{\textit{glycerol}}{HOCH_2CHCH_2OH}$$

Another group of alcohols derived from natural sources are steroid alcohols or **sterols,** present in all plant and animal tissues. The most important sterol is **cholesterol,** which is the major constituent of spinal cords. In certain disease conditions, cholesterol accumulates as bladder stones, and in atherosclerosis, it is deposited in blood vessels. Cholesterol is built up from isoprene units by way of the triterpenes **squalene** and **lanosterol** (the main ingredient of lanolin). Cholesterol is the parent compound of a large family of compounds called **steroids,** which have in common a structure with one five-membered and three six-membered rings fused together. Other important steroids are the cardiotonic drugs obtained from digitalis and the steroid hormones (Chapter 8).

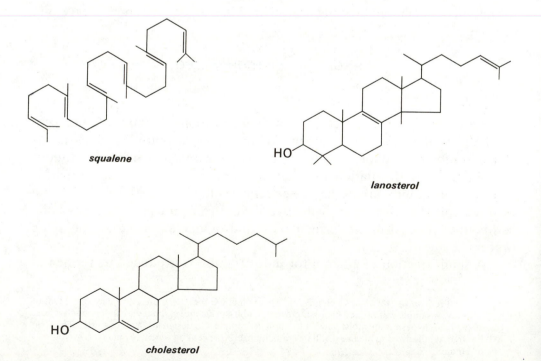

squalene

lanosterol

cholesterol

7.4 Synthesis of Ethers and Amines

As a class of compounds, ethers are not of major importance, and most of the occasions for their synthesis can be met by the first method indicated.

METHODS FOR PREPARATION OF ETHERS

1. **Reaction of alkyl halides and alkoxides.**

$$RCH_2X + R'ONa \longrightarrow RCH_2OR' + NaX$$

(Sects. 6.10 and 7.7)

2. **Symmetrical ethers from alcohols.**

$$2\ ROH \xrightarrow[140°]{H_2SO_4} ROR + H_2O$$

(Sect. 7.12)

The most reliable methods for preparing specific amines are those involving the reduction of compounds containing C—N multiple bonds and N—O bonds.

METHODS FOR PREPARATION OF AMINES

1. **Reaction of alkyl halides and ammonia or amines.**

$$RX + NH_3 \longrightarrow RNH_3{}^+X^- \xrightarrow{OH^-} RNH_2$$

(Sect. 7.7)

$$R'X + RNH_2 \longrightarrow R'NH_2R^+X^-$$

2. **Reduction of nitriles.**

$$RC{\equiv}N + 2\ H_2 \xrightarrow{Ni} RCH_2NH_2$$

(Sect. 10.13)

3. **Reduction of** $\diagup$ C=N **bonds.**

$$\diagup C{=}O + NH_2R \longrightarrow \diagup C{=}NR \xrightarrow{H_2,\ Ni} \diagup CH{-}NHR$$

(Sect. 8.14)

4. **Reduction of nitro compounds.**

$$ArNO_2 + 3\ H_2 \xrightarrow{Ni} ArNH_2 + 2\ H_2O$$

(Sect. 12.6)

7.5 Physical Properties

Alcohols with up to about 14 carbon atoms are liquids with sweetish odors; longer chain alcohols, like alkanes, are waxy solids. Amines and thiols have highly characteristic and unpleasant odors. The lower alkylamines have sharp odors resembling both ammonia and decaying fish. Thiols and sulfides are foul-smelling substances. Minute traces of thiols are added to domestic natural gas to provide a warning in case of a leak.

Several important properties of alcohols are due to the presence of the O—H bond, which is very similar in length and dipole moment to that in water. Two major effects of the OH group are the relatively high boiling points and the water solubility of alcohols. We can see the influence on boiling points by comparing the values for similar compounds of about the same molecular weight. Ethanol has a boiling point over 100° higher than that of the isomeric compound dimethyl ether.

CH_3CH_2OH CH_3OCH_3 $CH_3CH_2CH_3$ CH_3SH

ethanol   *dimethyl ether*   *propane*   *methanethiol*
mw 46   mw 46   mw 44   mw 48
bp 78°   bp −24°   bp −42°   bp 6°

The influence of the OH group is even more strikingly demonstrated by the properties of diols and triols. These compounds have very high boiling points; they are completely miscible with water, but are only slightly soluble in alkanes.

$HOCH_2CH_2OH$ $HOCH_2CHOHCH_2OH$

1,2-ethanediol   *1,2,3-propanetriol*
(ethylene glycol)   *(glycerol)*
bp 197°   bp 290°

The abnormally high boiling points of alcohols compared to those of the isomeric ethers indicate that alcohols are much more highly associated in the liquid state than are hydrocarbons or ethers. This extra attractive force is called **hydrogen bonding,** and is a specific interaction between the hydrogen of the OH group and an unshared electron pair on the oxygen atom of another molecule. Hydrogen bonding is usually represented by a dotted bond, as in Figure 7.1. This additional intermolecular bonding must be overcome when the liquid vaporizes, and a higher temperature is thus required than in the absence of the hydrogen bond. Hydrogen bonding depends on the dipole that exists between atoms of very different electro-

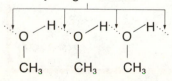

hydrogen bonds

FIGURE 7.1

Hydrogen bonding in methanol.

negativity, and is significant only with protons bonded to oxygen, nitrogen, and fluorine.

Problem 7.5   The strength of hydrogen bonding in alcohols depends on the extent to which the OH group is exposed and available for bonding to another molecule. The OH group is most exposed when the alcohol is primary, and is most shielded from hydrogen bonding in tertiary alcohols. With this in mind, and also the fact (discussed in Section 2.2) that chain-branching lowers the boiling point, write the structures of the four alcohols with formula $C_4H_{10}O$ and match the structures with the boiling points, which are 83°, 99°, 108°, and 118°.

Alcohols with three or fewer carbons are soluble in water in all proportions. Higher alcohols are not completely miscible, but they have much higher solubilities than do hydrocarbons or ethers of comparable size. We can readily account for the higher solubility of alcohols in water by hydrogen bonding between the OH group of the alcohol and the oxygen of water molecules, and *vice versa*. Ethers are intermediate in water solubility between alcohols and alkanes.

Problem 7.6   Diethyl ether has a solubility of 7 per cent in water, compared to 12 per cent for 2-butanol and 0.03 per cent for *n*-pentane. Suggest an explanation for this order of solubilities.

The physical properties of primary and secondary amines reflect hydrogen bonding of the $-NH_2$ and $-NH-$ groups. However, the $N-H---N$ hydrogen bond is weaker, as seen by comparing the boiling points of ethanol (bp 78°) and ethylamine (bp 16°). Amines of all types with up to four carbons are completely soluble in water because of hydrogen bonding to water molecules.

7.6   Acidic and Basic Properties

Alcohols.   The most fundamental chemical property of alcohols, like water, is the **amphoteric** behavior of the OH group, *i.e.*, the ability to react either as an acid or as a base.

The reaction of an alcohol as an acid gives the **alkoxide anion RO⁻**. Since alcohols are relatively weak acids, it follows that the alkoxides are strong bases. An alcohol can be converted to the alkoxide by reaction with metallic sodium. Hydrogen gas is evolved; with methanol or ethanol, care must be taken to avoid a fire. Tertiary alcohols react very slowly with sodium, and potassium is used instead to obtain the alkoxide in these cases. The sodium or potassium alkoxides can be isolated as solids by evaporation of excess alcohol. Alcohols are weaker acids than water, and alkoxides therefore react with water to give the alcohol plus hydroxide ion.

$$C_2H_5OH + Na \longrightarrow CH_3CH_2O^-Na^+ + \tfrac{1}{2}H_2$$
sodium ethoxide

$$C_2H_5ONa + H_2O \longrightarrow C_2H_5OH + NaOH$$

Problem 7.7   Write an equation for the reaction in which *tert*-butyl alcohol is converted to *tert*-butoxide anion.

Alcohols behave as bases toward sulfuric acid and other very strong acids. An unshared electron pair on oxygen accepts a proton, and the alcohol is converted to an alkyl oxonium ion, $\overset{+}{R}OH_2$, analogous to the formation of H_3O^+ when H_2SO_4 or HCl is added to water.

$$ROH + H_2SO_4 \longrightarrow ROH_2^+ + HSO_4^-$$
oxonium ion

Ethers have similar basic properties and are protonated by strong acids to give dialkyl oxonium ions, R_2OH^+.

$$ROR + H_2SO_4 \longrightarrow R_2OH^+ + HSO_4^-$$

Thiols are stronger acids than alcohols or water, and a **thiolate anion** RS^- is a weaker base than OR^- or OH^-. This means that a thiol can be converted to the thiolate by dissolving it in aqueous hydroxide solution.

$$RSH + NaOH \longrightarrow RS^-Na^+ + H_2O$$
sodium thiolate

Amines.   Just as the amphoteric properties of alcohols resemble those of water, the acidity and basicity of primary and secondary **amines** parallel those of ammonia. Since nitrogen is less electronegative than oxygen, an NH bond is much less acidic than OH. The anions RNH^- and R_2N^- can be obtained by adding metallic sodium or lithium to the amine, and are extremely strong bases. In water or alcohols, these amide anions are protonated, and hydroxide or alkoxide is formed.

$$RNH_2 + Li \longrightarrow RNH^- Li^+ + \tfrac{1}{2} H_2$$
$$RNH^- + R'OH \longrightarrow RNH_2 + R'O^-$$

The most important property of amines is their basicity, which is due to the readily available unshared electron pair on nitrogen. All classes of amines (RNH_2, R_2NH, and R_3N) are much stronger bases than are water or alcohols, and a solution of an amine in water thus has an alkaline pH. The most useful way to look at the basicity of amines is in terms of competition with water for a proton. In acid solutions, the equilibrium between an amine R_3N and the conjugate acid R_3NH^+ lies far to the right. Stated another way, the amine, which is more basic than water, neutralizes the acid H_3O^+ to give the ammonium ion and water.

$$R_3N: + H_3O^+ \rightleftharpoons R_3NH^+ + H_2O$$
amine   alkyl ammonium ion

Problem 7.8   In the equations in this chapter, unshared electron pairs are generally not shown, but they should be understood to be present as required by the formula and charge. Write Lewis electron structures, showing all unshared electrons, for the following molecules or ions: a) ROH, b) ROH_2^+, c) RS^-, d) RNH^-, e) NH_3, f) RNH_3^+.

The fact that an amine is converted to an ionic ammonium salt in acid solution means that amines can be dissolved in aqueous acid even though they may not be soluble in water. Thus, if the alkyl chains attached to nitrogen contain more than five or six carbons in all, the hydrocarbon character of the molecule becomes dominant and reduces the water solubility of the amine. However, the ionic ammonium salt will be completely soluble. We can take advantage of this behavior to separate an amine from a mixture with other organic compounds. When a solution of the mixture in an organic solvent is shaken with dilute hydrochloric acid, the amine dissolves in the water layer. After separation of the layers, the amine is recovered from the water solution by neutralizing the acid with hydroxide ion.

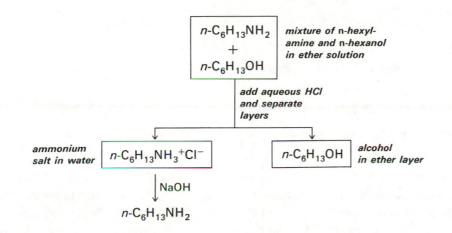

Problem 7.9   Arrange each of the following groups in order of *decreasing* acidity: a) CH_3OH, CH_3SH, CH_3NH_2; b) C_2H_5OH, $C_2H_5OH_2^+$, $C_2H_5NH_3^+$; c) H_2O, C_2H_5OH, H_3O^+.

7.7 Nucleophilic Properties of Alcohols, Amines, and Thiols

In reactions of alcohols, amines, and thiols as **nucleophiles**, an unshared electron pair from oxygen, nitrogen, or sulfur is donated to some electrophile other than a proton. The most important reactions are those in which carbon is the electrophilic center. In this and other chapters we will encounter several different functional groups that contain an electrophilic carbon atom. All of these groups react with alcohols, amines, and thiols in the same way—an electron pair from —OH, —NH—, or SH forms a σ bond to carbon. At some stage of the process a proton is usually detached from the nucleophilic O, N, or S atom. Several important examples are shown in the following chart, with RṄuH representing the alcohol, amine, or thiol.

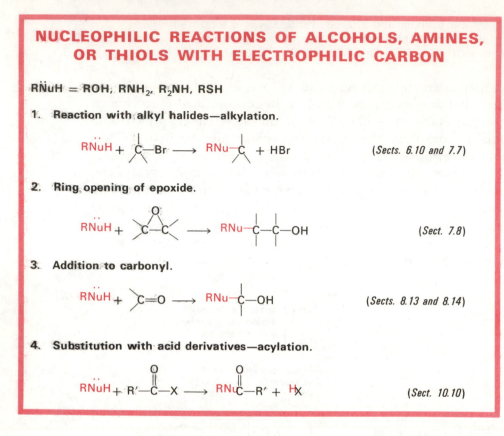

NUCLEOPHILIC REACTIONS OF ALCOHOLS, AMINES, OR THIOLS WITH ELECTROPHILIC CARBON

RN̈uH = ROH, RNH$_2$, R$_2$NH, RSH

1. **Reaction with alkyl halides—alkylation.**

   RN̈uH + C—Br ⟶ RNu—C + HBr   (*Sects. 6.10 and 7.7*)

2. **Ring opening of epoxide.**

   RN̈uH + C—C(O) ⟶ RNu—C—C—OH   (*Sect. 7.8*)

3. **Addition to carbonyl.**

   RN̈uH + C=O ⟶ RNu—C—OH   (*Sects. 8.13 and 8.14*)

4. **Substitution with acid derivatives—acylation.**

   RN̈uH + R′—C(=O)—X ⟶ RNu—C(=O)—R′ + HX   (*Sect. 10.10*)

Alcohols and Thiols.   In Section 6.10 we saw the reactions of alcohols, thiols, and amines as nucleophiles with alkyl halides. For the preparation of **ethers,** an alcohol is not sufficiently nucleophilic to react by an S$_N$2 substitution mechanism, and the first step is removal of a proton to give the alkoxide.

$$ \text{ROH} \xrightarrow{\text{Na}} \text{RO}^{:-}\text{Na}^+ \xrightarrow{\text{CH}_3\text{CH}_2\text{Br}} \text{ROCH}_2\text{CH}_3 + \text{NaBr} $$

Thiols are stronger bases than alcohols, and thiols and thiolate anions are also more reactive as nucleophiles than are the corresponding alcohols or alkoxides. **Sulfides** can be obtained by the S$_N$2 reaction of a thiol and an alkyl halide in the presence of sodium hydroxide.

$$ \text{RSH} \xrightarrow{\text{NaOH}} \text{RS}^{:-}\text{Na}^+ \xrightarrow{\text{CH}_3\text{CH}_2\text{Br}} \text{RSCH}_2\text{CH}_3 $$

Amines.   Amines are sufficiently nucleophilic as neutral molecules to react directly with alkyl halides and form dialkylammonium salts. These salts can then be neutralized with base to give a more highly substituted amine.

$$ \text{CH}_3\ddot{\text{N}}\text{H}_2 + \text{CH}_3\text{I} \xrightarrow[(\text{S}_N2)]{} \underset{\underset{\text{CH}_3}{|}}{\text{CH}_3\overset{+}{\text{N}}\text{H}_2} \text{I}^- \xrightarrow{\text{OH}^-} (\text{CH}_3)_2\text{NH} $$

methyl-   dimethyl-   dimethylamine   ammonium iodide   amine

A drawback to this reaction as a method for synthesizing amines is the fact that the starting amine also can act as a base toward the dialkylammonium salt to liberate the more substituted amine during the reaction. The new amine can then react as the nucleophile with the alkyl halide, and the substitution continues until all hydrogens on the nitrogen have been displaced to form a **quaternary ammonium salt.**

$$CH_3\overset{..}{N}H_2 + CH_3I \longrightarrow (CH_3)_2\overset{+}{N}H_2 \ I^- \underset{\xleftarrow{CH_3\overset{..}{N}H_2}}{\overset{CH_3\overset{..}{N}H_2}{\rightleftarrows}} (CH_3)_2\overset{..}{N}H + CH_3\overset{+}{N}H_3 \ I^-$$

$$\Big\backslash CH_3I$$

$$\underset{\xrightarrow{\overset{|}{\underset{|}{}}N:}}{}$$

$$(CH_3)_4\overset{+}{N} \ I^- \xleftarrow{CH_3I} (CH_3)_3N: \xleftarrow{} (CH_3)_3\overset{+}{N}H \ I^-$$

tetramethyl-
ammonium iodide

trimethyl-
amine

(a quaternary
ammonium salt)

Because of the subsequent acid-base reactions, substitution reactions of amines are of limited usefulness as a preparative method. If a quaternary ammonium salt is the desired product, the reaction with the alkyl halide is carried out in the presence of an inorganic base to neutralize the intermediate ammonium ions at each step.

$$RNH_2 + 3 \ CH_3Cl + K_2CO_3 \longrightarrow R\overset{+}{N}(CH_3)_3 \ Cl^- + 2 \ KCl + CO_2 + H_2O$$

The quaternary ammonium salts which are the end product in these substitution reactions are "permanent" cations. Since there is no proton that can be removed from the ammonium nitrogen, the quaternary ammonium ion forms stable, neutral salts $R_4N^+X^-$, analogous to Na^+X^-. Quaternary ammonium salts with long alkyl chains have surfactant and bactericidal properties, and "alkyl dimethyl benzyl ammonium chlorides" with alkyl chains 14 to 16 carbons long are used in small amounts in most germicidal soaps and cleansers.

Problem 7.10 A typical cationic surfactant is

$$\left[\begin{array}{c} CH_3 \\ | \\ C_6H_5CH_2\overset{}{N}-(CH_2)_{13}CH_3 \\ | \\ CH_3 \end{array} \right]^+ \ Cl^-$$

Show how the compound could be prepared from $(CH_3)_2N(CH_2)_{13}CH_3$.

7.8 Ring Opening of Epoxides

Reactions of alcohols and amines with epoxides lead to some very useful products. An epoxide ring is highly strained because the normal bond angles of 105° to 110° for sp^3 carbon and oxygen are compressed to about 60°. The ring is therefore very susceptible to the attack of a nucleophile at carbon. Unlike ethers in general, which are quite unreactive, a C—O bond in the epoxide breaks, and the three-membered ring opens up. As the reaction progresses, a proton is transferred to the oxygen. A typical example is the reaction of ethylene oxide with water to give ethylene glycol, which is used in enormous volumes as a permanent antifreeze and as a component in polyester fiber.

$$H_2O: \overset{+}{} CH_2—CH_2 \longrightarrow HO\,CH_2CH_2OH$$
$$\textit{ethylene oxide} \qquad \textit{ethylene glycol}$$

With an alcohol as the nucleophile, the product is the 2-alkoxyethanol. If the reaction is carried out with a large molar excess of the oxide, the newly formed OH group can attack another mole of oxide, and a chain of any desired length with repeating —(CH_2CH_2O)— units can be obtained.

$$ROH + CH_2—CH_2 \longrightarrow RO\,CH_2CH_2OH \xrightarrow{n\ CH_2—CH_2} RO\,CH_2CH_2O(CH_2CH_2O)_nH$$
$$\textit{2-alkoxyethanol}$$

Problem 7.11   Write structures for the products that would be obtained in the reaction of ethylene oxide (one mole) with a) diethylamine, b) methanethiol.

Problem 7.12   2-Methoxyethanol is used as a jet fuel deicer. Suggest a synthesis of this compound.

The ring opening of epoxides with good nucleophiles such as amines is a rapid, exothermic reaction. Advantage is taken of this reaction in **epoxy resins** and adhesives. The formation of an epoxy resin is a two-step operation. First, a low molecular weight polymer is prepared with an epoxide group at each end. Epichlorohydrin is usually used in this step, taking advantage of the fact that the primary chloride readily undergoes S_N2 displacement. In the final step, the epoxy polymer is mixed with a triamine to cross-link the polymer by opening the epoxide rings.

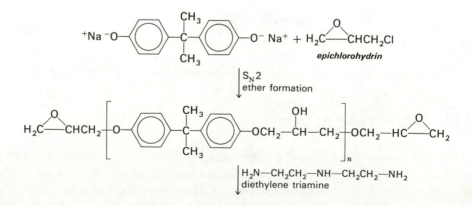

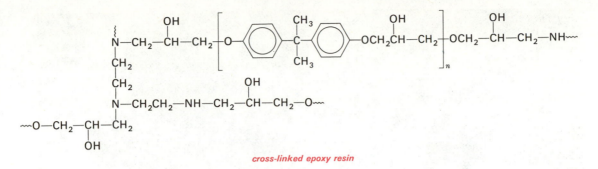

cross-linked epoxy resin

7.9 Esters of Inorganic Acids

The products formed by splitting out water between an alcohol and an acid are called **esters**. Alkyl esters of sulfuric, nitric, and phosphoric acids are all important compounds in a variety of ways.

$$R-OH + H-A \longrightarrow R-A + H_2O \qquad \text{where } A = HSO_4^-, NO_3^-, H_2PO_4^-$$

alcohol   *acid*   *ester*

Sulfuric acid, with two OH groups, forms both **monoalkyl sulfates** **(ROSO$_2$OH)** and **dialkylsulfates (ROSO$_2$OR)**. The mono esters can be neutralized to give sodium alkyl sulfates. With a long alkyl chain, these compounds are effective detergents and are more readily biodegradable than alkylbenzene sulfonates. Sodium dodecyl (lauryl) sulfate is a component of shampoos, toothpaste, and many other cleansing agents.

$$CH_3(CH_2)_{10}CH_2OSO_3^- Na^+$$

sodium lauryl sulfate

The **nitrate esters** of glycerol and a few other polyhydroxy compounds are the main ingredients in high explosives. Glyceryl trinitrate, commonly called nitroglycerine, is a very dangerous substance; but when it is absorbed in sawdust or clay it is a safe and dependable explosive used in dynamite (Figure 7.2). Another major component of dynamite is ammonium nitrate.

$$\underset{\text{nitroglycerine}}{HOCH_2-\underset{\underset{OH}{|}}{CH}-CH_2OH + H\,NO_3 \xrightarrow{H_2SO_4} O_2NOCH_2-\underset{\underset{ONO_2}{|}}{CH}-CH_2\,ONO_2}$$

An effective explosive for demolition, mining, or quarrying must decompose very rapidly with liberation of a large amount of heat and a large

FIGURE 7.2

The Ripple Rock blast. These photographs show the removal of a barely submerged peak of rock in the inland shipping channel to Alaska in Seymours Narrows. This blast, the largest nonatomic explosion fired, used 2.7 million pounds of dynamite to remove a navigation hazard which had destroyed over 100 vessels.

volume of gas. To achieve this, the ratio of carbon to nitrogen to oxygen is usually close to a composition corresponding to self-combustion:

$$4\ C_3H_5N_3O_9 \longrightarrow 12\ CO_2 + 10\ H_2O + 6\ N_2 + O_2$$

nitroglycerine

Other requirements for a useful explosive are safety in handling and sensitivity to detonation under reproducible conditions. High explosives are detonated by a blasting cap which contains a small amount of a highly sensitive primer such as lead azide, $Pb(N_3)_2$.

Esters of phosphoric acid, **alkyl phosphates,** are central to every reaction that occurs in a living cell, because these compounds permit the storage and transfer of energy. The compounds involved in this function are esters of mono-, di-, and triphosphoric acid.

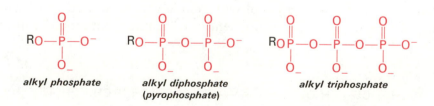

alkyl phosphate   *alkyl diphosphate*   *alkyl triphosphate*
                (pyrophosphate)

In the cell at pH 7 to 8, the acidic OH groups attached to phosphorus are neutralized. The R group in these energy-rich phosphates is a complex sugar derivative, discussed in Section 15.2. Dialkyl phosphate esters, $ROPO_2OR$, provide the links between unit of nucleic acids (Chapter 16).

7.10 Substitution and Elimination Reactions of Alcohols

In this section we will look at reactions of alcohols in which the C—O bond is broken in substitution or elimination reactions. These are the same processes that were discussed in Chapter 6 with alkyl halides, but now it is the OH group of an alcohol that is displaced or lost by elimination.

An essential requirement for substitution or elimination is a good leaving group. The group must be a stable anion or molecule with low basicity, or, to put it another way, the conjugate base of a strong acid. In reactions of alcohols, hydroxide ion cannot function as a leaving group since it is a very strong base. To use the alcohol as an electrophile in substitution or elimination reactions, therefore, it must be converted to some other form.

One way to provide the driving force for breaking the C—O bond is to take advantage of the basicity of the OH group and protonate the alcohol to give the oxonium ion. In this form, a water molecule (the conjugate base of H_3O^+) is the actual leaving group. Loss of water from the oxonium ion leads to the carbocation, which can then undergo either substitution or elimination.

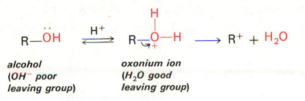

Another way to create a good leaving group is to convert the alcohol to an ester such as the dialkyl sulfate. A similar and more convenient ester is a **sulfonate,** prepared by reaction of the alcohol with toluenesulfonyl chloride.

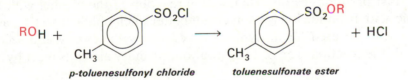

In these esters, the sulfate or sulfonate anions meet our requirements for a leaving group, and these derivatives can be used like alkyl halides in substitution reactions with nucleophiles.

$$(CH_3)_3N + CH_3OSO_2C_7H_7 \longrightarrow (CH_3)_4N^+ \ ^-OSO_2C_7H_7$$

trimethyl-   methyl toluene-   tetramethylammonium
amine   sulfonate   toluenesulfonate

In biochemical reactions, the phosphate or diphosphate group is the leaving group for substitution and elimination reactions, as seen for example in the biosynthesis of terpenes (p. 76).

7.11 Preparation of Alkyl Halides

The most useful and flexible method for preparing alkyl halides is replacement of the OH group of an alcohol by halogen. Secondary and tertiary alcohols react very readily with concentrated hydrochloric acid to give the alkyl chloride. With these alcohols, the reaction is usually an S_N1 substitution; the alcohol is protonated, the oxonium ion loses water, and the carbocation is captured by chloride ion.

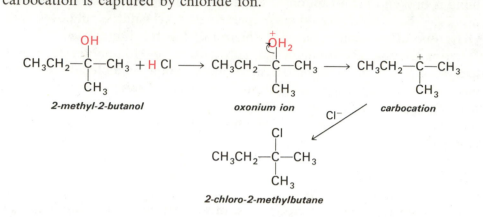

A better method for converting a primary or secondary alcohol to the chloride is by reaction with thionyl chloride, $SOCl_2$. An intermediate chlorosulfite ester is formed, and this then breaks down, with displacement of the C—O bond by Cl and loss of SO_2.

$$CH_3CH_2CH_2CH_2OH + SOCl_2 \longrightarrow CH_3CH_2CH_2CH_2O\overset{\overset{O}{\|}}{S}Cl + HCl \longrightarrow$$

n-butyl alcohol   *n-butyl chlorosulfite*

$$CH_3CH_2CH_2CH_2Cl + SO_2$$

n-butyl chloride

Alkyl bromides can be obtained by reaction of the alcohol with hydrobromic acid. Bromide ion is a somewhat better nucleophile than Cl^-, and this process can be used for all types of alcohols. With tertiary and secondary alcohols the reaction is an S_N1 substitution; primary alcohols react by an S_N2 mechanism.

> **Problem 7.13** Write out the steps in the preparation of a) 2-bromopentane and b) 1-bromopentane by the reaction of the alcohols with HBr.

Another way to obtain alkyl bromides is the reaction of the alcohol with phosphorus tribromide. In this method, the alkyl phosphite is the intermediate which undergoes substitution by bromide ion.

$$3 ROH + PBr_3 \longrightarrow (RO)_3P + 3 HBr \longrightarrow 3 RBr + P(OH)_3$$

7.12 Dehydration of Alcohols to Alkenes and Ethers

The elimination of water from an alcohol is called **dehydration;** this reaction is a useful method for preparing alkenes. Dehydration of secondary and tertiary alcohols occurs readily by the E1 process when the alcohol is heated with sulfuric or phosphoric acid. The anions of these acids are poor nucleophiles, and the carbocation that is formed by loss of water undergoes elimination without competing substitution. The major product is the alkene with the double bond between the most highly substituted carbons.

Dehydration is the reverse of the acid-catalyzed addition of water to alkenes described in Chapter 3. The same carbocation is involved in both reactions; the cation can either lose a proton or undergo nucleophilic attack by water, depending on the conditions. Treatment of the alcohol with anhydrous acid causes dehydration to the alkene; this reaction is promoted by raising the temperature. In the reverse process, the alcohol is formed by absorbing the alkene in aqueous acid.

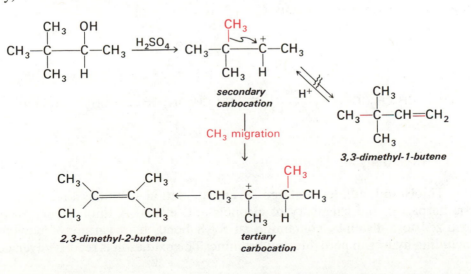

In any reaction involving carbocations, rearrangement of the carbon skeleton may occur if a more stable cation can be formed by a shift of a hydrogen atom or an alkyl group (see p. 55). For example, in the dehydration of 3,3-dimethyl-2-butanol, a secondary carbocation is formed by loss of water. An electron pair from an adjacent C—CH$_3$ bond moves into the vacant p orbital, leading to the *tertiary* carbocation. Loss of a proton gives 2,3-dimethylbutene in high yield. Rearrangement of the secondary carbocation also occurs if the isomeric 3,3-dimethyl-1-butene (prepared in another way) is treated with sulfuric acid.

Problem 7.14   Write structures of the alkenes that would be obtained by dehydration of a) cyclohexanol, b) 1-phenyl-2-propanol, and c) 2,2-dimethyl-1-propanol.

Dehydration of alcohols is often accomplished by passing the alcohol vapor over an acidic catalyst such as aluminum oxide (alumina). The alumina acts as a Lewis acid, coordinating the OH group and absorbing water as it is eliminated.

$$\text{OH H} \qquad \overset{Al_2O_3}{\underset{250°}{\longrightarrow}} \qquad \text{C}=\text{C} + Al_2O_3 \cdot H_2O$$

A third compound that can enter the picture in the acid-catalyzed interconversion of alcohols and alkenes is the dialkyl ether. Conditions can be adjusted in the dehydration so that a second mole of alcohol acts as a nucleophile for the carbocation. The resulting dialkyloxonium ion can then lose a proton to give the ether.

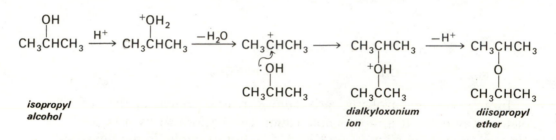

isopropyl alcohol   *dialkyloxonium ion*   *diisopropyl ether*

With primary alcohols, dehydration and ether formation also occur, but the ether is formed by an S_N2 mechanism. Diethyl ether can be obtained from either ethanol or ethylene. In the preparation of the ether from the alcohol, ethanol is added to concentrated sulfuric acid at 130° and the low boiling ether is continuously distilled out. At 180°, ethylene is formed by elimination, but the dehydration is more easily carried out by vapor phase dehydration over a solid catalyst.

$$CH_3CH_2OH$$
$$\downarrow H_2SO_4$$

$$H_3O^+ + CH_3CH_2OCH_2CH_3 \xleftarrow[130°]{CH_3CH_2OH} CH_3CH_2\overset{+}{O}H_2 \underset{}{\overset{180°}{\rightleftharpoons}} CH_2\!=\!CH_2 + H_3O^+$$

7.13   Reactions of Thiols

Thiols and sulfides undergo several oxidation reactions that have no counterpart in the chemistry of alcohols and ethers. A thiol is very easily oxidized to a **disulfide,** containing an S—S bond, by a variety of reagents including hydrogen peroxide and bromine. The oxidation is readily reversed, and interconversions of —SH and —S—S— groups play an important role in the chemistry of proteins in biological systems.

$$2 \text{ CH}_3\text{CH}_2\text{SH} \underset{[H]}{\overset{[O]}{\rightleftharpoons}} \text{CH}_3\text{CH}_2\text{—S—S—CH}_2\text{CH}_3$$

ethanethiol   *diethyl disulfide*

The —S—S— bonds in the proteins of wool and hair form cross-links that give these fibers their springiness. In cold permanent waving of hair, these bonds are broken by adding a reducing lotion; after the limp fiber is curled, the hair is rinsed with a neutralizer that contains peroxide to reform the —S—S— bonds.

A series of oxidation products containing S—O bonds exists because sulfur, in the second row of the periodic table, can utilize 3d orbitals in bonding. A few of the possibilities can be seen in the following compounds:

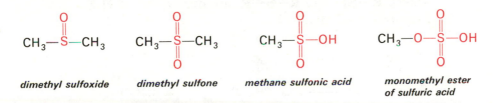

dimethyl sulfoxide   *dimethyl sulfone*   *methane sulfonic acid*   *monomethyl ester of sulfuric acid*

Another important aspect of organosulfur compounds is the cleavage of C—S bonds by hydrogenolysis (breaking by means of hydrogen). Thiols and sulfides can be **desulfurized** by reaction with metal catalysts that are impregnated with hydrogen. The overall reaction is replacement of C—S bonds by C—H bonds; the sulfur is converted to H_2S, which is retained by the metal.

$$(\text{CH}_3\text{CH}_2\text{CH}_2)_2\text{S} \xrightarrow[\text{Ni}]{\text{H}_2} 2 \text{ CH}_3\text{CH}_2\text{CH}_3$$

This reaction has some utility in laboratory synthesis, but its major significance is in petroleum chemistry. As noted in Chapter 2, crude petroleum contains sulfur compounds, in amounts ranging up to 6 per cent sulfur by weight. The amount of sulfur varies with the location—Venezuelan and some United States crudes are high in sulfur; mid-Eastern crudes have a relatively low sulfur content.

The presence of sulfur in crude petroleum leads to two problems. Sulfur compounds in the more volatile gasoline fractions act as poisons toward the metal catalysts used for isomerization and aromatization steps in the refining process. These volatile sulfur compounds are present in small amounts and are fairly easily removed by extraction methods.

The sulfur content is enriched in the heavy residual oils that remain after distillation of the gasoline fraction. These high-boiling residues are used for boiler fuel, and in the past they represented a major source of atmospheric SO_2. In many areas air-quality legislation has set strict limits on the maximum permissible sulfur content in fuel oils, however. The removal of sulfur from residual fractions has become a major challenge in refinery operations, particularly since the sulfur compounds in these higher molecular weight fractions are aromatic and relatively unreactive. The main approach to the problem is hydrodesulfurization, in which the oil is treated with hydrogen and catalysts containing cobalt and molybdenum oxides, and the sulfur is converted to H_2S.

SUMMARY AND HIGHLIGHTS

1. **Alcohols** contain an OH group: RCH_2OH (primary), R_2CHOH (secondary), R_3COH (tertiary).

Ethers have the general structure **R—O—R.**

Amines are derivatives of NH_3: **R—NH₂** (primary), **R₂NH** (secondary), **R₃N** (tertiary).

Thiols, R—SH and **R—S—R,** are derivatives of H_2S.

2. **Preparation of Alcohols**

$$RCH{=}CH_2 + H_2O \xrightarrow{H^+} RCHOHCH_3$$

$$RCH{=}CH_2 \xrightarrow{BH_3} (RCH_2CH_2)_3B \xrightarrow[+OH^-/H_2O]{H_2O_2} RCH_2CH_2OH$$

$$CO + 2H_2 \xrightarrow[\text{cat.}]{Ni} CH_3OH$$

$$\text{glucose} \xrightarrow{\text{yeast}} C_2H_5OH$$

3. **Hydrogen bonding** of alcohols and amines causes high boiling points and solubility in water.

4. **Amphoteric** properties of alcohols, thiols, and amines:

$ROH + Na \longrightarrow RO^-Na^+$ alkoxide ion—strong base

$ROH + H_2SO_4 \longrightarrow ROH_2^+$ alkyloxonium ion—strong acid

$RSH + NaOH \longrightarrow RS^-Na^+$ thiolate ion

$RNH_2 + H_3O^+ \longrightarrow R\overset{+}{N}H_3$ ammonium ion

5. **Nucleophilic substitutions** with alcohols or thiols are carried out with alkoxide or thiolate: $RO^-Na^+ + R'X \longrightarrow R{-}O{-}R'$ or $RS^-Na^+ + R'X \longrightarrow R{-}S{-}R'$. Amines are alkylated to mixtures of di-, tri-, and quaternary ammonium salts:

$$RNH_2 \xrightarrow{R'X} (RNH_2R')^+ \longrightarrow RNHR' \longrightarrow (RNHR'_2)^+ \longrightarrow RNR'_2 \longrightarrow (RNR'_3)^+X^-$$

6. **Epoxides** react with alcohols and amines as nucleophiles to undergo ring opening:

$$ROH + H_2C\overset{O}{\overset{\triangle}{-}}CH_2 \longrightarrow ROCH_2CH_2OH$$

7. **Alkyl esters** of inorganic acids: $ROSO_3^-$, $RONO_2$, $ROPO_3^=$ are important as detergents, explosives, and biochemical intermediates, respectively.

8. **Substitution** and **elimination** reactions of alcohols occur *via* oxonium ion (ROH_2^+) or alkyl ester.

9. **Alkyl halides** are prepared from alcohols with HCl, $SOCl_2$, HBr, or PBr_3.

10. **Dehydration** of alcohols gives alkenes; rearrangement may occur:

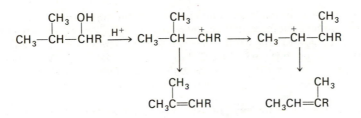

11. **Ethers** can be formed by intermolecular dehydration:

$$2\ RCH_2OH \xrightarrow{H^+} RCH_2OCH_2R$$

12. **Thiols** are reversibly oxidized to disulfides; C—S bonds can be broken by hydrogen:

$$R-SH \underset{red}{\overset{ox}{\rightleftharpoons}} R-S-S-R$$

$$R-S-R' \xrightarrow{H_2,\ Ni} RH + R'H$$

ADDITIONAL PROBLEMS

7.15 Write the structures of the following compounds:

a) 2-fluoroethanol

b) 3-amino-1-propanol

c) di-*n*-propyl ether

d) dibenzyl sulfide

e) 2-aminoethanethiol

f) 1-phenylethanol

g) methylethylamine

h) 3-methyl-2-cyclohexenol

i) tetra-*n*-butylammonium hydroxide

j) ethylene glycol

k) dimethyl sulfate

l) 2,4-dimethylbenzylamine

7.16 Write names for the following compounds:

a) $(CH_3)_2CHNH_2$

b) $HOCH_2CH_2CH_2OH$

c) $CH_3(CH_2)_5CH(OCH_3)CH_2NH_2$

d) $CH_3CH_2CH_2SCH_2CH_2CH_3$

e) $CH_3CH_2CH_2CH_2ONO_2$

f)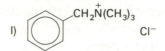

g) $CH_3CH=CHCH_2OH$

h) $HOCH_2CH_2N(CH_2CH_3)_2$

i) $HC\equiv CCH_2OC_2H_5$

j) $C_6H_5OCH_3$

k) $CH_3(CH_2)_{17}CH_2NH_2$

l)

7.17 Suggest an explanation for the fact that dimethylamine (bp 7°) has a higher boiling point than either methylamine (bp $-7°$) or trimethylamine (bp 3°).

7.18 There are four stable compounds with the formula $C_3H_8O_2$ [as will be seen in Chapter 8, compounds with the structural unit $R_2C(OH)_2$ or $R_2C(OH)(OCH_3)$ are *not* stable, but those with $C(OCH_3)_2$ are]. The compounds will be designated **A, B, C,** and **D.** The boiling points of the four compounds are **A,** 41°; **B,** 124°; **C,** 187°; **D,** 214°. Write the four possible stable structures, and on the basis of the boiling points, state which is **A, B, C,** and **D.** (Hint: primary alcohols are generally higher boiling than secondary alcohols.)

7.19 Write equations for any reactions that would occur when the following compounds and reagents are combined as indicated. If no net reaction would take place, write N.R.

a) sodium added to 1-propanol

b) diethylamine added to aqueous hydrochloric acid

c) aqueous KOH solution mixed with ethanol

d) concentrated H_2SO_4 added to 1-butanol

e) water added to a solution of lithium diethylamide $[(CH_3CH_2)_2N^-Li^+]$ in diethylamine

f) solutions of methylammonium bromide ($CH_3NH_3^+Br^-$) and sodium methoxide, both in methanol, mixed together (a white precipitate appears)

g) potassium *tert*-butoxide dissolved in water

h) butylamine added to a solution of sodium ethoxide in ethanol

7.20 Write structures of the products that would be formed in the following reactions:

a) 1-chloropentane + sodium ethoxide $\longrightarrow$

b) 2-butanol + conc. HCl $\longrightarrow$

c) $CH_2=CHCH_2Cl + (CH_3)_3N \longrightarrow$

d) $CH_3-CH\overset{O}{\overbrace{}}CH-CH_3 + CH_3OH \xrightarrow{H^+}$

e) $C_6H_5CH_2SH + CH_3I \xrightarrow{NaOH}$

f) 1-hexanol $\xrightarrow[\text{heat}]{Al_2O_3}$

g) 1-butene + diborane, then $H_2O_2/^-OH/H_2O \longrightarrow$

h) *tert*-butyl chloride + methanol $\longrightarrow$

i) $(CH_3CH_2)_2NH + CH_3OSO_2OCH_3 \longrightarrow$

j) cyclohexanol + $SOCl_2 \longrightarrow$

7.21 Write structures and give names of *all* the isomeric amines with the formula $C_4H_{11}N$. (Note that the general formula of saturated amines is $C_nH_{2n+3}N$, *i.e.*, an additional hydrogen in the formula C_nH_{2n+2} is needed for each nitrogen, as can be seen by comparing CH_3CH_3 and CH_3NHCH_3.)

7.22 A basic compound **A** with formula $C_4H_{11}N$ reacts with excess CH_3I in the presence of K_2CO_3 to give a neutral salt **B**, $C_6H_{16}N^+I^-$. Consider the reaction that occurs, as shown by the change in molecular formula, and your answer to problem 7.20, and write all possible structures for compound **A**.

7.23 Dehydration of 2-methyl-1-butanol with sulfuric acid gives a trace of 2-methyl-1-butene; the major product is an isomeric alkene. Suggest the structure of the main product and show the pathway by which it is formed.

7.24 Describe a simple test, which could be carried out quickly in a test tube with a small sample, that would tell whether a compound was tri-*n*-propylamine or di-*n*-butyl ether. Indicate the basis for the test and what observation would be made.

7.25 Show how the following compounds could be prepared from the starting substances indicated, plus any other compounds required. More than one step may be needed; show all steps and necessary reagents.

    a) cyclohexyl bromide from cyclohexanol

    b) $(CH_3)_2NCH_2CH_2OH$ from dimethylamine

    c) $CH_3CH_2CH_2SCH_3$ from $CH_3CH_2CH_2SH$

    d) di-*n*-propyl ether from *n*-propanol

    e) 1,2-dibromopropane from 2-propanol

    f)

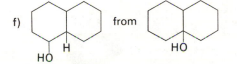

7.26 Ethyl *n*-butyl ether is only slightly soluble in water, but it dissolves readily in concentrated (50 per cent) aqueous HBr. After the solution is allowed to stand for some time and is then diluted with water, the ether separates out and can be recovered unchanged. However, when ethyl *tert*-butyl ether is dissolved in 50 per cent HBr, a liquid layer rapidly separates from the concentrated acid solution. Explain the observations, with equations, in each case, and account for the difference in behavior between the two ethers.

7.27 An *optically active* compound **A** has the formula $C_5H_{12}O$. Passage of the vapors over heated aluminum oxide granules gave a product **B**, which reacted rapidly with bromine. Ozonolysis of **B** gave $CH_3CH_2CH=O$ and $CH_3CH=O$. Write the structures of **A** and **B**.

7.28 Compound **A**, C_8H_9Br, gives a precipitate quite rapidly with NaI in acetone, but very slowly with alcoholic silver nitrate. Reaction of **A** with NaSH in alcohol gives compound **B**, $C_8H_{10}S$. Bubbling a stream of oxygen through **B** converts it to compound **C**, $C_{16}H_{18}S_2$. Treatment of **B** or **C** with hydrogen in the presence of nickel gives ethylbenzene. Suggest structures for compounds **A**, **B**, and **C**.

8
ALDEHYDES
AND KETONES

The compounds that we will see in this chapter contain a **carbonyl group,** $C=O$, which consists of one σ and one π bond between carbon and oxygen. The carbon atom is sp^2 hybridized, and the geometry around a carbonyl group is planar, as in an alkene. Two unshared electron pairs occupy nonbonding orbitals on oxygen. Because of the much greater electronegativity of oxygen, a dipolar resonance structure contributes significantly to the π bond.

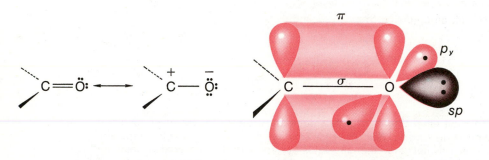

The carbonyl group is a very stable system, and compounds with an OH group and another electronegative group X such as OH, OR, or Cl on the same carbon are usually unstable with respect to the carbonyl compound. Thus, with a few exceptions that we will see later, the following equilibria lie far to the right.

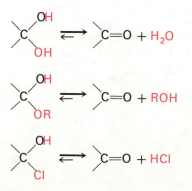

8.1 Structures and Names

Aldehydes contain the group $-\overset{\overset{\text{O}}{\|}}{\text{C}}-\text{H}$, which is often written as $-\text{CHO}$. The important distinction between aldehydes and ketones is that aldehydes have a hydrogen attached to the carbonyl carbon. In **ketones** the carbonyl group is flanked by two carbons, which can be either alkyl or aryl groups; the carbonyl carbon can also be part of a ring.

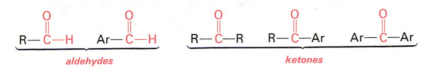

The systematic names of aliphatic aldehydes are formed by replacing the **-ane** ending of the parent alkane by **-al**. The chain is numbered with the CHO group as 1; thus, the carbonyl group takes precedence over an OH group or a double bond. A number of aldehydes have trivial names derived from those of the corresponding acids (Chapter 10); the trivial names are always used for the two simplest compounds, formaldehyde and acetaldehyde. Compounds with the $-\text{CHO}$ group attached to a benzene ring are named as derivatives of the parent benzaldehyde.

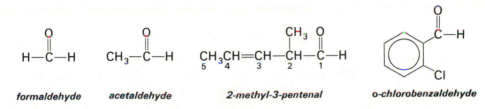

In the system developed by IUPAC, ketones are named by replacing **-ane** with **-one**; the chain is numbered to give the CO group the lowest number. The trivial name **acetone** is always used for the first member of the series. Another type of name is based on the names of the two groups attached to the carbonyl, with the separate word "ketone." If a compound contains two functional groups, the prefix **oxo** can be used to designate a carbonyl group.

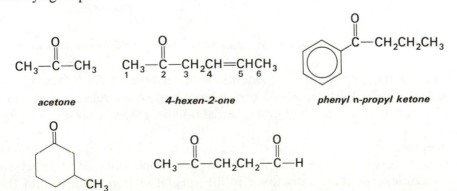

Problem 8.1   Write structures for a) 2-chlorohexanal, b) 4-methyl-2-pentanone, c) phenylacetone, d) 2-butenal, e) dicyclopropyl ketone.

8.2   Properties and Occurrence

Aldehydes and ketones have higher boiling points than alkanes or ethers of comparable molecular weight because the polarity of the carbonyl group leads to electrostatic attraction between molecules in the liquid.

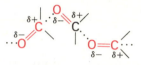

This type of interaction is weaker than hydrogen bonding, however, and the boiling points of aldehydes and ketones are therefore lower than those of the corresponding alcohols.

$$CH_3CH_2CH_2OH \qquad CH_3CH_2CHO \qquad CH_3\overset{OH}{\underset{|}{C}}HCH_3 \qquad CH_3\overset{O}{\overset{||}{C}}CH_3$$

1-propanol   *propanal*   *2-propanol*   *acetone*
bp 97°   **bp 49°**   **bp 82°**   **bp 56°**

Aldehydes and ketones are very commonly encountered, both in the laboratory and as naturally occurring compounds. Aldehydes are the principal constituents of a number of essential oils used as fragrances and flavors, such as those from lemon, hyacinth, cumin, and cinnamon.

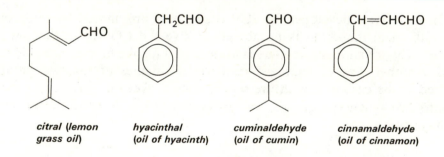

citral (lemon grass oil)   *hyacinthal (oil of hyacinth)*   *cuminaldehyde (oil of cumin)*   *cinnamaldehyde (oil of cinnamon)*

Problem 8.2   Write systematic names for the four aldehydes shown.

One of the important components of the aroma of fresh bread is *n*-butanal; the palatibility of stored bread decreases as the release of butanal slows. As we will see in Chapter 11, the aldehyde group is present in most sugars.

Ketones are also widely distributed in nature. One of the best known terpenes is camphor; menthone from mint and irone from oil of violet are other ketonic terpenes. As discussed in the special section at the end of this chapter, the important steroid hormones are complex cyclic ketones.

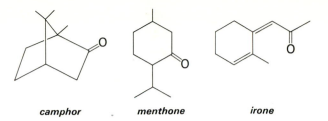

camphor   menthone   irone

8.3 Preparation of Aldehydes and Ketones

The major methods for preparing aldehydes and ketones are outlined in the following chart.

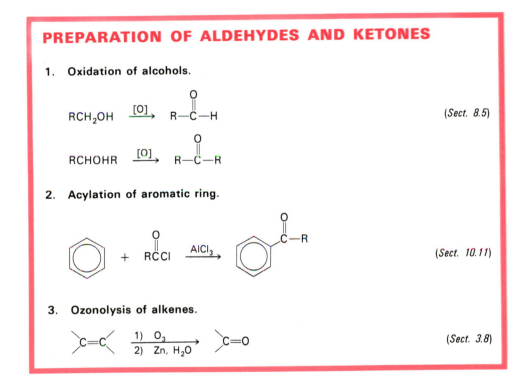

PREPARATION OF ALDEHYDES AND KETONES

1. Oxidation of alcohols.

$$RCH_2OH \xrightarrow{[O]} R-\overset{O}{\overset{\|}{C}}-H$$

(*Sect. 8.5*)

$$RCHOHR \xrightarrow{[O]} R-\overset{O}{\overset{\|}{C}}-R$$

2. Acylation of aromatic ring.

$$\bigcirc + R\overset{O}{\overset{\|}{C}}Cl \xrightarrow{AlCl_3} \bigcirc\overset{O}{\overset{\|}{C}}-R$$

(*Sect. 10.11*)

3. Ozonolysis of alkenes.

$$>C=C< \xrightarrow[2)\ Zn,\ H_2O]{1)\ O_3} >C=O$$

(*Sect. 3.8*)

8.4 Oxidation and Reduction in Organic Chemistry

In the following sections we will encounter several examples of oxidation and reduction in the chemistry of aldehydes and ketones. Before going into specific reactions, let us first recall the meanings of these terms. The basic definition of oxidation is the loss of electrons; reduction is the gain of electrons. Oxidation and reduction are always coupled together; the electrons are actually *transferred,* and when we balance an equation we must have the same number of electrons in reactants and in products. Electron transfer processes are at the heart of many important biochemical reactions. We obtain energy from food by a complex series of oxidation-reduction reactions called an **electron transport chain** (Section 15.4) which permits the controlled combustion of organic compounds in the body by molecular oxygen.

$$-CH_2- + \tfrac{3}{2}O_2 \longrightarrow CO_2 + H_2O + 156 \text{ kcal}$$

In reactions of organic compounds, oxidation and reduction usually involve changes in the relative electronegativities of the groups attached to carbon. Thus, the replacement of hydrogen in a C—H bond by a more electronegative element such as O or Cl amounts to oxidation of the carbon. We can see that replacement of H by OH is equivalent to the loss of electrons by considering the relationship between carbanions and carbocations, and the behavior of each.

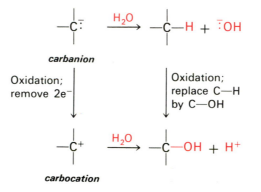

The conversion of a carbanion to a carbocation by removing an electron pair clearly corresponds to an oxidation by our definition. Now let us compare the reactions of the two ions with water. The carbanion behaves as a base, and a C—H bond is formed. The carbocation, on the other hand, is a Lewis acid, and the product with water is an alcohol. These acid-base reactions with water do not involve oxidation or reduction, since only protons and no electrons are transferred. In the overall replacement of a C—H bond by C—OH, the carbon has been **oxidized,** or converted to a higher **oxidation level.** We can write the successive oxidation levels of carbon in a CH_3 group simply by replacing each H in turn with an OH group, and removing water when there are two OH groups on the same carbon.

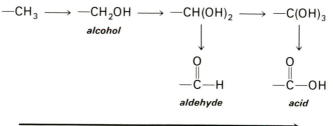

PLATE 1

A. Hyacinth blossom.

B. Hyacinths clogging an irrigation ditch.

C. Hyacinth infestation in cypress-wooded stream area.

Water hyacinth, *Eichhornia crassipes,* has invaded streams, swamps, and canals in Florida and elsewhere in the southern U.S. As a weed, these plants interfere with navigation, provide breeding sites for mosquitos, and interfere with fish and waterfowl populations. The hyacinth can be controlled with 2,4-D. These prolific plants have been suggested as a possible biomass for use as an energy source (p. 46).

B

C

PLATE 2

A

B

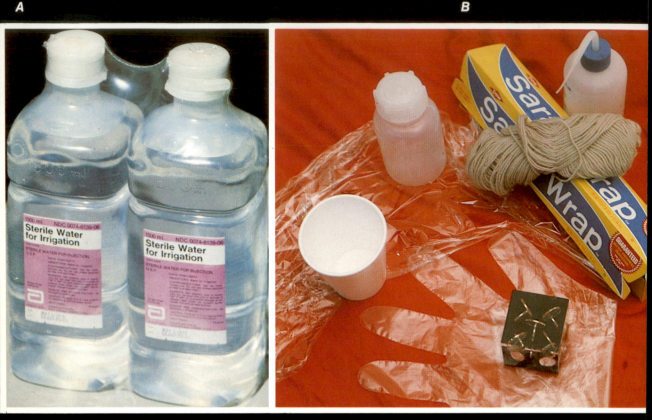

A. Shrink wrapping provides a durable sanitary barrier for a variety of products, including sterile solutions used in operating rooms. The wrap is a polyethylene film which has been stretched in two directions to provide biaxial orientation. After the film is fitted over the object as a loose sleeve, it is heated to give a form-fitting wrap. (Photo courtesy Du Pont Company.) B. Some familiar objects made of vinyl plastics are disposable cup (polystyrene), screw-cap bottle (polypropylene), squeeze bottle (polyethylene), Saran Wrap (polyvinyl-polyvinylidene chloride), yarn (polyacrylonitrile), transparent paperweight (polymethylmethacrylate), and glove (polyethylene). C. Plastics in surgical reconstruction. An ankle joint with polyethylene insert in the upper surface (see p. 71). (Photo courtesy Richards Manufacturing Co., Inc.)

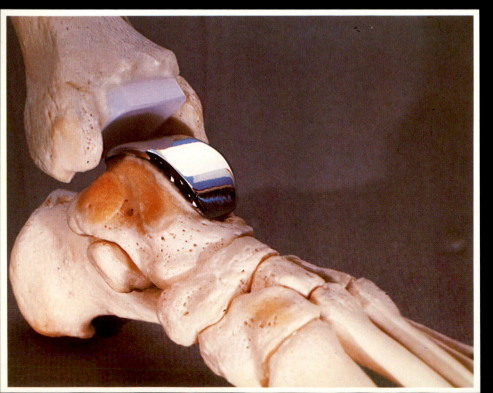

C

PLATE 3

Three types of molecular models. The orange and black models are the "space-filling" type and most closely approximate the overall molecular shape. Wire models on the right in each photo show bonds only, and are useful in visualizing spatial relationships.

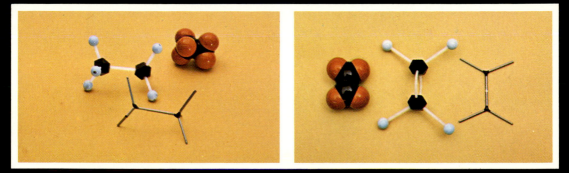

A. Ethane

B. Ethylene

C. Acetylene

D. *trans*-1,3-Dichlorocyclohexane

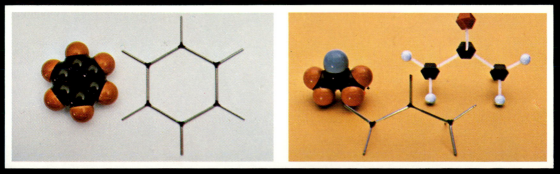

E. Benzene

F. Acetone

PLATE 4

A. The Hope Diamond.

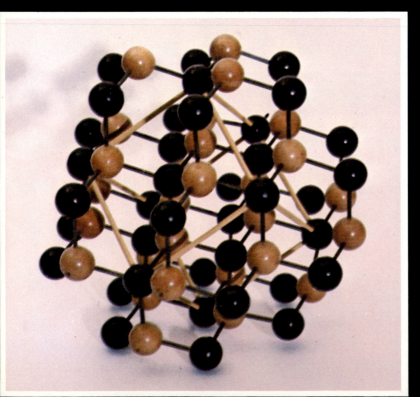

B. Model of the diamond crystal lattice. All atoms represent carbon; the black and tan spheres show alternating planes made up of chair cyclohexane rings.

A

PLATE 5

Defoliating insects susceptible to insecticides (see pp. 101–107). *A.* Female gypsy moths with egg cases; each mass contains up to 600 eggs. *B.* Gypsy moth larvae. *C.* Cankerworm has a characteristic looping movement as it feeds on leaf. *D.* Saddled caterpillars are responsible for extensive damage to oaks, maples, and other hardwoods. *E.* Leaf miner has caused this damage in birch twig by tunneling through leaf in maggot stage.

B

C

D

E

A

PLATE 6

Agricultural insects can be controlled by proper choice of insecticides (p. 104). *A.* Mexican bean beetle—eggs, larva (spiny), pupa, and adult. *B.* Japanese beetle on soybean leaf. *C.* Corn earworm, *Heliothis armigera*—three color phases. *D.* Corn cutworm and cut stalk. *E.* Corn earworm also attacks tomatoes.

B

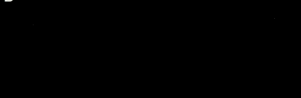

C

D

E

PLATE 7

A. Cross dyeing of nylon. These carpet samples contain four different types of nylon. The top band is cationic dyeable, and the next three bands below have progressively increasing capacity for anionic dyes. All four types are dyed equally by disperse dyes. From right to left the strips were dyed with a red anionic dye, a blue cationic dye, a yellow disperse dye and a bath containing all three dyes. B. Cationic dyes on Dacron. A single cationic dye was used for the two patches on the left of each group; the orange, purple and green patches are dyed with a mixture of the two dyes at the left (see pp. 252 and 295–297).

A

PLATE 8

A. Corn plot with a normal growth of pigweed and giant foxtail. (Courtesy of Dr. Larry Hawf.)

B. This corn plot was sprayed, prior to emergence of the corn, with a solution of atrazine (p. 306). (Courtesy of Dr. Larry Hawf.)

C. Test plot of fescue, in which the center strip (in front of sign) was treated with maleic hydrazide to control the height of the grass (p. 308). (Courtesy of Dr. Larry Hawf.)

B

C

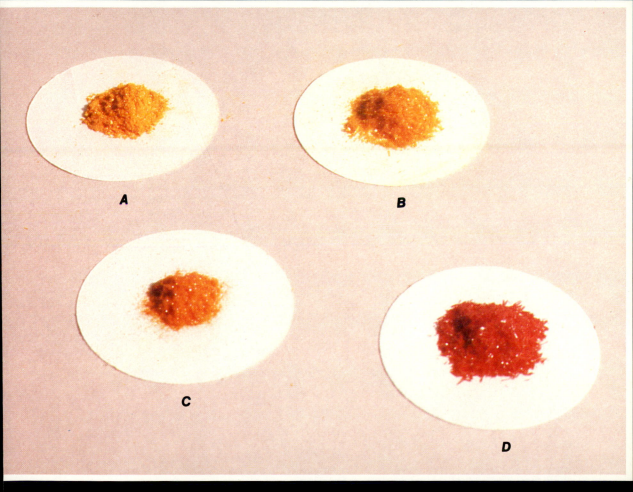

PLATE 9

2,4-Dinitrophenylhydrazones of aldehydes and ketones (p. 189). The crystals are derivatives of A) acetaldehyde, B) cyclohexanone, C) benzaldehyde, D) methyl 1-cyclohexenyl ketone. The deep red color of the methyl 1-cyclohexenyl ketone derivative is typical of dinitrophenylhydrazones of conjugated unsaturated ketones.

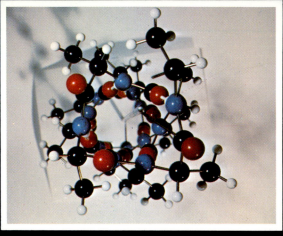

A

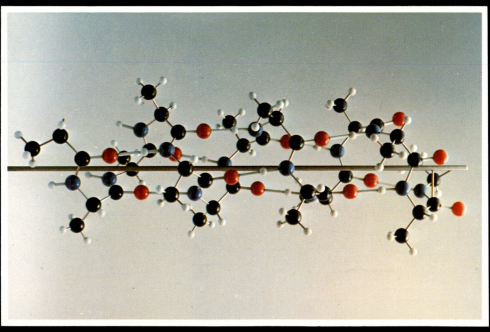

B

PLATE 10

The α-helix protein structure. *A*. End view looking down helical axis. *B*. Side view of helix. Black spheres are carbon, red are oxygen, blue are nitrogen and white are hydrogen. The amino acid units are alanine. White vertical tie lines are hydrogen bonds (p. 338).

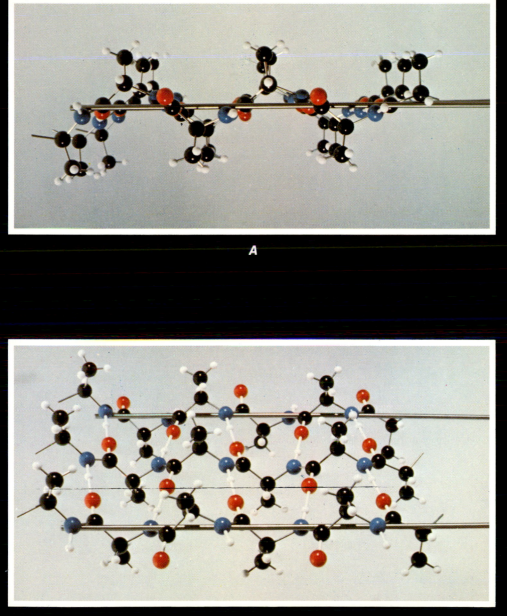

A

B

PLATE 11

The β-pleated sheet protein structure (p. 338). *A*, Side view of sheet. *B*, Top view of sheet

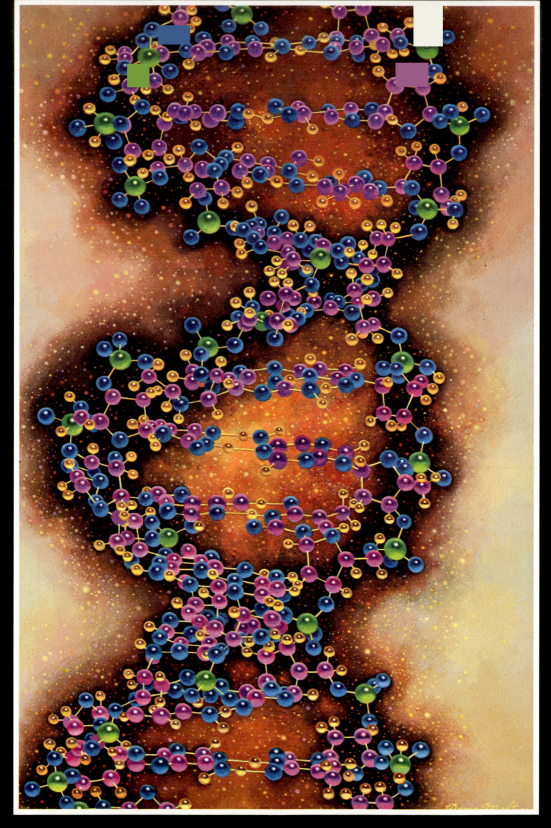

Artist's rendering of a portion of the DNA molecule (p. 362). The colors represent: purple, carbon; blue, oxygen and nitrogen; green, phosphorus; gold, hydrogen. Illustration of DNA molecule reproduced with permission of EM Laboratories, Inc., Elmsford, NY, associate of E. Merck, Darmstadt, W. Germany.

8.5 Oxidation of Alcohols to Aldehydes and Ketones

Oxidation of alcohols is a very general method for preparing aldehydes and ketones. Primary alcohols are converted to aldehydes and secondary alcohols to ketones. Tertiary alcohols cannot undergo oxidation unless a C—C bond breaks; therefore, these compounds do not react under the usual oxidizing conditions.

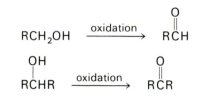

One of the most useful reagents for alcohol oxidation is chromic acid, H_2CrO_4, which is obtained by the reaction of sodium dichromate with acid or of chromic oxide with water. The oxidation proceeds by way of the chromate ester of the alcohol. In the oxidation step, chromium changes oxidation state from Cr^{+6} to Cr^{+4}. The latter is unstable and undergoes disproportionation to give Cr^{+3}. Regardless of the details of the subsequent change from Cr^{+4} to

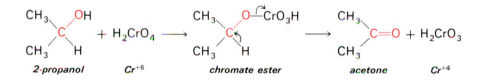

Cr^{+3}, we can write an equation for the overall oxidation with a two-electron change for the alcohol and a three-electron change for chromium.

$$3\ CH_3CHOHCH_3 + 2\ H_2CrO_4 + 3\ H_2SO_4 \longrightarrow 3\ CH_3COCH_3 + Cr_2(SO_4)_3 + 8\ H_2O$$

orange   green

Oxidation with chromic acid is the basis for a rapid analysis of ethanol used in breath analyzers for law enforcement. The analysis depends on the color change that occurs on going from orange Cr^{+6} to greenish-blue Cr^{+3} when the alcohol is oxidized. A sample of expired air from deep in the lungs is passed through chromic acid solution, and the change in color detected by the instrument can be directly correlated with the amount of ethanol in the lungs and bloodstream.

8.6 Oxidation Reactions of Aldehydes

An important difference in the properties of aldehydes and ketones is the susceptibility of aldehydes to oxidation. In the chromic acid oxidation of a primary alcohol, for example, the aldehyde is easily oxidized further to a carboxylic acid by breaking the C—H bond of the —CHO group. To isolate the aldehyde, it must be removed by distillation from the reaction as fast as it is formed. Ketones, however, are quite stable to chromic acid.

$$RCH_2OH \xrightarrow{H_2CrO_4} R-\overset{\overset{\displaystyle O}{\|}}{C}-H \xrightarrow{H_2CrO_4} R-\overset{\overset{\displaystyle O}{\|}}{C}-OH$$

Aldehydes are also oxidized by other reagents. Silver and copper ions in alkaline solution are specific oxidizing agents for aldehydes and are particularly useful for the identification and determination of a —CHO group. A solution of $Ag(NH_3)_2OH$, called Tollens' reagent, is reduced by the aldehyde to give metallic silver, which is deposited as a mirror on the wall of the flask. This reaction, using formaldehyde, is the basis of the manufacturing process for mirrors.

$$R-\overset{\overset{\displaystyle O}{\|}}{C}-H + 2\ Ag(NH_3)_2{}^+ + 3\ OH^- \longrightarrow R-\overset{\overset{\displaystyle O}{\|}}{C}-O^- + 2\ Ag + 4\ NH_3 + 2\ H_2O$$

mirror

An alkaline solution of copper ion, called Benedict's reagent, is widely used for the determination of sugars, which are aldehydes, in biological fluids (Section 11.1). The blue Cu^{+2} ion is reduced to red Cu_2O.

$$R-\overset{\overset{\displaystyle O}{\|}}{C}-H + 2\ Cu^{+2} + 5\ OH^- \longrightarrow R-\overset{\overset{\displaystyle O}{\|}}{C}-O^- + Cu_2O + 3\ H_2O$$

*blue
solution* *red
precipitate*

Problem 8.3   Write the structures of the products that would be obtained in any of the following reactions that would take place. Identify those cases in which no reaction would occur.

a) 3-methyl-2-butanol plus $Na_2Cr_2O_7$-H_2SO_4
b) propanal plus $Na_2Cr_2O_7$-H_2SO_4
c) ethanol plus $Ag(NH_3)_2OH$
d) acetone plus $Na_2Cr_2O_7$-H_2SO_4
e) 4-hydroxybutanal plus alkaline Cu^{+2}

ADDITION REACTIONS OF ALDEHYDES AND KETONES

8.7 General Mechanism of Nucleophilic Addition

In the next few sections we will see reactions of aldehydes and ketones which lead to a number of quite different kinds of products. It will be helpful in understanding and remembering these reactions to recognize at the beginning that all of these products arise by the *same process*—addition of a nucleophile to the carbonyl double bond.

The carbonyl group has a strong dipole, with the electrons in the π bond shifted toward the more electronegative oxygen atom. The carbonyl carbon is therefore electron deficient, and unlike the C=C bond in an alkene, the carbonyl group is easily attacked by a nucleophile (Nu⁻). As a new bond is formed from the nucleophilic center to the carbon of the C=O group, the carbonyl oxygen gains an unshared electron pair. The addition is completed by transfer of a proton to the oxygen, and the overall reaction is thus the addition of Nu—H.

$$Nu\overset{..}{:}\,\curvearrow\,\overset{\delta+}{\underset{}{C}}=\overset{\delta-}{\underset{..}{O}}\,\colon\;\longrightarrow\;Nu-\underset{|}{\overset{|}{C}}-\overset{..}{\underset{..}{O}}\colon\;\xrightarrow{H^+}\;Nu-\underset{|}{\overset{|}{C}}-\overset{..}{O}H$$

Nucleophilic addition is assisted by coordination of the electron pair on oxygen by a proton or other Lewis acid; *i.e.*, carbonyl addition is an acid-catalyzed process. When the addition is carried out in the presence of strong acid, nucleophilic attack occurs on the conjugate acid of the aldehyde or ketone. The protonated carbonyl compound is actually a resonance-stabilized carbocation, and is of course much more strongly electrophilic than the carbonyl group.

$$\overset{}{\underset{}{C}}=\overset{..}{\underset{..}{O}}\colon\;\xrightarrow{H^+}\;\left[\overset{}{\underset{}{C}}=\overset{+}{\underset{..}{O}}-H\;\longleftrightarrow\;\overset{+}{\underset{}{C}}-\overset{..}{\underset{..}{O}}-H\right]\;\xrightarrow{\overset{..}{N}uH}\;Nu-\underset{|}{\overset{|}{C}}-OH\;+\;H^+$$

The relative reactivities of aldehydes and ketones toward nucleophilic addition depend on two factors: a) the electronic influence of the groups attached to the carbonyl carbon, and b) the steric bulk of the groups. The more electron-releasing the groups, the less electron deficient is the carbonyl carbon, and the less reactive toward nucleophiles. Alkyl groups are electron-releasing relative to hydrogen and are also much bulkier. Thus, for both electronic and steric reasons, ketones—with the C=O group flanked by *two* carbons—are in general less reactive than aldehydes.

A final point about the general carbonyl addition process is the stereochemical course of the reaction. In many carbonyl additions, an achiral aldehyde or ketone becomes a chiral molecule when the nucleophile adds to the carbonyl carbon. We must keep in mind that in such reactions, addition occurs at both sides of the planar carbonyl group at exactly equal rates, and both enantiomers are formed in exactly the same amounts; *i.e.*, the product is a racemic mixture.

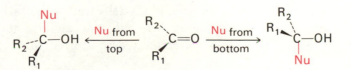

8.8 Addition of Hydrogen

Aldehydes and ketones are reduced to alcohols by the addition of hydrogen to the C=O double bond. The reaction can be carried out by catalytic hydrogenation under the same conditions used for addition of hydrogen to an alkene (Section 3.4); if a C=C bond is present in the same molecule, it is, of course, also reduced.

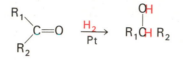

A more selective method for reducing a carbonyl group is by transfer of a **hydride ion** (H$\bar{\ }$). The source of hydride ion for this reaction is sodium borohydride, $NaBH_4$, or lithium aluminum hydride, $LiAlH_4$. The former is usually more convenient. All four hydride ions are available, and thus one mole of $NaBH_4$ reduces four moles of aldehyde or ketone. As the hydride ions are transferred to the carbonyl carbon, the oxygens form a borate complex. The OH proton is supplied by hydrolysis of the borate with water.

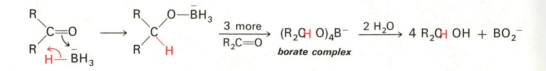

Since a C=C bond is not susceptible to nucleophilic attack, sodium borohydride permits the selective reduction of a carbonyl group in a compound containing both C=C and C=O groups.

Problem 8.4 Write the structures of the products that would be obtained with (1) H_2 plus catalyst and (2) $NaBH_4$ with the following compounds; identify any cases in which no reaction would occur: a) propene, b) 2-pentanone, c) cyclohexanone, d) 3-hexene-2-one.

The simplest and most common reductions in biological systems occur by a hydride transfer very similar to that accomplished by $NaBH_4$. These biochemical reactions involve enzymes that contain a group which is designated NADH in its reduced form and NAD$^+$ in the oxidized form (Section 15.4). NADH contains a dihydropyridine ring, and transfer of a hydride anion from NADH occurs readily because it gives NAD$^+$ with an aromatic (six π electron) pyridinium ion (Section 13.2). Since the reaction takes place on a chiral enzyme surface, the hydride ion is delivered to the carbonyl group *stereospecifically;* as shown in Figure 8.1, only one enantiomer is formed when a chiral —CHOH— group is generated in the reaction.

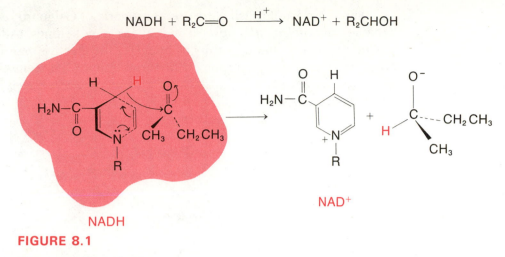

FIGURE 8.1

Stereospecific reduction.

8.9 Addition of Organometallic Compounds

In organometallic compounds, carbon is bonded directly to a metal. Among the most important examples are organomagnesium halides, which are readily obtained by reaction of alkyl or aryl halides with magnesium. These compounds are usually called **Grignard reagents** in honor of their discoverer, the French chemist Victor Grignard. The reaction of the halide and metal is carried out in the presence of ether, which acts as a Lewis base to coordinate the metal atom; the Grignard reagent is not isolated, but is used as an ether solution.

$$CH_3Br + Mg: \xrightarrow{\text{ether}} \overset{\delta-}{CH_3}-\overset{\delta+}{Mg}Br$$

methylmagnesium bromide

Since carbon is more electronegative than the metal, the carbon-metal bond is polarized very strongly, with *carbon as the negative end* of the dipole. As a result of this polarity, Grignard reagents have two important and distinctive properties. First, since the carbon is anionic, Grignard reagents are very *strong bases* and react with any compounds containing OH or NH groups. The product is the hydrocarbon, *i.e.*, the conjugate acid formed by protonation of the carbanion.

$$CH_3CH_2MgCl + H_2O \longrightarrow CH_3CH_3 + MgClOH$$

ethylmagnesium chloride   *ethane*

This type of acid-base reaction is a problem in working with Grignard reagents, since the solvent and reactants must be completely free of water, alcohols, or any other acidic substances.

The second and much more useful property of Grignard reagents is their behavior as *nucleophiles* toward aldehydes and ketones. Addition of the organomagnesium halide occurs by attack of the electron pair in the C—Mg bond on the carbonyl carbon, with formation of a magnesium alkoxide. Subsequent treatment of the alkoxide with water and acid gives the alcohol plus magnesium salts.

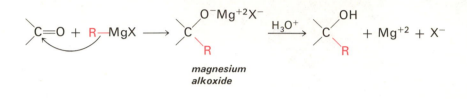

magnesium alkoxide

In this Grignard addition reaction, a new C—C bond is created, together with a hydroxyl group for use as a "handle" in further transformations. Since the process can be used with any Grignard reagent and any aldehyde or ketone, it is one of the most useful and flexible methods for creating a new carbon skeleton. Subsequent reactions of the alcohol (such as dehydration to an alkene, conversion to a halide, or oxidation) can provide routes to many possible products, with functional groups at any desired point.

Examples of the Grignard synthesis of various types of alcohols are given in the following chart. Primary alcohols are obtained by using formaldehyde as the carbonyl compound, or by reaction of the Grignard reagent with ethylene oxide. Higher aldehydes give secondary alcohols, and addition to ketones gives tertiary alcohols.

GRIGNARD SYNTHESIS OF ALCOHOLS

1. Primary alcohols, RCH$_2$OH.

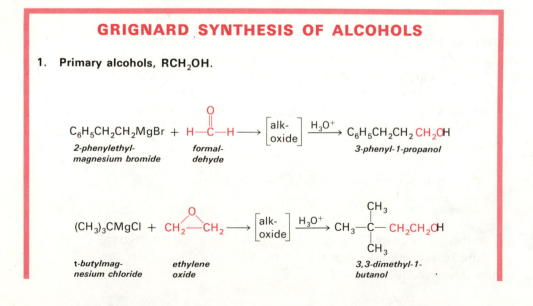

2. Secondary alcohols, RCHOHR

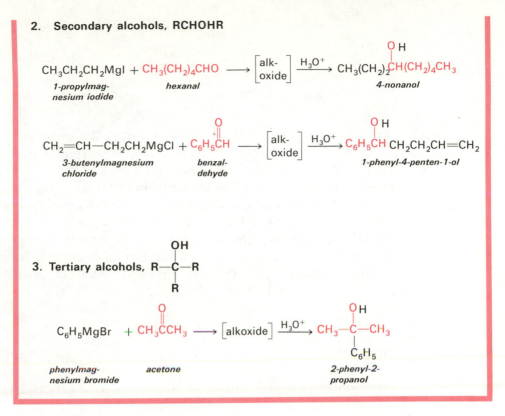

CH$_3$CH$_2$CH$_2$MgI + CH$_3$(CH$_2$)$_4$CHO $\longrightarrow$ $\begin{bmatrix}\text{alk-}\\\text{oxide}\end{bmatrix}$ $\xrightarrow{\text{H}_3\text{O}^+}$ CH$_3$(CH$_2$)$_2$CH(CH$_2$)$_4$CH$_3$

1-propylmag-   *hexanal*   *4-nonanol*
nesium iodide

CH$_2$=CH—CH$_2$CH$_2$MgCl + C$_6$H$_5$CH $\longrightarrow$ $\begin{bmatrix}\text{alk-}\\\text{oxide}\end{bmatrix}$ $\xrightarrow{\text{H}_3\text{O}^+}$ C$_6$H$_5$CH CH$_2$CH$_2$CH=CH$_2$

3-butenylmagnesium   *benzal-*   *1-phenyl-4-penten-1-ol*
chloride   *dehyde*

3. Tertiary alcohols, R—C—R (with OH above C and R below)

C$_6$H$_5$MgBr + CH$_3$CCH$_3$ $\longrightarrow$ [alkoxide] $\xrightarrow{\text{H}_3\text{O}^+}$ CH$_3$—C—CH$_3$ (with OH above and C$_6$H$_5$ below)

phenylmag-   *acetone*   *2-phenyl-2-*
nesium bromide   *propanol*

In the preparation of secondary and tertiary alcohols, more than one combination of Grignard reagent and carbonyl compound can be used to obtain the same alcohol. Thus, 3-methyl-2-butanol, for example, could be prepared from either isopropylmagnesium halide plus acetaldehyde or the methyl Grignard reagent plus 2-methylpropanal. The choice between the two

CH$_3$CHMgX + CH$_3$CHO $\longrightarrow$ $\xrightarrow{\text{H}_3\text{O}^+}$ CH$_3$CHCHCH$_3$ $\xleftarrow{\text{H}_3\text{O}^+}$ $\longleftarrow$ CH$_3$CHCHO + CH$_3$MgX
(CH$_3$ below first carbon; OH and CH$_3$ on middle structure; CH$_3$ below last)

possibilities would depend mainly on the relative availabilities of the starting materials. The following examples illustrate how this kind of question can be approached.

Example 1.   Prepare 1-phenyl-3-pentanol by a Grignard reaction.

This secondary alcohol could be obtained in either of two ways. To determine the starting components, we must work back from the structure of the product. The —CHOH— group represents the carbonyl group of an aldehyde, and either of the bonds attached to the CHOH could be formed in the addition.

$$OH$$
$$C_6H_5CH_2CH_2\overset{|}{C}HCH_2CH_3$$

$$C_6H_5CH_2CH_2MgBr + H\overset{O}{\overset{||}{C}}CH_2CH_3 \qquad C_6H_5CH_2CH_2\overset{O}{\overset{||}{C}}H + BrMgCH_2CH_3$$

Example 2.   Prepare 3-heptanol, using as the only organic starting materials *alcohols* containing no more than four carbon atoms.

In this case, the problem must be worked back a few more steps. Starting again with the structure of the product, there are two ways to obtain the alcohol by Grignard addition. One of these involves a 2-carbon fragment plus a 5-carbon fragment, and the other involves 3-carbon and 4-carbon fragments. The second combination fits the restriction on starting materials. We must then prepare the aldehyde and the Grignard reagent from alcohols. These steps involve simple reactions that have been discussed previously.

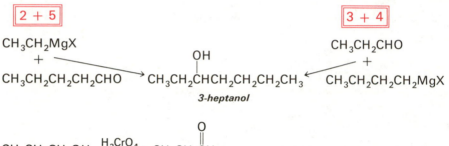

$$\boxed{2 + 5} \qquad\qquad\qquad\qquad \boxed{3 + 4}$$

$$CH_3CH_2MgX \qquad\qquad\qquad\qquad CH_3CH_2CHO$$
$$+ \qquad\qquad\qquad\qquad\qquad\qquad +$$
$$CH_3CH_2CH_2CH_2CHO \longrightarrow CH_3CH_2\overset{OH}{\overset{|}{C}}HCH_2CH_2CH_2CH_3 \longleftarrow CH_3CH_2CH_2CH_2MgX$$

3-heptanol

$$CH_3CH_2CH_2OH \xrightarrow{\text{H}_2\text{CrO}_4} CH_3CH_2\overset{O}{\overset{||}{C}}H$$

$$CH_3CH_2CH_2CH_2OH \xrightarrow{\text{HBr}} CH_3CH_2CH_2CH_2Br \xrightarrow[\text{ether}]{\text{Mg}} CH_3CH_2CH_2CH_2MgBr$$

Limitations.   In devising a synthesis that makes use of a Grignard reaction, it must be remembered that a Grignard reagent reacts as a very strong base with all types of acidic hydrogen. Therefore, it is not possible to prepare a Grignard reagent from a halide that contains an OH, SH, or NH group, such as $HOCH_2CH_2CH_2Br$. If the *carbonyl* compound contains an active hydrogen group, two moles of Grignard reagent must be used, since one mole is consumed in the acid-base reaction:

$$HOCH_2CH_2CH_2CH_2CHO + 2\ CH_3MgBr \longrightarrow$$

$$\qquad\qquad\qquad\qquad\qquad\qquad\qquad\qquad\qquad OMgBr$$
$$CH_4 + BrMgOCH_2CH_2CH_2{-}\overset{|}{C}H{-}CH_3$$

Problem 8.5   Show, with structural formulas, reactions that could be used to prepare the following alcohols by the Grignard reaction.

a)   3-pentanol
b)   2-phenylethanol
c)   1-methylcyclohexanol
d)   2,5-dimethyl-2-hexanol

Problem 8.6   Show how the following compounds could be prepared by Grignard condensation followed by dehydration or oxidation of the alcohol, using starting materials with no more than four carbons.

a)   2-methyl-2-pentene
b)   3-heptanone
c)   3-methyl-1-butene
d)   [structure: CH₃–C(=O)– attached to cyclobutane ring]

8.10   Addition of Acetylides

As we saw in Section 6.9, acetylide anions (HC≡C⁻ or RC≡C⁻) are obtained from the reactions of 1-alkynes with a very strong base, and can be used as nucleophiles in S_N2 reactions. These anions can also be used in the same way that Grignard reagents are used in carbonyl addition reactions. The initial addition product is an alkoxide, and protonation with aqueous acid then gives the acetylenic alcohol. The products obtained in these reactions are difunctional compounds which undergo reactions of alkynes such as partial hydrogenation to *cis*-alkenes (Section 3.9) as well as reactions typical of alcohols.

$$RC\equiv CH \xrightarrow{NaNH_2} RC\equiv C^{-}\,Na^{+} \xrightarrow{C=O} RC\equiv C-\underset{|}{\overset{|}{C}}-O^{-}Na^{+} \xrightarrow{H_3O^{+}} RC\equiv C-\underset{|}{\overset{|}{C}}-OH$$

Problem 8.7   Write the structures of the intermediates and products designated by bold letters.

a)   acetaldehyde + HC≡CNa ⟶ **A** $\xrightarrow{H_3O^{+}}$ **B** (C₄H₆O)

b)   1-propyne $\xrightarrow{NaNH_2}$ **A** $\xrightarrow[\text{2)}\ H_3O^{+}]{\text{1)}\ \text{acetone}}$ **B** $\xrightarrow[\text{cat}]{H_2}$ **C** (C₆H₁₂O)

Problem 8.8   One of the main structural features of oral contraceptive drugs (p. 198) is the acetylenic alcohol group attached to a five-membered ring, as shown in structure **B**. Show how **B** could be obtained from the alcohol **A**; indicate steps and intermediates.

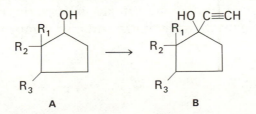

8.11 Addition of Cyanide

Addition products of hydrogen cyanide with aldehydes and ketones are called **cyanohydrins.** These compounds are useful intermediates in several synthetic methods for acids (Section 10.6) and carbohydrates (Section 11.3). Unlike the stable alkoxides formed by addition of hydride or Grignard reagents to carbonyl groups, the alkoxide intermediate formed by addition of cyanide tends to undergo the reverse reaction and lose cyanide ion.

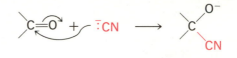

To obtain the cyanohydrin, the pH of the reaction must be kept within the range from about 8 to 11. In more acidic solution (pH below 8), cyanide ion is almost completely protonated to give HCN, and there is insufficient nucleophile for the addition. In strongly basic solution (pH above 11), protonation of the alkoxide intermediate is prevented, and the equilibrium is shifted to the left.

$$\text{HCN} \underset{\text{pH} < 8}{\overset{\text{OH}^-}{\rightleftharpoons}} \text{CN}^- + \text{CH}_3\text{CHO} \rightleftharpoons \underset{\overset{|}{\text{CN}}}{\overset{\overset{\text{O}^-}{|}}{\text{CH}_3{-}\text{C}{-}\text{H}}} \underset{\text{pH} > 11}{\overset{\text{H}^+}{\rightleftharpoons}} \underset{\overset{|}{\text{CN}}}{\overset{\overset{\text{OH}}{|}}{\text{CH}_3{-}\text{C}{-}\text{H}}}$$

acetaldehyde   *acetaldehyde*
                                                                             cyanohydrin

8.12 Addition of Water

As we saw earlier in this chapter, the addition products of water or alcohols to a carbonyl group are generally not stable compounds. Attack by a water molecule and proton transfer with the solvent occur readily to give a hydrate, but the reactions are reversible and the equilibrium lies on the side of the aldehyde or ketone. Evidence that the addition products are indeed formed is obtained by carrying out the reaction in water labeled with the "heavy" isotope ^{18}O. The addition product from H_2^{18}O contains a normal ^{16}OH group and an ^{18}OH group, either of which can be lost at almost the same rate. When ^{16}OH is lost, the resulting carbonyl group retains the ^{18}O, and the carbonyl oxygen is "exchanged" by the isotope. In the presence of ^{18}O-enriched water, aldehydes and ketones undergo this exchange until the isotope concentration is the same in the carbonyl compound and the water.

$$\text{C}{=}^{16}\text{O} + \text{H}_2^{18}\text{O} \rightleftharpoons \underset{^{18}\text{OH}}{\overset{^{16}\text{O H}}{\text{C}}} \rightleftharpoons \text{C}{=}^{18}\text{O} + \text{H}_2^{16}\text{O}$$

Problem 8.9 The rate of exchange of carbonyl oxygen with $H_2{}^{18}O$ is greatly increased in the presence of either acid or base. Suggest an explanation for the effect in each case. (Hint: consider the actual nucleophile and electrophile involved.)

The position of the equilibrium between carbonyl compound and hydrate depends on the groups or atoms that are adjacent to the carbonyl group. Because of the greater electron deficiency of the carbonyl carbon in aldehydes, the hydrate is more favored with aldehydes than with ketones. In a few special cases such as formaldehyde, the equilibrium lies far on the side of the hydrate. Another well known example is chloral hydrate. In trichloroacetaldehyde (chloral), the presence of the three electron-withdrawing chlorine atoms makes the carbonyl group so electron deficient that the hydrate obtained by addition of water is a stable compound.

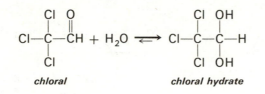

Chloral hydrate is a hypnotic or sleep-producing agent, and was one of the earliest drugs used for this purpose. The combination of chloral hydrate with ethanol is referred to as "knock-out drops" because of its very rapid action.

8.13 Addition of Alcohols

The addition of alcohols to the carbonyl group closely resembles the reaction with water. The addition product of an alcohol with an aldehyde is called a **hemiacetal.** The addition is reversible, and hemiacetals readily dissociate into the aldehyde and alcohol. Further reaction of the hemiacetal with excess alcohol in the presence of acid under conditions in which water is removed leads to the **acetal** containing the $-CH(OR)_2$ group. The corresponding products from a ketone are called hemiketal and ketal, respectively; these compounds are more difficult to obtain, and we will be concerned here only with the aldehyde derivatives.

Overall acetal formation

$$CH_3\overset{O}{\overset{\|}{C}}H + ROH \rightleftharpoons CH_3\overset{OH}{\underset{}{C}HOR} \xrightarrow{ROH} CH_3\overset{OR}{\underset{}{C}HOR} + H_2O$$

hemiacetal   acetal

Formation of an acetal is catalyzed by acid, and occurs by a series of proton-transfer and addition-elimination steps. Protonation of the hydroxyl group in the hemiacetal permits the loss of water to give an alkoxycarbo-cation, with the charge stabilized by an unshared electron pair on oxygen. Addition of alcohol to this intermediate leads to the acetal. If the acetal is treated with aqueous acid, the reverse series of steps takes places; alcohol is lost, water is added to the alkoxycation, and elimination of a second mole of alcohol gives the aldehyde.

Mechanism:

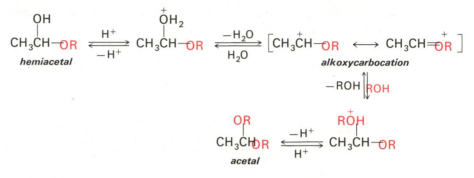

Problem 8.10   Write equations, with structural formulas and arrows showing electron shifts, for the following reactions:

a)   Addition of methanol to propanal to give $C_4H_{10}O_2$.
b)   Reaction of pentanal plus excess ethanol and sulfuric acid to give $C_9H_{20}O_2$.

8.14   Addition of Amines

The addition of an amine to the carbonyl group of an aldehyde or ketone is analogous to the formation of hydrates or hemiacetals, and is readily reversible. The addition product from a primary amine can undergo a further reaction, however, in which water is eliminated with formation of a C=N bond. The overall reaction is an example of a very common process, called **condensation,** in which two organic molecules combine with the loss of water. The general name for the product from this type of amine-carbonyl condensation is **imine** or **Schiff base.**

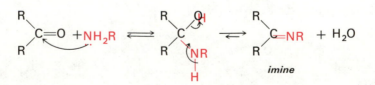

Several special amines are used to produce crystalline derivatives of carbonyl compounds. These reagents and the names of the products are illustrated by the following examples.

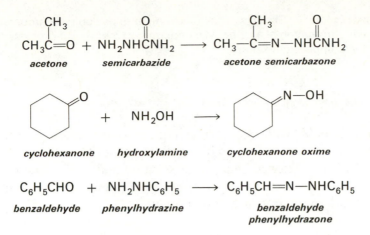

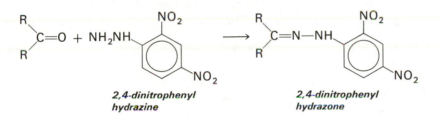

A particularly useful reagent is 2,4-dinitrophenylhydrazine, which gives beautifully crystalline orange derivatives with aldehydes and ketones (Color Plate 9). Formation of a crystalline 2,4-dinitrophenylhydrazone by treatment of an unknown compound with a solution of this reagent is a reliable test for an aldehyde or ketone.

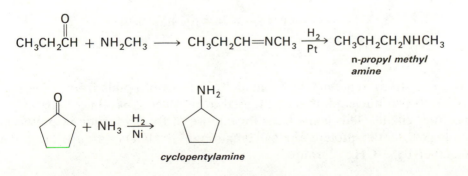

The condensation products of carbonyl compounds with ammonia and simple primary amines readily undergo hydrolysis by addition of water to the C=N bond, and are difficult to isolate. Although these compounds are not particularly important in themselves, the condensation of carbonyl compounds with NH_2 groups in proteins is a key step in many biochemical reactions.

When the condensation of an aldehyde or ketone with an amine is carried out in the presence of hydrogen and a metal catalyst, the imine is reduced to a saturated amine. This process is called **reductive amination,** and is a useful general method for preparing amines.

Problem 8.11   Show how the following amines could be prepared by reductive amination of a carbonyl compound: a) methyl-*iso*-propyl amine, b) ethylaminocyclohexane.

8.15   Reactivity and Uses of Formaldehyde

Formaldehyde, with the carbonyl group flanked by two hydrogens, is literally in a class by itself, and its reactions are quite distinct from those of other aldehydes or ketones. In aqueous solution (called "formalin"), formaldehyde exists entirely as the hydrate plus polymeric structures in which a number of aldehyde units are added together.

$$\underset{\text{HCH}}{\overset{\text{O}}{\|}} + H_2O \rightleftharpoons HOCH_2OH \underset{\xrightarrow{n\,HCHO}}{\rightleftharpoons} HOCH_2(OCH_2)_nOH$$

With gaseous formaldehyde, a high molecular weight linear polymer is obtained. To prevent depolymerization, the ends of the chain are "capped" by replacing the H with R, and the product is a tough, stable resin called Delrin, which is useful in gears and other machine parts and in electrical equipment.

$$ROCH_2(OCH_2)_nOR$$

Delrin

Formaldehyde also condenses very readily with amines. Any —NH group is converted by formaldehyde to —NCH$_2$OH, called an *N*-methylol group. In the reaction with a primary amine, the condensation product RN=CH$_2$ is not obtained; this group undergoes addition of another mole of amine to give a diaminomethane, which may then react with more formaldehyde. If an alcohol is present, addition of alcohol across the C=N bond can occur.

$$RNH_2 + HCHO \longrightarrow RNH\,CH_2OH \underset{H_2O}{\overset{-H_2O}{\rightleftharpoons}} [RN{=}CH_2] \xrightarrow{R'OH} RNH\,CH_2OR'$$

N-*methylol* amine

$$\underset{\underset{H\ \ OCH_2}{|}\quad\underset{CH_2OH}{|}}{RN{-}CH_2{-}NR} \overset{HCHO}{\longleftarrow} RNH\,CH_2NHR \quad \Big\downarrow RNH_2$$

The strong tendency of formaldehyde to form bonds from a —CH$_2$— group to two nucleophiles can be used to produce **cross-links** between two polymer chains. This is the basis for the use of formaldehyde in preserving biological tissue; protein chains containing —NHCO— groups are linked together by —CH$_2$— groups.

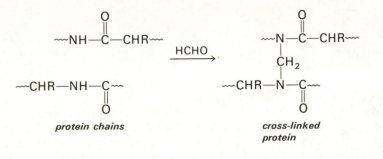

protein chains   cross-linked
                                   protein

SPECIAL TOPIC: Formaldehyde and Fabrics

A specialized use of formaldehyde is in finishes for cotton fabrics. This application is quite evident in the sharp, irritating smell of formaldehyde that is often noticeable in textile stores, particularly when a new bolt of fabric is opened. Cotton is cheap, strong, and comfortable to wear, but cotton garments wrinkle easily and do not hold their shape well. To overcome the latter drawbacks, a resin finish is applied to most cotton fabric. The purpose of this treatment is to form cross-links between hydroxyl groups on neighboring cellulose chains. These cross-links stabilize the fibrous structure and impart a resilience to the fiber. In this process the cotton becomes resistant to wrinkling and also holds a pleat or fold that is pressed into the fabric before the cross-linking is done. The finish also markedly reduces shrinkage of the fabric.

Most of the reagents used for durable press cotton finishes contain a dimethylolurea unit, which provides a cross-link of the proper length by acetal formation with the cellulose. "Wash-wear" cotton contains a permanent press finish. Several variations have been developed to achieve the best performance and to overcome the complications that arise owing to the presence of the finish.

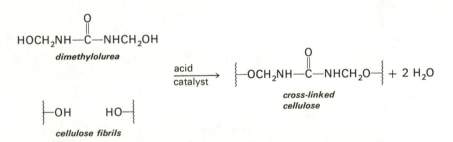

One of the most satisfactory processes for preparing durable press garments is **deferred curing** (Figure 8.2). The fabric is impregnated with a cross-linking agent and an acid catalyst, and garments are then cut, finished, and pressed before a final heat-treatment to "set" the resin. The smooth, permanently creased finish is obtained at the expense of some brittleness and loss of strength of the fabric. In turn, these factors are compensated for by incorporating in the fabric synthetic polymers that restore the soft feel and reduce abrasion. Finally, synthetic fibers, particularly polyesters (Section 10.21), are frequently blended with the cotton to improve strength and wear resistance. The final product is a far cry from the starched native cotton of a generation ago.

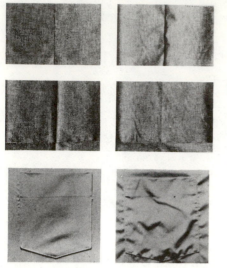

FIGURE 8.2

Comparison of untreated and permanent press garment after laundering. The samples of seam, cuff and pocket on the left have been finished by the deferred cure process. Control samples without this treatment, laundered in the same way, are shown on the right. (Courtesy of Mr. H. Goldstein, Chester, South Carolina.)

8.16 Enolization of Aldehydes and Ketones

The numerous addition reactions of aldehydes and ketones that we have seen in the preceding sections occur because of the strongly electron deficient carbonyl carbon. Let us now examine a second consequence of the electron deficient carbonyl group. This is the fact that hydrogens attached to the carbon atoms adjacent to the carbonyl group are considerably *more acidic* than those in other C—H bonds. These carbons are designated as α relative to the carbonyl group, and the acidic hydrogens are called **α-hydrogens.**

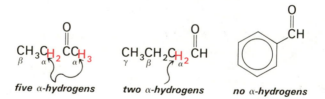

five α-hydrogens   *two α-hydrogens*   *no α-hydrogens*

The acidity of the α-hydrogens in an aldehyde or ketone can be explained in terms of the attraction of electrons in the $H_\alpha \rightarrow C_\alpha \rightarrow C{=}O$ bonds away from the α-hydrogen by the electron deficient carbonyl group. The same effect can be stated another way in terms of the stabilization of the negative charge when the α-hydrogen is removed by a base. The carbanion is a resonance hybrid in which the charge resides mostly on the oxygen atom.

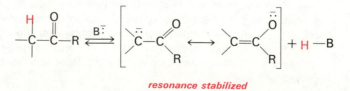

resonance stabilized anion

Since the negative charge in the anion is shared by both carbon and oxygen, a proton can be replaced on either atom. If the anion is reprotonated on carbon, the original carbonyl compound is obtained. (Or, if the reaction is carried out in "heavy water," *i.e.*, D_2O, the carbonyl compound with a D atom on the α-carbon is produced.) Protonation of the anion on the oxygen gives an isomer with a C=C bond and an —OH group, called the **enol.**

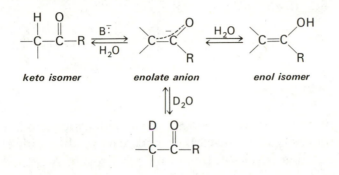

keto isomer   enolate anion   enol isomer

Thus we see that the anion, usually called the **enolate** anion, is actually the conjugate base of the two acids—the keto and enol isomers. The term "tautomerism" is used to refer to this special type of isomerization, and keto and enol forms are called **tautomers.** Since proton transfer is rapid, an aldehyde or ketone containing α-hydrogens is always in equilibrium with the tautomeric enol. With simple carbonyl compounds the keto tautomer is the most stable one, and the amount of enol is less than 1 per cent. If a C—H bond is α to *two* carbonyl groups (as in 2,4-pentanedione, for example), the acidity of the α-H is much higher because of the increased stabilization of the negative charge in the anion by two oxygens. The amount of enol is also larger, and the enol may in fact be the major tautomeric form.

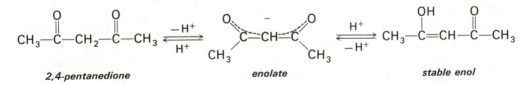

2,4-pentanedione   enolate   stable enol

Keto and enol tautomers have the same *conjugate acid* as well as having the same conjugate base. This means that the interconversion of the two isomers is catalyzed by acid just as it is by base, and reactions that depend upon enolization can be carried out in acid.

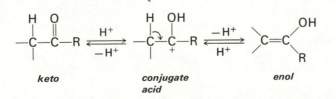

keto   conjugate   enol
         acid

Another form of the enolized carbonyl group is the **enamine** or amino-alkene, in which the oxygen of the enol is replaced by nitrogen. The reaction of an aldehyde or ketone with a secondary amine leads to an addition product analogous to that obtained with a primary amine (Section 8.13), but since no proton is available on nitrogen, the final dehydration involves loss of an α-hydrogen.

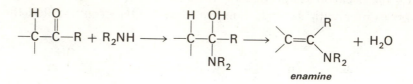

8.17   Enols as Nucleophiles

In the form of the enol or enolate anion, aldehydes and ketones have a high electron density at the α-carbon, and react readily with electrophiles. The overall reaction is substitution of an α-hydrogen. A typical example is the bromination of a ketone. In the presence of acid the reaction occurs by way of the enol, and an α-bromoketone is the product.

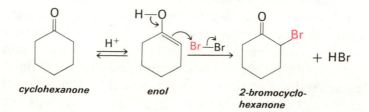

In *base*, halogenation of the enolate anion is very rapid; with a methyl ketone, all three α-hydrogens in the CH_3 group are substituted by halogen. The resulting trihalomethyl ketone then undergoes addition of hydroxide ion. Although this type of addition to a ketone usually leads only to a reversible equilibrium, in this case the electron-withdrawing effect of the three halogens is so great that the CX_3 group can be eliminated as an anion. In a final step, proton exchange gives CHX_3, called a **haloform,** and the anion of the carboxylic acid. When iodine is used, the yellow solid iodoform, CHI_3, is obtained, and isolation of this product is a test for the presence of a CH_3 group adjacent to the C=O group in the original ketone.

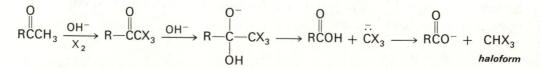

8.18   Aldol Condensation

When an aldehyde containing α-hydrogen atoms is treated with base, the carbonyl compound (which can undergo nucleophilic addition) and the enolate (which is a nucleophile) are present together. We might therefore expect a reaction between the two, and this in fact occurs. The enolate attacks the carbonyl group of the original aldehyde, and a new C—C bond is formed. The product is a β-hydroxyaldehyde, and the overall process is called the **aldol condensation.**

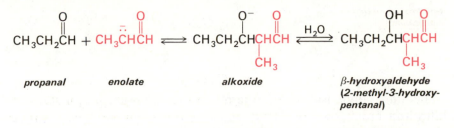

propanal *enolate* *alkoxide* **β-hydroxyaldehyde** *(2-methyl-3-hydroxy-pentanal)*

The aldol condensation is a useful reaction with simple aldehydes. The products can be reduced to 1,3-diols or dehydrated to unsaturated aldehydes, and the latter can then be reduced to give alcohols. The aldol reaction cannot be used with a mixture of two different aldehydes which both contain α-hydrogens, since a mixture of products derived from the two enolates and the two different carbonyl groups would result.

Problem 8.12   Write equations showing the formation of all possible aldol condensation products that would be obtained from the reaction of acetaldehyde and butanal in base.

"Mixed" aldol condensations are practical if only one reactant contains α-hydrogens. Aromatic aldehydes, for example, react smoothly with a second carbonyl compound that contains a —CH$_2$CO— or CH$_3$CO— group. The product in this case is always the unsaturated carbonyl compound resulting from loss of water from the initial aldol compound. Dehydration occurs readily in basic solution because the proton removed is α to the carbonyl group.

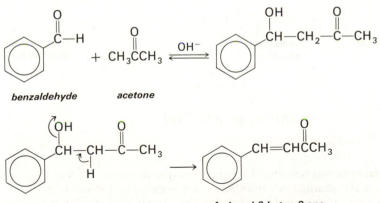

benzaldehyde   *acetone*

4-phenyl-3-buten-2-one

Formation of an aldol condensation product is an equilibrium process, and the position of the equilibrium is usually unfavorable when the carbonyl compound is a ketone. Thus, if a β-hydroxyketone is prepared indirectly under non-equilibrium conditions and treated with base, the alkoxide intermediate undergoes loss of enolate anion, and two moles of ketone are formed.

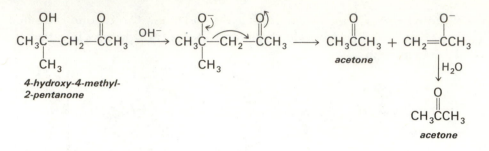

4-hydroxy-4-methyl-2-pentanone

The aldol condensation and the reverse reaction are major steps in the buildup and breakdown of sugars and in energy storage. The compounds involved are somewhat more elaborate than those we have been looking at, but the same reactions occur. The condensation, as always in biochemical processes, is controlled by enzymes which direct the course and the stereochemistry of the reaction.

The nucleophile is an enamine formed by the combination of the phosphate ester of dihydroxyacetone with a primary amine. The carbonyl compound which undergoes addition is the phosphate of glyceraldehyde (2,3-dihydroxypropanal). After condensation, the resulting imine is hydrolyzed to give fructose 1,6-diphosphate.

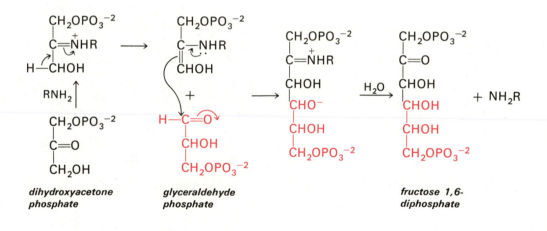

dihydroxyacetone phosphate **glyceraldehyde phosphate** **fructose 1,6-diphosphate**

SPECIAL TOPIC: Steroid Hormones and Oral Contraceptives

Hormones are compounds secreted by endocrine glands which regulate metabolism, growth, and reproduction. Hormones vary widely in structure and include amines, peptides (Chapter 14), and steroids (Section 7.3). Some typical examples of steroid hormones are the male sex hormone **testosterone,** the ovarian follicle hormone **progesterone,** and **hydrocortisone** from the adrenal cortex. Although the structures of these compounds are very similar, the specific physiological action depends on minor features such as the number of and location of hydroxy groups. The regulatory function of the hormones is maintained by minute amounts of the compounds, and endocrine insufficiency can be overcome by administration of the hormones obtained from another animal source such as beef glands.

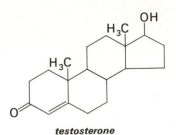

testosterone
(male—testis)

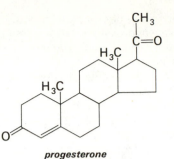

progesterone
(ovarian follicle)

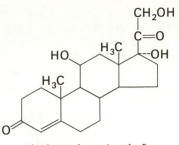

hydrocortisone (cortisol)
(11,17,21-trihydroxy-4-
pregnene-3,20-dione)
(adrenal cortex)

A major development in medicine was brought about with the discovery in 1949 that certain of the adrenal steroids, particularly hydrocortisone, relieve the inflammation and pain of rheumatoid arthritis. This finding created a need for synthesis of the compounds, since amounts available from beef or hog adrenals were far too limited.

Two general approaches can be considered in a problem of this kind. One is *partial* synthesis starting with other steroids; the other is *total* synthesis in which the ring system is built up from smaller molecules. Given an adequate supply of a more abundant steroid, partial synthesis is a simpler task, and this was the first route used. Reactions were developed for the conversion of steroids obtained from yams or soybeans to progesterone.

The further transformation of progesterone to hydrocortisone requires the introduction of hydroxy groups at carbon atoms 11, 17, and 21. The two latter positions are α to a ketone group, and introduction of substituents was readily accomplished by way of enolization. The OH group at carbon 11 posed a more difficult problem, however, since this position is not specifically activated for any type of substitution reaction. This problem was solved by **microbiological hydroxylation.** Many common yeasts and molds contain oxidase enzymes that can bring about oxidation of a C—H bond to C—OH; this is one of the first

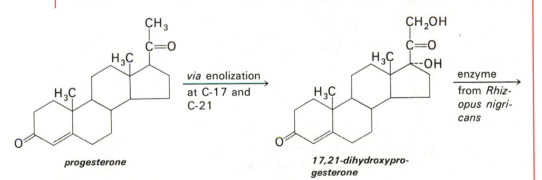

progesterone

via enolization at C-17 and C-21

17,21-dihydroxypro-gesterone

enzyme from *Rhiz-opus nigri-cans*

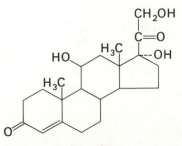

hydrocortisone

reactions involved in the decomposition of organic compounds in sewage. By adaptation of the proper organism, hydroxylation of a given molecule can be carried out selectively and efficiently. With this approach, a practical synthetic source of corticosteroids was in hand.

The availability of adrenal cortex steroids provided for the first time a means to relieve the crippling inflammation of arthritis. The fact that these compounds are potent hormones limits their utility as drugs, however, since side effects can arise from hormone oversupply. To overcome this problem, modifications of the structure were explored and the introduction of additional substituents was found to enhance selectively the anti-inflammatory activity, thus permitting far smaller dosage. Alkyl groups, additional double bonds, and, surprisingly, fluorine substituents are effective. Hydrocortisone derivatives with one or more of these modifications are now in wide use for alleviating inflammation in such diverse conditions as skin eruptions and mononucleosis.

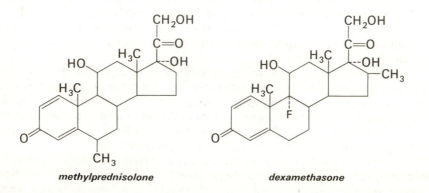

methylprednisolone   dexamethasone

Oral Contraceptives.   Another development in the use of steroids that has had a profound impact in recent years is the availability of orally administered drugs for preventing pregnancy. Ovulation and the implantation of an embryo are under the control of the female sex hormones. Following ovulation, progesterone is secreted by the ovarian follicle during the second half of the menstrual cycle; cessation of progesterone output causes menstruation. Maintenance of the progesterone level throughout the cycle leads to an endocrine situation resembling that of pregnancy, and further ovulation is inhibited. The effect of steroid contraceptive drugs is related primarily to their progestational activity.

Practical contraceptive drugs were developed from the discovery of modified steroids that are highly active progestational agents when taken by mouth. The most effective compounds have structures which contain a 17-hydroxy-17-acetylenic side chain, and hydrogen in place of the C-19 methyl group. Two of these compounds, **norethynodrel** and **norethisterone**, originally prepared by partial synthesis from progesterone, are widely used in oral contraceptive pills.

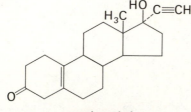

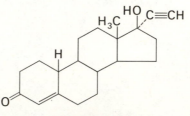

norethynodrel   norethisterone

The success of norsteroids* in controlling fertility has created a demand for these compounds at low cost and in amounts much larger than could be obtained by synthesis from other steroids. To meet this need, methods for total synthesis have been worked out in which the tetracyclic ring system is assembled in just a few steps from two fragments.

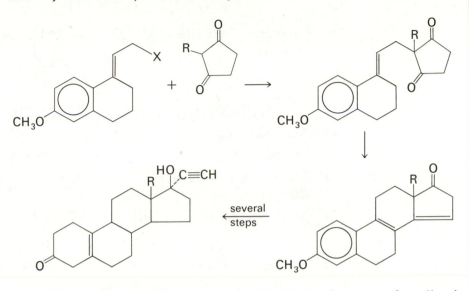

This type of synthesis has made steroid contraceptive agents from "coal tar" chemicals available for mass use, and has provided one approach for coping with the problem of global overpopulation.

*The prefix *nor-* means the absence of a carbon atom.

SUMMARY AND HIGHLIGHTS

1. **Aldehyes, R—C̈—H,** and **ketones, R—C̈—R,** are compounds containing a **carbonyl group, C=O.**

2. **Oxidation** in organic reactions involves replacement of C—H by C—OH; successive **oxidation levels** are:

$$-CH_2OH \xrightarrow{[O]} -CH(OH)_2 \quad \text{or} \quad -\overset{O}{\overset{\|}{C}}H \xrightarrow{[O]} -C(OH)_3 \quad \text{or} \quad -\overset{O}{\overset{\|}{C}}-OH$$

3. Aldehydes and ketones are prepared by oxidation of primary alcohols, RCH_2OH, and secondary alcohols, $RCHOHR$, respectively, with H_2CrO_4. Aldehydes can be further oxidized to acids; Cu^{+2} and Ag^+ are specific reagents for aldehyde oxidation.

4. **Addition reactions of aldehydes and ketones:**

a) Hydrogen: $R_2C=O + H_2 \xrightarrow{Pt} R_2CHOH$

$+ NaBH_4 \longrightarrow (R_2CHO)_4B^- \xrightarrow{H_2O} 4\,RCH_2OH$

b) Grignard reagents:

$RCHO + R'MgX \longrightarrow RCHOHR'$ (2° alcohol)

$R_2C=O + R'MgX \longrightarrow R_2C(OH)R'$ (3° alcohol)

$H_2C\overset{O}{\diagup\!\!\diagdown}CH_2 + R'MgX \longrightarrow R'CH_2CH_2OH$ (1° alcohol)

c) Acetylides: $R_2C=O + R'C\equiv C^{\overline{\cdot}}\ Na^+ \longrightarrow R_2C(OH)C\equiv CR'$

d) Cyanide: $R_2C=O + \overline{\cdot}CN \longrightarrow R_2C(OH)CN$ (cyanohydrin)

e) Water: $R_2C={}^{16}O + H_2{}^{18}O \longrightarrow R_2C={}^{18}O$ (oxygen exchange)

f) Alcohols: $RCHO + R'OH \longrightarrow RCH(OH)OR'$ (hemiacetal)

$RCHO + 2\,R'OH \xrightarrow{-H_2O} RCH(OR')_2$ (acetal)

g) Amines: $R_2C=O + NH_2X \longrightarrow R_2C=NX$ (hydrazones, imines)

$R_2C=O + NH_2R \xrightarrow{H_2 \atop Pt} R_2CHNHR$

5. **Formaldehyde** is exceptionally reactive; forms polymer $-(CH_2O)_n-$ and R_2NCH_2OH derivatives with amines; forms $-CH_2-$ cross-links between protein chains.

6. **Keto** and **enol tautomers** exist when **α-hydrogens** are present; equilibrium is catalyzed by acid or base:

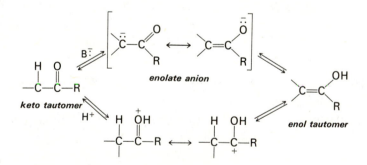

7. **Enols** and **enolates** and **nucleophiles:**

8. **Haloform reaction:** $RCCH_3 + 3 X_2 + OH^- \longrightarrow RCO^- + CHX_3$

9. **Aldol condensation** occurs between two moles of aldehyde in base; with aromatic aldehydes, the β-hydroxyaldehyde undergoes dehydration.

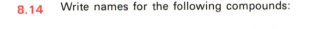

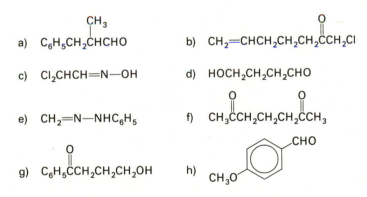

ADDITIONAL PROBLEMS

8.13   Write the structures of the following compounds:

a)   3-octanone

b)   *o*-methoxybenzaldehyde

c)   2-methylcyclopentanone

d)   1,1,3-trichloroacetone

e)   *n*-butylmagnesium bromide

f)   hexanal diethyl acetal

g)   diethyl ketone semicarbazone

h)   3-penten-2-one oxime

i)   acetaldehyde 2,4-dinitrophenyl hydrazone

j)   acetone cyanohydrin

8.14   Write names for the following compounds:

a) $C_6H_5CH_2\overset{\overset{\displaystyle CH_3}{|}}{C}HCHO$

b) $CH_2{=}CHCH_2CH_2CH_2\overset{\overset{\displaystyle O}{||}}{C}CH_2Cl$

c) $Cl_2CHCH{=}N{-}OH$

d) $HOCH_2CH_2CH_2CHO$

e) $CH_2{=}N{-}NHC_6H_5$

f) $CH_3\overset{\overset{\displaystyle O}{||}}{C}CH_2CH_2CH_2\overset{\overset{\displaystyle O}{||}}{C}CH_3$

g) $C_6H_5\overset{\overset{\displaystyle O}{||}}{C}CH_2CH_2CH_2OH$

h) CH_3O— (phenyl ring with CHO substituent)

8.15   Write structures and give names for all compounds of molecular formula $C_5H_{10}O$ that contain a carbonyl group.

8.16   Write equations, with structural formulas, showing how acetone could be converted to the following products. More than one step may be involved. Show reagents. (Other organic compounds may be needed.)

a)   acetone oxime

b)   2-propanol

c)   chloroform

d)   *tert*-butyl alcohol

e)   2-methyl-1-buten-3-yne

f)   *N*-ethyl isopropylamine

g)   2-methyl-2-hexene

h)   4-phenyl-3-buten-2-one

8.17 Write equations, with structural formulas, showing how acetaldehyde could be converted to the following products. Show all reagents and any other compounds needed.

a) $CH_3\overset{\overset{\displaystyle O}{\|}}{C}OH$ b)   1,1-dimethoxyethane

c)   ethanol   d)   1-phenylethanol

e)   ethylamine   f)   2-butenal

8.18 Write structures of the products that would be formed in the following reactions:

a) $CH_3CH_2MgBr + 3,5\text{-dimethoxybenzaldehyde} \longrightarrow \xrightarrow{H_3O^+}$

b) $CH_3CHOHCH_2Cl + H_2CrO_4 \xrightarrow{H^+}$

c) $CH_3CH_2CHO + Ag_2O \xrightarrow{NH_3}$

d) $C_6H_5MgBr + \text{ethylene oxide} \longrightarrow \xrightarrow{H_3O^+}$

e) $CH_3MgI + D_2O \longrightarrow$

f) $CH_3(CH_2)_4CHO + CH_3OH \xrightarrow{H^+} (C_8H_{18}O_2)$

g) $CH_2{=}CHCH_2\overset{\overset{\displaystyle O}{\|}}{C}CH_3 + NaBH_4 \longrightarrow$

h) $C_6H_5CHO + HCN \xrightarrow{CN^-}$

i) $C_6H_5\overset{\overset{\displaystyle O}{\|}}{C}CH_3 + C_6H_5CHO \xrightarrow{OH^-} (C_{15}H_{12}O)$

j)   3-penten-2-one $+ H_2 \xrightarrow{Pt}$

k) $(CH_3)_2CH\overset{\overset{\displaystyle O}{\|}}{C}H + (CH_3CH_2)_2NH \longrightarrow (C_8H_{17}N)$

8.19 Show how the following syntheses could be carried out from the starting materials indicated and any other compounds and reagents that may be needed. Several steps will be required.

a)   4-octanone starting with alcohols with no more than four carbon atoms
b)   2-methylpentane-1,3-diol from propanal
c)   1-methyl-1-cyclohexanol from cyclohexanol
d) $C_6H_5CHOHCH{=}CH_2$ from benzaldehyde
e)   di-*n*-propylamine from propanal
f)   2-ethyl-1-hexanol from a four-carbon compound

g)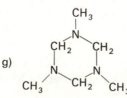

from compounds with no more than one carbon atom.

8.20 The following structures represent unstable substances that cannot be isolated. Write the structure of a stable compound that would be formed spontaneously from each.

a) $CH_3CH{=}CHOH$ b) $C_6H_5\overset{\overset{\displaystyle Cl}{|}}{C}H{-}OH$

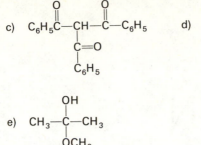

c)
$$C_6H_5\overset{O}{\underset{}{C}}-\underset{\underset{C_6H_5}{\overset{|}{\underset{}{C}=O}}}{\overset{|}{CH}}-\overset{O}{\underset{}{C}}-C_6H_5$$

d) $H_2N-NHCH_2CH_2CH_2CHO$

e) $CH_3-\underset{\underset{OCH_3}{|}}{\overset{\overset{OH}{|}}{C}}-CH_3$

8.21 The reaction of butanal with ethylene glycol ($HOCH_2CH_2OH$) in the presence of sulfuric acid gives a product $C_6H_{12}O_2$. Infrared data (Chapter 9) showed that the compound contained no C=O, C=C, or OH groups. Suggest the structure of the product.

8.22 A low-boiling compound **A** gave a silver mirror with Tollens' reagent. Compound **A** reacted with 2,4-dinitrophenylhydrazine to give a yellow-orange product with the composition $C_8H_8N_4O_4$. Deduce the structure of compound **A**.

8.23 Describe simple chemical tests which would distinguish between the following pairs of compounds; state what observations would be made.

a) $C_6H_5\overset{O}{\overset{\|}{C}}CH_3$ and $C_6H_5CH_2CH_2OH$

b) CH_3CH_2CHO and $CH_3CH_2\overset{O}{\overset{\|}{C}}CH_3$

c) $CH_3CH(OCH_3)_2$ and $CH_3OCH_2CH_2OCH_3$

8.24 In order to prepare hexanedial, a student treated cyclohexene with ozone. After reductive hydrolysis of the ozonide, the reaction was inadvertently made basic before distilling the product. The main compound isolated had molecular formula C_6H_8O. Suggest the structure of this compound and trace its formation from cyclohexene.

9
SPECTRAL IDENTIFICATION AND STRUCTURE DETERMINATION OF ORGANIC COMPOUNDS

Structural formulas are the language of organic chemistry. Throughout this book or any book on organic chemistry, compounds are represented by structural formulas, and their reactions and properties are rationalized in terms of the structural features of the molecules. Already you may have asked, "How do I know that these are actually the structures of the molecules?" In this chapter we will briefly describe the most important instrumental methods by which a chemist may deduce the structure of an organic molecule. While the information in this chapter is insufficient to solve the structures of any but very simple molecules, you will obtain a "feel" for the ways in which it is normally done.

The structures of some simple, long-known compounds evolved directly from the concepts of bonding together with basic chemical reactions. Thus, long before the advent of modern instrumental techniques, ethanol was found to have the molecular formula C_2H_6O. Since the compound reacted with sodium to liberate hydrogen, it was reasoned by analogy with water that ethanol must contain an —OH group. Given the requirement of tetracovalent carbon, along with the observed chemistry, it was concluded that ethanol must have the structure CH_3CH_2OH.

Going a step further, it was known that ethanol could be oxidized to a compound called acetaldehyde, C_2H_4O, and further oxidation of acetaldehyde gave acetic acid, $C_2H_4O_2$. Long before a structural theory was available, it was inferred by indirect and laborious methods that all three compounds, ethanol, acetaldehyde, and acetic acid, contained a CH_3 unit which remained unchanged when these compounds were chemically interconverted. Thus, the structures could be represented as shown here.

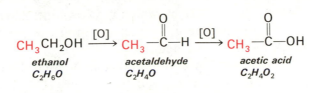

In the case of these long-known compounds, the structures literally grew out of the development of structural theory. They are consistent with a variety of related facts. However, there was no single, clearly identifiable point when it could be stated that the structure of any of these compounds was "proven."

As new compounds were chemically characterized, they were fitted into the existing pattern of structural theory. Until about 1940, the structure of a newly discovered compound was based almost entirely on its molecular formula and chemical reactions. A compound was transformed into one which was already known by a series of specific reactions, or it was synthesized from smaller known compounds by specific reactions. Such structural identifications were extremely time consuming, and were subject to considerable uncertainty since they were only as reliable as the understanding of the reactions that were carried out.

Organic chemistry has been revolutionized in the past 30 years by the use of **spectroscopic methods** for analysis and structure determination. Practically all functional groups and structural features can be detected quickly with very small amounts of sample. An added bonus is that the same sample can often be used for several spectroscopic analyses, since most of these methods are nondestructive. Analytical methods for the quantitative determination of organic compounds in air, water, soil, and body fluids and tissue also depend almost exclusively upon spectroscopic measurements.

9.1 Electromagnetic Radiation

The term **spectroscopy** refers to the dispersion of **electromagnetic radiation** into a spectrum of different energies. The application of spectroscopy to chemistry depends upon the interaction of molecules with radiation of specific energy.

The most familiar form of electromagnetic radiation is ordinary white light, which can be dispersed by a prism into a spectrum of colors. This visible light represents only a small part of the entire electromagnetic spectrum, which extends from high-energy cosmic rays to low-energy radio or radar waves. Electromagnetic radiation can be described either as a stream of energetic particles called **photons** or as a *wave motion*. The fundamental equations of electromagnetic radiation are

$$E = h\nu \qquad \text{and} \qquad \nu\lambda = c$$

where E is the energy of the radiation, h is Planck's constant, ν is the frequency of the radiation (waves per second), λ is the wavelength of the radiation, and c is the velocity of light, which is a constant. Since ν and λ are related by a constant (c), the energy of the radiation can be described by either the frequency or the wavelength.

The wavelength regions of most importance in organic chemistry are the ultraviolet (uv), visible (vis), and infrared (ir) (Figure 9.1). The ultraviolet

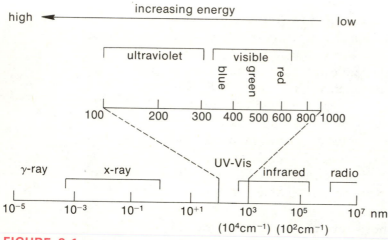

FIGURE 9.1

The electromagnetic spectrum.

and visible regions are adjacent, and the boundary is somewhat arbitrary since it depends upon human perception of light. The usual unit for wavelength in this region is the **nanometer** (1 nm = 10^{-9} meter = 0.1 Ångstrom). The infrared region is at longer wavelengths (lower energy, since $E = hc/\lambda$); in infrared spectroscopy it is customary to designate a band position in the spectrum in terms of the *reciprocal* of the wavelength in centimeters. Since $1/\lambda = \nu/c$, this quantity (called a wavenumber) is proportional to the frequency of the radiation. The units of cm^{-1} (1/cm) are called *reciprocal centimeters*.

Electromagnetic radiation can be absorbed by a molecule *only* when the energy of the radiation corresponds to the energy difference between two energy levels of the compound; *i.e.*, energy absorption is *quantized*. In the **uv-vis** region the energies correspond to transitions of *electrons* from one orbital to a higher energy orbital. When a molecule has been raised to a higher energy state by absorption of energy, the molecule is said to be in an "excited state." The larger the gap between the ground and excited states, the higher the energy (shorter wavelength, $E = hc/\lambda$) required to produce the transition. In the **infrared** region, the energy levels correspond to certain vibrations of the *atoms* in the molecule.

9.2   Measurement of Spectra

The **absorption spectrum** of a compound is a graph of the wavelength or frequency of the light impinging on the sample versus the fraction of the light absorbed by the sample. Spectra can be recorded automatically in a few minutes, using a solution containing only a few milligrams of compound, with a spectrometer such as that shown in Figure 9.2. The instrument consists of a source of radiation in the desired wavelength region, a prism or similar device to disperse the radiation into beams of very narrow wavelength spread, and an electronic system to detect and record the difference in intensities of the beam passing through the sample and a second beam which passes through the pure solvent.

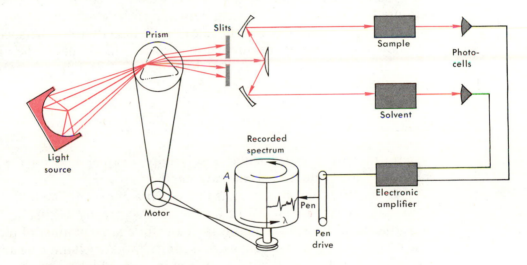

FIGURE 9.2

Schematic diagram of recording spectrometer.

9.3   Ultraviolet and Visible Spectra

Ultraviolet (uv) and visible (vis) spectra arise from the absorption of light which "excites" an electron in a compound to a higher energy level. Organic compounds that absorb light in this region are those with C=C and C=O bonds. The spectral range which is accessible with usual instruments is from 200 to 700 nm, and this entire region is often referred to as the ultraviolet spectrum although it also includes the visible region. The wavelength of the highest point of the absorption band is indicated by the symbol λ_{max}, and can be read directly from the horizontal axis of the spectrum (Figure 9.3).

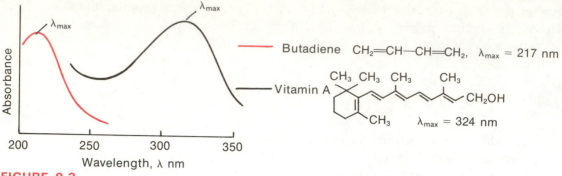

FIGURE 9.3

Ultraviolet absorption spectra.

The vertical axis of the spectrum measures **absorbance** (A), which is the logarithm of the ratio I_0/I. The term I_0 is the intensity of the light beam which strikes the sample, and I is the intensity of the beam after it has passed through the sample. Obviously, if the sample does not absorb any of the light, $I_0/I = 1$ and $\log I_0/I = 0 = A$. In practice, very dilute solutions of the compound in a non-absorbing solvent (*e.g.*, water, ethanol, or cyclohexane) are used in a quartz cell of accurately known length.

The absorbance (A), which we can read directly from the spectrum, is directly proportional to the number of absorbing molecules in the light path. Thus, A is dependent upon the length of the cell, l, and the concentration of the solution, C. The exact relationship is

$$\log I_0/I = A = \epsilon \cdot C \left(\frac{\text{moles}}{\text{liter}}\right) \cdot l \text{ (cm)}$$

where the term ϵ, called molar absorptivity, is a constant for a particular absorbing molecule at a given wavelength.

If the molar absorptivity, ϵ, is known for a pure compound, the concentration of the compound can be quickly and accurately calculated from the absorbance measurement. This is the basis for a large number of the analytical measurements in biochemical and clinical work, including the automated analyses carried out in clinical analyzers in medical laboratories. An example, using the strong smelling, bicyclic ketone camphor (used as a moth repellant and in embalming fluids, among other things) is shown here.

Example: For a solution of camphor in hexane in a 10 cm cell, A at 295 nm was found to be 2.52. What is the concentration of camphor?

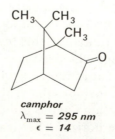

camphor
$\lambda_{max} = 295$ *nm*
$\epsilon = 14$

Answer: $2.52 = (14)(10) C$
$C = 1.8 \times 10^{-2}$ moles/liter

Nonconjugated carbonyl compounds have a very weak ($\epsilon = 10$ to 30) absorption band in the 200 to 300 nm region. This band arises from excitation of one of the nonbonding electrons (from an unshared pair) to the next highest energy level or orbital.

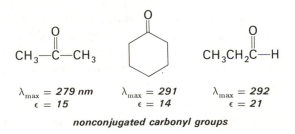

$CH_3-\overset{\displaystyle O}{\overset{\|}{C}}-CH_3$ $CH_3CH_2\overset{\displaystyle O}{\overset{\|}{C}}-H$

$\lambda_{max} = 279$ nm $\lambda_{max} = 291$ $\lambda_{max} = 292$
$\epsilon = 15$ $\epsilon = 14$ $\epsilon = 21$

nonconjugated carbonyl groups

All nonconjugated alkenes have an *intense* absorption which occurs below 200 nm and is therefore inaccessible to most commonly used ultraviolet spectrometers. For example, ethylene has $\lambda_{max} = 166$ nm with $\epsilon = 10,000$. This absorption comes from the light-induced promotion of a π-electron to the next higher energy level.

The most important structural information which can be obtained from the uv spectrum is whether or not two unsaturated groups are in conjugation. When two or more C=C and/or C=O units are in the same molecule, but are not conjugated, the uv spectrum is simply that expected of the isolated absorbing groups. However, when the two groups are conjugated, the difference between the electronic ground state and the excited states, ΔE, becomes smaller and results in a shift of λ_{max} to longer wavelengths. Thus, it is a simple task to distinguish between the two isomeric dienes 1,5-hexadiene and 2,4-hexadiene from the relative positions of λ_{max}.

$CH_3CH{=}CH-CH{=}CHCH_3$ $CH_2{=}CH-CH_2-CH_2-CH{=}CH_2$

$\lambda_{max} = 227$ nm $\lambda_{max} = 178$ nm
conjugated   *vs.*   *unconjugated*

Conjugation of a carbon-carbon double bond and a carbonyl group shifts the λ_{max} of both groups to longer wavelengths. This effect is dramatically demonstrated by a comparison of the following three molecules:

$CH_3-\overset{\displaystyle CH_3}{\overset{|}{C}}{=}CH-CH_2CH_3$ $CH_3-\overset{\displaystyle CH_3}{\overset{|}{C}}H-CH_2-\overset{\displaystyle O}{\overset{\|}{C}}-CH_3$ $CH_3-\overset{\displaystyle CH_3}{\overset{|}{C}}{=}CH-\overset{\displaystyle O}{\overset{\|}{C}}-CH_3$

$\lambda_{max} = 180$ nm $\lambda_{max} = 283$ nm $\lambda_{max} = 230$ ($\epsilon = 12,600$)(from C=C)
$\epsilon = 11,000$ $\epsilon = 20$ $\lambda_{max} = 327$ ($\epsilon = 40$)(from C=O)

As the number of double bonds in conjugation increases, ΔE for the excitation of an electron continues to get smaller and λ_{max} therefore increases. With five conjugated double bonds, as in vitamin A (Figure 9.3), the absorption peak is now moved into the visible part of the spectrum. Dyes (see Chapter 12) owe their colors to conjugated systems which have strong absorption in the visible region.

9.4   Infrared Spectra

Infrared (ir) radiation ranging from about 10,000 to 100 cm^{-1} is absorbed and converted into energy of *molecular vibration*. Only certain discrete vibrational energy levels are available to the molecule, and thus the absorption of infrared light corresponds to a transition between two vibrational energy levels. Most compounds will produce a rather complex ir spectrum because for a molecule of n atoms there are $3n - 6$ allowed vibrations, although exactly this number of transitions will not be observed in the region accessible to the usual ir spectrophotometer, which is from 4000 to 666 cm^{-1}.

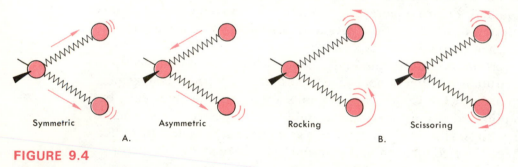

Symmetric   Asymmetric   Rocking   Scissoring

A.   B.

FIGURE 9.4

Some molecular vibrations: (A) stretching, (B) in-plane bending.

A good working analogy for a vibrating molecule is an assembly of balls (atoms) held together by springs (bonds). Most of the absorption peaks in the ir region are due to two types of vibrations—**stretching** and **bending** (Figure 9.4). Stretching vibrations are those which cause an increase or decrease in the distance between the two atoms involved. Bending vibrations, sometimes descriptively classified as "wagging," "rocking," and "scissoring," refer to vibrations which result in a change in bond angle.

The position of an ir band depends upon the relative masses of the atoms involved in the vibration and the strengths of the bonds. Thus, we find that the stretching frequencies of C—H, O—H, C—C, C—O, C=C, C=O, and many other bonds appear in quite different regions of the ir spectrum. The positions of some typical vibrational frequencies are presented in Table 9.1. From tabulations such as this, we can find where to expect a certain group to absorb in the infrared and then check our spectrum to see if this absorption is present.

TABLE 9.1   INFRARED VIBRATIONAL FREQUENCIES

Frequency ν cm^{-1}	Vibration	Frequency ν cm^{-1}	Vibration
Stretching		*Stretching*	
3600–3300	O—H	1650–1500	C=C
3500–3100	N—H	1200–1000	C—O
3300	C≡C—H		
3100–3000	Ar—H		*Bending*
3000–2800	RC—H	1500–1300	—OH
3000–2500	O—H···O—	1470–1420	—CH$_2$—
2200	C≡N	960–900	—C=C—H
2100	C≡C	1600–1500	Aromatic ring
1830–1650	C=O	900–700	Ar—H

The relative intensities of peaks caused by the vibrations of various groups are dependent upon the change in bond dipole moment occurring during the vibration. Since this bond moment is the product of the amount of charge separation times the distance, the more polar groups (such as C—O or C=O) produce much stronger peaks than do the vibrations of less polar units, such as C—C or C=C.

The main value of ir spectra for structure determination is in revealing the presence of specific functional groups such as O—H or C=O. Peaks due to these groups are readily identified, as can be seen in the spectra in Figures 9.5 and 9.6. A C=O peak is easily identified by both its position and its

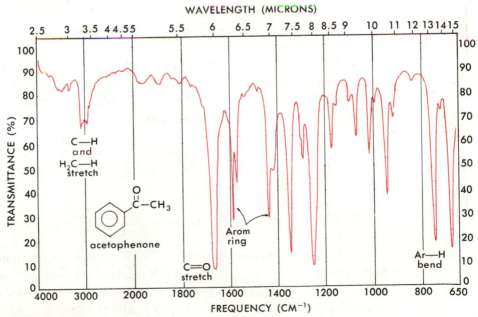

FIGURE 9.5

Infrared spectrum of acetophenone.

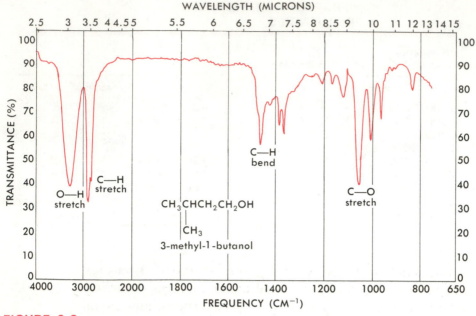

FIGURE 9.6

Infrared spectrum of 3-methyl-1-butanol.

intensity. Although a C=O peak always appears in the same general region of the spectrum, the exact frequency depends on the groups attached to the carbon atom. For example, an aliphatic ketone or aldehyde absorbs at 1710

to 1735 cm^{-1}, while an acid chloride $(R-\overset{\overset{\text{O}}{\|}}{C}-Cl)$ has a C=O stretching bond at *ca.* 1800 cm^{-1}. No one tries to assign *all* of the bands in the spectrum of an organic molecule, but the presence of major groups is readily recognized from a quick inspection.

In addition, there are a number of other peaks in the lower frequency region of both spectra. These are due to various bending vibrations of C—C bonds which are characteristic of the particular molecule as a whole, and provide a "fingerprint" of the compound. The infrared spectra of two different compounds always show at least minor differences. Thus, a perfect matching of the ir spectra of a known and an unknown compound can provide rapid identification of the unknown molecule.

9.5 Nuclear Magnetic Resonance (NMR) Spectra

A third type of transition that absorbs radiation is experienced by nuclei which possess a magnetic dipole. To the organic chemist, the most important nucleus that has a magnetic dipole is the proton. In a magnetic field a proton can align itself so that its nuclear spin is either parallel to (more stable) or against (less stable) the direction of the applied field; thus, the two spin states correspond to different dipole orientations. In an nmr spectrometer, the

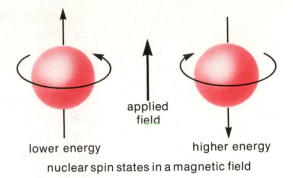

applied field

lower energy — higher energy

nuclear spin states in a magnetic field

sample is placed between the poles of a powerful magnet and irradiated with radio-frequency energy. At a particular combination of frequency and magnetic field strength, energy is absorbed and the nucleus is "flipped" from one spin state to another. This absorption of energy is recorded as a peak or "resonance signal" in the nmr spectrum.

A major value of proton nmr spectroscopy lies in the fact that the position of the resonance signal for a particular proton depends upon the magnetic field immediately surrounding that proton. This field will not be exactly the same as the applied magnetic field of the instrument, because it is influenced by the electron density around the proton and is therefore dependent upon the type of bonding in which the proton is involved. A general rule is that high electron density around the proton tends to **shield** the nucleus from the applied magnetic field, so that a stronger field must be applied to obtain resonance at a constant irradiation frequency. Naturally, the opposite effect occurs if electron density is removed from the vicinity of the proton, for now a weaker field will be required to reach resonance for such "deshielded" hydrogens. Let us briefly examine how such effects can

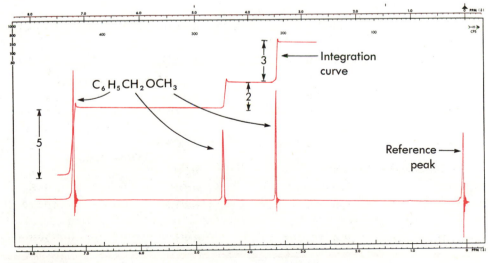

FIGURE 9.7

NMR spectrum of methyl benzyl ether.

provide us with a great deal of information about the *structural environment* of the absorbing protons.

A typical proton spectrum, that of methyl benzyl ether, is shown in Figure 9.7. The horizontal axis is calibrated in parts per million (ppm) of the applied magnetic field relative to the standard reference peak of tetramethylsilane, $(CH_3)_4Si$, which is taken as zero. The three lines in the spectrum correspond to the absorptions of three different types of protons in the ether, CH_3, CH_2, and C_6H_5. Furthermore, we find that the relative areas underneath these peaks correspond exactly to the number of protons that are present in each group and are absorbing at that region of the spectrum. Integration of these peak areas is done automatically by the nmr spectrometer, and is shown in Figure 9.7 by an integration curve which reveals the relative areas to be 5:2:3.

The position of a peak on the horizontal axis (called **chemical shift**, δ, in ppm) depends on the electronic environment of the carbon to which the protons are attached. For instance, protons on a carbon attached to oxygen,

TABLE 9.2   CHEMICAL SHIFTS OF PROTON GROUPINGS*

Methyl Groups CH_3-		Methylene Groups RCH_2-	
Group	*Shift, δ ppm*	*Group*	*Shift, δ ppm*
CH_3-R	0.8–1.2	$R-CH_2-R$	1.1–1.5
$CH_3-CR=C\big<$	1.6–1.9	$R-CH_2-Ar$	2.5–2.9
CH_3-Ar	2.2–2.5		
		$R-CH_2-\overset{\overset{\displaystyle O}{\|\|}}{C}-R$	2.5–2.9
$CH_3-\overset{\overset{\displaystyle O}{\|\|}}{C}-R$	2.1–2.4	$R-CH_2OH$	3.2–3.5
$CH_3-\overset{\overset{\displaystyle O}{\|\|}}{C}-Ar$	2.4–2.6	$R-CH_2OAr$	3.9–4.3
$CH_3-\overset{\overset{\displaystyle O}{\|\|}}{C}-OR$	1.9–2.2	$R-CH_2-\overset{\overset{\displaystyle O}{\|\|}}{O}CR$	3.7–4.1
$CH_3-\overset{\overset{\displaystyle O}{\|\|}}{C}-OAr$	2.0–2.5	$R-CH_2-Cl$	3.5–3.7
$CH_3-N\big<$	2.2–2.6	**Methine Groups R_2CH-**	
CH_3-OR	3.2–3.5	R_3CH	1.4–1.6
CH_3-OAr	3.7–4.0	R_2CHOH	3.5–3.8
$CH_3-O\overset{\overset{\displaystyle O}{\|\|}}{C}R$	3.6–3.9	Ar_2CHOH	5.7–5.8
Unsaturated Groups		**Other Groups**	
		ROH	3–6
$RCH=C\big<$	5.0–5.7	$ArOH$	6–8
$Ar-H$	6.0–7.5	RCO_2H	10–12
$R-\overset{\overset{\displaystyle O}{\|\|}}{C}-H$	9.4–10.4	$RNH-$	2–4

*In this table, R = saturated carbon $\left(CH_3-,\ -CH_2-,\ -\overset{\textstyle |}{\underset{\textstyle |}{C}}H,\ -\overset{\textstyle |}{\underset{\textstyle |}{C}}-\right)$; Ar = aromatic ring.

such as CH_3—O—R, are found in nearly the same position (δ) in the spectrum of any compound containing this grouping, regardless of the nature of R. The characteristic positions of some common groups are given in Table 9.2. It is interesting to note that the protons of our example CH_3OR are found considerably further **downfield** (larger δ) than those in an alkane, CH_3—R. This is understandable, since the electronegative oxygen tends to pull electron density away from the protons, therefore **deshielding** them from the applied magnetic field. We find, therefore, that hydrogens on carbon attached to other electronegative groups, such as halogen, appear much further downfield than do those of alkanes. Changing the hybridization of the carbon to which the proton is attached from sp^3 to sp^2 also brings about a downfield shift. For example, compare in Table 9.2 the chemical shift of CH_3—R with

the chemical shifts of CH_3—CR=C$\diagdown$ and CH_3—$\overset{\overset{\displaystyle O}{\parallel}}{C}$—R. Thus, a lot can be learned about the environment of a proton simply from its chemical shift; as a result, much is learned about the molecule.

Another important feature of nmr spectra can be seen in Figure 9.8. The signals due to the CH_3 and CH_2 protons in ethyl bromide are **split** into three and four lines, respectively. As shown in Figure 9.9, the CH_3 protons appear as a **triplet** because of interaction with the three possible spin states of the CH_2 protons with respect to the magnetic field. Since the CH_3 protons can adopt four possible spin arrangements with respect to the magnetic field, the adjacent CH_2 protons interact with four different spin arrangements and thus appear as a **quartet** of four lines. It is important to note that protons that have identical chemical shifts will not split each other—see, for example, the CH_3 group in Figure 9.7, where three protons produce only a single peak. Protons that are more remote from each other than those on adjacent carbons, as in CH_3OCH_2—, do not generally show splitting.

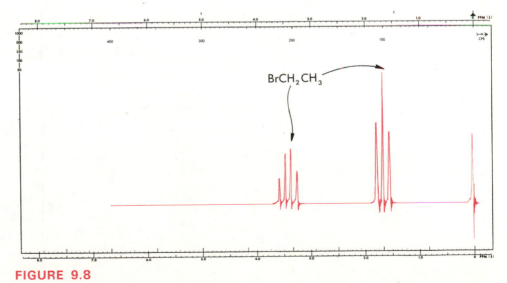

$BrCH_2CH_3$

FIGURE 9.8
NMR spectrum of ethyl bromide.

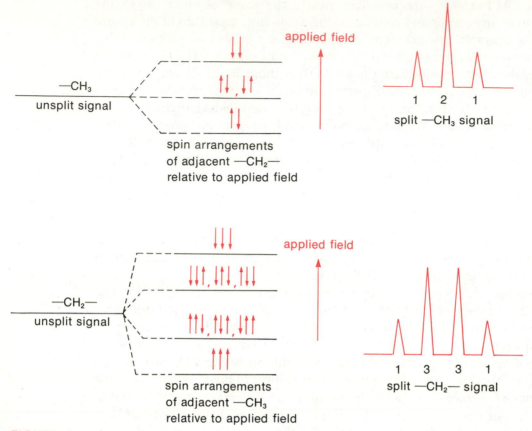

FIGURE 9.9

Signal splitting by protons on adjacent carbons. Symbol ↑ indicates spin alignment with, and ↓ against, the direction of the applied field.

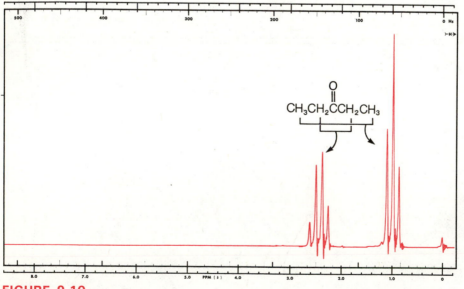

FIGURE 9.10

NMR spectrum of 3-pentanone.

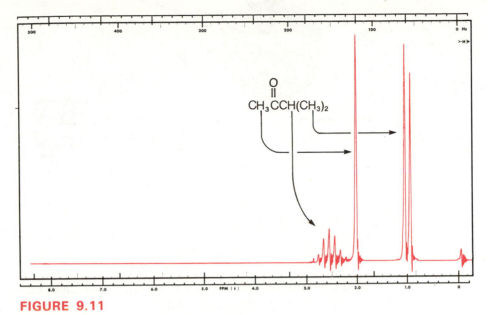

FIGURE 9.11

NMR spectrum of 3-methyl-2-butanone.

When the chemical shifts of the interacting protons are well separated, the signal for a given proton grouping (*e.g.*, CH_3, CH_2, or CH) will be split into $n + 1$ peaks by **n** protons on the next carbon atom. Thus, the two CH_2 protons are split by the adjacent CH_3 signal into a quartet ($n = 3$).

A good example of this signal splitting by protons on adjacent carbons is found in Figure 9.11. In the case of the isopropyl group, $(CH_3)_2CH—$, the six equivalent (same δ) methyl hydrogens split the resonance signal due to the single tertiary proton into $(6 + 1)$ lines, or a **septet.** In turn, the lone tertiary hydrogen splits the methyl peak into a **doublet.**

The nmr spectra of the two ketones in Figures 9.10 and 9.11 illustrate how the nmr method allows rapid distinction between two isomers. In Figure 9.10 the expected quartet (CH_2 split by CH_3) clearly distinguishes 3-pentanone from 3-methyl-2-butanone, whose spectrum (Figure 9.11) contains a sharp singlet (no H's on C adjacent to one —CH_3) and the doublet and septet of the isopropyl group.

9.6   Mass Spectra

In a mass spectrometer (Figure 9.12), molecules in the gas phase are bombarded by a stream of very high energy electrons. The resulting collisions impart considerable energy to the molecules, which in turn emit electrons to produce positively charged ions. These cations possess so much excess energy that they often fragment through various bond cleavages to produce new cations and neutral fragments. The resulting stream of cations is accelerated

$$\text{molecule} \xrightarrow[\text{electron beam}]{\text{high energy}} e^- + \text{molecule}^{+\cdot} \xrightarrow{\text{fragmentation}}$$

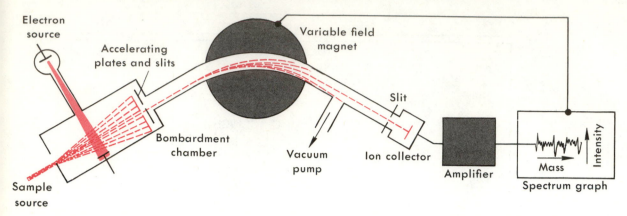

FIGURE 9.12

Schematic diagram of mass spectrometer.

out of the bombardment chamber by negatively charged grids and passes through a region of strong magnetic field. The path in the magnetic field region taken by any given cation depends upon its mass. Thus, the mass spectrometer separates the cations on the basis of their relative masses. Finally, the beams of separated ions pass through an exit slit and the relative abundance of ions of each mass is recorded. The mass spectrum of a molecule can be obtained in a few seconds with a tiny fraction of a gram of material. The mass spectrum is usually presented as a graph of ion abundance versus mass, as shown in Figure 9.13.

The ion formed by removing one electron from the starting molecule is a radical cation, since it now must have one unpaired electron, and it is usually represented as $M^{+\cdot}$. This **molecular ion** is the most important ion, since its mass is the molecular weight of our compound. Fragmentation of the molecular ion usually gives a new, smaller cation and a neutral fragment.

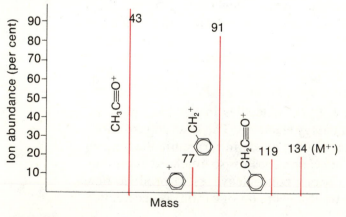
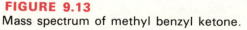

FIGURE 9.13

Mass spectrum of methyl benzyl ketone.

Only the cations are accelerated by the negatively charged plates, and therefore only these ions appear in the mass spectrum.

The mass spectrum of methyl benzyl ketone, a metabolite extracted from urine, is shown in Figure 9.13. The molecular ion, corresponding to the molecular weight, is quite distinct at mass 134. This ion apparently fragments through homolytic cleavage of either of the two carbon bonds to the carbonyl group (paths *a* and *b*). Further fragmentation of the ion derived from loss of $\cdot CH_3$ (ion of mass 119) occurs through loss of carbon monoxide to give the very stable benzyl ion (mass 91).

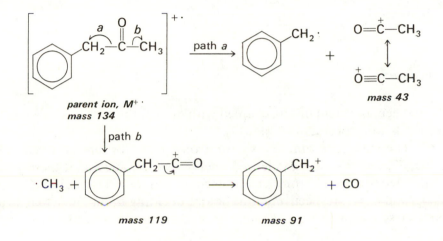

The interpretation of a mass spectrum is much like putting together a jigsaw puzzle. Often there are several ways we can connect the various pieces, but other spectral methods (ir, uv, nmr) will place enough restrictions on our game that only one answer will fit *all* the data.

9.7 Illustration of a Structure Determination from Spectroscopic Data

Now that we have looked briefly at the spectroscopic methods that are used by the organic chemist, let us see how the structure of an unknown compound could be solved with these methods. We will gather various pieces of information about our unknown molecule and finally put them all together.

1. **Combustion analysis** gives the percentages of C and H in the molecule by burning a weighed sample and measuring the amounts of CO_2 and H_2O formed. Our unknown compound was found to contain 72.0% C and 6.7% H by weight. If we assume that the remainder of the molecule is oxygen, the percentage of O can be obtained by subtraction ($100 - 72.0 - 6.7 = 21.3\%$).

2. In the **mass spectrum** (Figure 9.14), the peak of largest mass ($M^{+\cdot}$) is

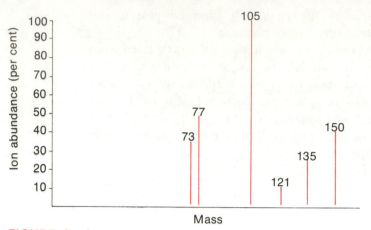

FIGURE 9.14

Mass spectrum of unknown.

150. A molecular weight of 150, coupled with our combustion analysis results, corresponds to a formula of $C_9H_{10}O_2$.*

3. The mass spectrum also shows strong peaks for ions of masses 135, 121, 105, 77, and 73. An experienced chemist would guess that these peaks corresponded to losses from the molecular ion of CH_3, C_2H_5, C_2H_5O, $C_3H_5O_2$, and C_6H_5, respectively. But before deciding upon the actual structures of these fragment ions, the nmr, ir, and uv spectra must be examined.

---

*The formula is found by reasoning as follows: Since one molecule contains a total of 150 mass units, 72% of this—or 108 mass units—is due to carbon atoms. Since each carbon atom accounts for 12 mass units, there must be $108 \div 12 = 9$ carbon atoms in the molecule. This process is repeated for hydrogen and oxygen.

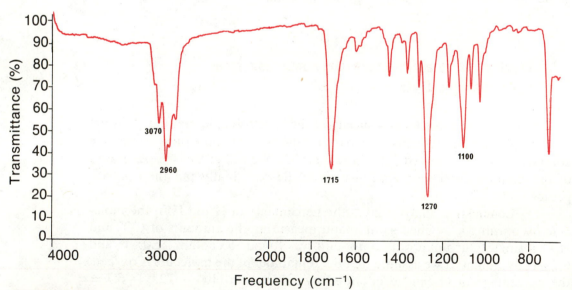

FIGURE 9.15

Infrared spectrum of unknown.

4. The **infrared spectrum** (Figure 9.15) shows strong bands at 3070, 2960, 1715, 1270, and 1100 cm^{-1} which correspond to the usual regions for C—H(Ar), C—H(RH), C=O, C—O(asym.), and C—O(sym.) stretching vibrations.

5. The **nmr spectrum** (Figure 9.16) has a complicated set of peaks (area 5) in the region for aromatic protons (δ 7.2–8.1). Both the chemical shift and the integration make it almost certain that this absorption is due to a phenyl group (C_6H_5—). We also observe a quartet (area 2) centered at δ 4.4. The area indicates this to be a —CH_2— group; the splitting into four peaks suggests that it is attached to a methyl group (—CH_2—CH_3, $n + 1$ peaks where $n = 3$); and the position of the absorption (δ 4.4) suggests that the CH_2 group is attached directly to an oxygen atom (—O—CH_2—CH_3). The presence of a methyl group attached to the —CH_2— is confirmed by the triplet at δ 1.4, which integrates correctly for three protons.

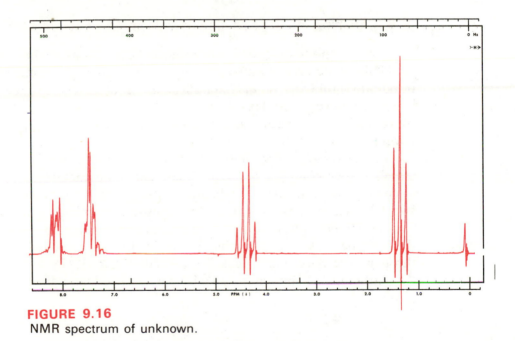

FIGURE 9.16
NMR spectrum of unknown.

6. Now we can put it all together. The nmr tells us that we have the —O—CH_2—CH_3 unit, and this is confirmed by the mass spectrum through the loss of this fragment (150 − 45 = 105). The C—O bond is also seen in the ir as the strong bands at 1270 and 1100 cm^{-1}. The nmr also says that we have a phenyl group, and the mass spectrum confirms this both through the loss of mass 77 from M$^{+\cdot}$ to produce a peak at mass 73, and from observation of a peak for $C_6H_5^+$ at mass 77. The ir spectrum says that we also have a carbonyl C=O unit, and the only place left for us to put this unit is between

the C_6H_5— and —OCH_2CH_3 groups. Thus, we conclude that our unknown has the structure shown below, that of ethyl benzoate.

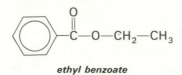

ethyl benzoate

It should be strongly emphasized that this simple illustration is intended *only* to demonstrate the type of analysis that a chemist uses to determine a molecular structure. This type of analysis normally requires much more information than can be included in a course of this length. However, you should now be aware of the power of these spectroscopic tools; this is the source of confidence in the organic structures you have seen and will be seeing in this book.

9.8 Analysis of Organic Compounds in the Environment

The methods described so far have been discussed in terms of the molecular structures of individual compounds. However, some of the most important problems in the identification of organic compounds involve substances in very complex mixtures. Methods for separation and purification by crystallization, fractional distillation, and column chromatography are covered in the laboratory part of the course and in the laboratory text.

An extremely important tool in the analysis of complex mixtures of closely related compounds is **vapor phase chromatography** (vpc), particularly in conjunction with spectroscopic methods. In vpc, a sample of a mixture is swept by helium gas through a narrow tube containing an adsorbent. The various components are separated because of different relative affinities for the adsorbent. As each compound emerges from the vpc column, it passes through a detector and a peak is recorded (Figure 9.17).

Individual compounds can also be collected and then identified by

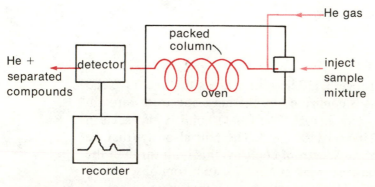

FIGURE 9.17
Schematic representation of a gas chromatograph.

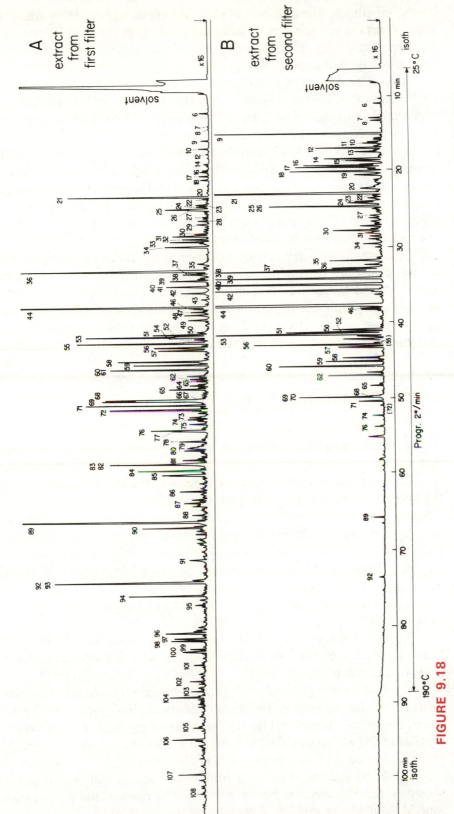

FIGURE 9.18

Vapor phase chromatogram of organic compounds in air sample. (Courtesy Dr. K. Grob, University of Zürich, Zürich, Switzerland; and Journal of Chromatography, *62*:1, 1971.)

spectral methods. The amounts of material are often extremely small, but the vapor from the column can be passed directly into a mass spectrometer. The mass spectra of the different components are then recorded sequentially as they leave the column.

The effectiveness of combined vpc–mass spectral analysis is illustrated by the results shown in Figure 9.18. This chart is a chromatogram of the organic compounds present in an air sample taken from a green park in Zurich, Switzerland, 1000 feet from streets or buildings. The air was drawn through an absorbent filter, and the organic compounds were then extracted from the filter and passed through the instrument. From the total of over 400 peaks in the chromatogram, 108 compounds were distinctly separated, and 74 of these were fully identified with specific structures by means of their mass spectra. About 90 per cent of the compounds in the sample were found to be hydrocarbons emitted by oil burners.

Similar analyses are carried out in many laboratories to study environmental problems. Insecticide residues in food, air pollution from automobile exhausts, and industrial effluents in rivers all present complex mixtures of organic compounds in trace amounts. Identification of the individual compounds is an important step in determining the precise nature of the problem and the measures required to overcome it.

SUMMARY AND HIGHLIGHTS

1. Spectroscopic methods are the major tools in structure determination.

2. The energy (E) of electromagnetic radiation is related to frequency (v) and wavelength (λ) by the equations $E = hv = hc/\lambda$.

3. Absorption spectra are produced by molecules absorbing discrete amounts of energy corresponding to quantized energy level changes in the particular molecules.

4. **Ultraviolet-visible** spectra are produced by **electronic transitions.** The longer the conjugation, the lower the energy; polyenes absorb in the visible region and are colored. The amount of light absorbed is proportional to concentration.

5. **Infrared** absorption is due to molecular **vibrations;** different types of bonds have characteristic absorption frequencies which reveal the presence of such functional groups as —OH, $\diagup C{=}C \diagdown$, and $\diagup C{=}O$.

6. **NMR** absorption arises from **nuclear spin** transitions of protons. The position of the resonance, known as chemical shift, indicates the bonding environment of protons, while the number of protons of each type can be measured from the area under the peaks, and the number of neighboring protons is shown by splitting of the peaks.

7. **Mass spectra** arise from molecular cations and resulting fragment ions produced by electron bombardment. The mass of the parent molecular ion, $M^{+\cdot}$, gives the molecular weight of the compound.

PROBLEMS

9.1 State which of the major spectroscopic methods—uv-visible, infrared, nmr, or mass spectroscopy—would be best suited for solving each of the following problems. In some cases, an answer could be obtained by more than one method; nearly all the structural questions could be solved by nmr or mass spectra. However, for purposes of this question, note the following very approximate data and be guided accordingly. Thus, if a clear answer can be gained from infrared, then mass spectra would probably not be method of choice.

	uv-vis	infrared	nmr	mass spectroscopy
cost of instrument	$2000–4000	$2000–4000	$15,000–30,000	$30,000–60,000
time required to obtain spectrum	5–10 min.	5–10 min.	10–20 min.	20–40 min.

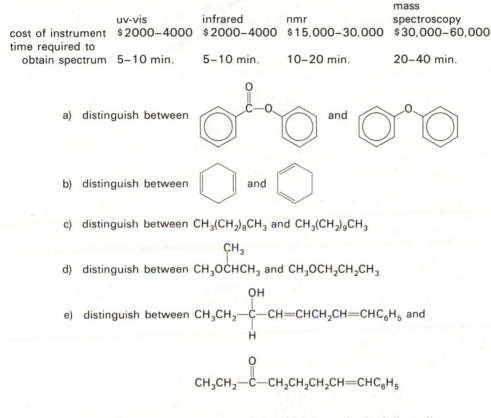

a) distinguish between [structures] and [structures]

b) distinguish between [structure] and [structure]

c) distinguish between $CH_3(CH_2)_8CH_3$ and $CH_3(CH_2)_9CH_3$

d) distinguish between CH_3OCHCH_3 (with CH_3 group) and $CH_3OCH_2CH_2CH_3$

e) distinguish between $CH_3CH_2-\overset{OH}{\underset{H}{C}}-CH=CHCH_2CH=CHC_6H_5$ and

$$CH_3CH_2-\overset{O}{C}-CH_2CH_2CH_2CH=CHC_6H_5$$

f) determine the concentration of vitamin A in crude shark liver oil
g) determine whether a substance obtained in a synthesis is the same compound as one previously obtained from a different source
h) determine the molecular weight of an unknown compound
i) determine whether a sample of an alcohol contains a carbonyl compound as an impurity

9.2 A 5.0×10^{-5} M solution of 1,2-diphenylethylene in a 1 cm ultraviolet spectrometer cell has $A = 1.60$. Calculate ϵ.

9.3 If a solution of pure compound of concentration 1.4×10^{-2} M has $A = 0.84$, what is the concentration of this compound in a solution which has $A = 0.42$?

9.4 The nmr absorption of the circled proton is split into two peaks. Which atom in this molecule causes this spin-spin splitting?

$$\underset{Cl \quad Br}{Cl-\overset{\textcircled{H} \quad Br}{\underset{|}{C}}-\overset{|}{C}-H}$$

9.5 How many absorptions (signals with different chemical shifts, disregarding spin-spin splitting) would there be in the nmr spectra of the following compounds?

a) *p*-dimethylbenzene   b) 2-methyl-4,4,4-trichloro-1-butene

c) 2,2-dimethylpentane   d) 2-butanone

9.6 State how many peaks would be observed in the nmr spectrum of 1,1,2-trichloro-ethane for the absorption of H_A.

$$\begin{array}{cc} H_A & H_B \\ | & | \\ Cl-C- & -C-H_B \\ | & | \\ Cl & Cl \end{array}$$

What would be the H_A/H_B ratio for the areas under their absorptions in the nmr spectrum?

9.7 State how many lines would be seen for the absorption in the CH_3 region of the nmr spectrum (σ 1–2.5) of the following compounds:

a) CH_3CHCl_2 b) $CH_3\overset{\displaystyle O}{\overset{\displaystyle \|}{C}}C_6H_5$

c) CH_3CH_2Cl d) $(CH_3)_3COH$

9.8 Using the information in Table 9.2, interpret the following data and suggest structures for the compounds.

a) A compound $C_2H_4Cl_2$ has an nmr spectrum containing one peak at δ 3.7.

b) A compound $C_5H_{12}O$ has an nmr spectrum with two singlet peaks in an area ratio of 3:1 at δ 1.1 (three) and δ 3.2 (one). The ir spectrum shows no peak above 3000 cm^{-1}.

c) A hydrocarbon C_8H_{10} has an nmr spectrum with three peaks in an area ratio of 3:2:5 at δ 1.2 (three, triplet), δ 2.6 (two, quartet), and δ 7.1 (five, singlet).

9.9 Using the characteristic chemical shift (δ) values from Table 9.2, tell how you would be able to distinguish between our example compound in Section 9.7 (A) and the isomer (B).

(A) $C_6H_5-\overset{\displaystyle O}{\overset{\displaystyle \|}{C}}-OCH_2CH_3$,   and   (B) $C_6H_5-O-\overset{\displaystyle O}{\overset{\displaystyle \|}{C}}-CH_2CH_3$.

9.10 Geraniol, a fragrant liquid isolated from flower petals, undergoes ozonolysis to yield a mixture of products from which one compound is easily distilled. The ir spectrum of this compound contains a strong band at 1720 cm^{-1}. In the mass spectrum of this compound, the highest significant peak has mass 58; a stronger peak appears at mass 43. Write the structure of this volatile product and suggest a structure for the ion with mass 43.

$$CH_3-\underset{\underset{\displaystyle CH_3}{|}}{C}=CH-CH_2CH_2-\underset{\underset{\displaystyle CH_3}{|}}{C}=CH-CH_2OH \xrightarrow{\;O_3\;} ?$$

geraniol

CARBOXYLIC ACIDS AND DERIVATIVES

Carboxylic acids have the general formula RCO_2H. The carboxyl group,

$$\overset{O}{\underset{\|}{-C}}-OH$$, is formally a combination of carbonyl and hydroxyl groups, but the properties of carboxylic acids are distinct from the properties of either of these groups. The $-CO_2H$ group is acidic, and gives rise to the name of the series.

10.1   Names

The systematic IUPAC names of aliphatic compounds containing the $-CO_2H$ group are formed by changing the **-e** in the parent alkane name to **-oic acid.** When other functional groups are present in the molecule, the $-CO_2H$ group takes precedence and the carbon of this group is numbered 1.

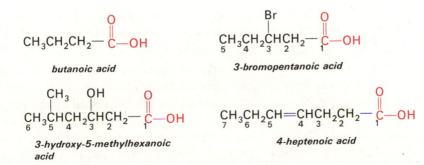

butanoic acid

3-bromopentanoic acid

3-hydroxy-5-methylhexanoic acid

4-heptenoic acid

Problem 10.1   Draw structures for the following compounds: a) 4-methyloctanoic acid, b) 3-butenoic acid, c) 2-phenylpentanoic acid.

227

A number of the carboxylic acids were known long before the IUPAC nomenclature rules were devised. The **trivial names** for the first three acids of the series are always used, and one often encounters trivial names for the C_4 to C_{10} acids as well as the lower dibasic acids. When a trivial name is used, the positions in the chain are indicated by Greek letters rather than numbers. The position next to the CO_2H carbon is called α; further carbons are designated β, γ, and so on.

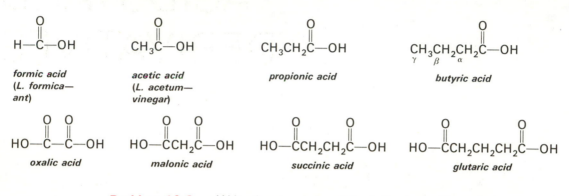

formic acid
(L. formica—
ant)

acetic acid
(L. acetum—
vinegar)

propionic acid

butyric acid

oxalic acid

malonic acid

succinic acid

glutaric acid

Problem 10.2   Write the structures of the following acids and give an alternate name: a) α-bromopropionic acid, b) β-hydroxybutyric acid.

When the carboxyl group is attached to a saturated ring, the acid is named as a **cycloalkanecarboxylic acid.** Aromatic acids are named as derivatives of **benzoic acid.**

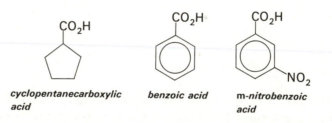

cyclopentanecarboxylic
acid

benzoic acid

m-nitrobenzoic
acid

10.2   Properties and Occurrence of Acids

Aliphatic acids with up to about eight carbons are liquids with pungent, unpleasant odors. The odor of rancid butter is, in part, due to butanoic acid; hence its trivial name, butyric acid from the Latin *butyrum,* or butter. Two of the compounds associated with the odor of goats are hexanoic and octanoic acids, whose trivial names recognize this relationship—caproic and caprylic acids from the Latin *caper,* or goat. Aromatic acids are usually high-melting solids.

Acetic acid, CH_3CO_2H, the sour component of vinegar, is a typical carboxylic acid. The boiling point of 118° is 20° higher than that of $CH_3CH_2CH_2OH$, which has the same molecular weight, showing that the acid is much more highly associated than the alcohol. Even in the gas phase the acid molecules are associated by *two* hydrogen bonds.

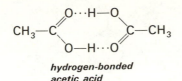

*hydrogen-bonded
acetic acid*

Acetic acid is a key biochemical building block and is widely distributed in nature. The acetic acid in vinegar is formed by enzymatic oxidation of ethanol in cider or wine. Higher straight-chain acids with even numbers of carbons occur in large amounts in the form of fats, as discussed in a later section.

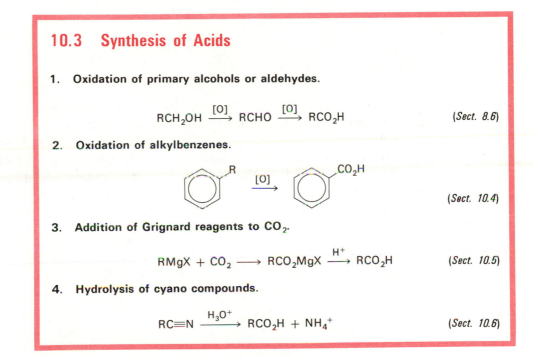

10.3   Synthesis of Acids

1. **Oxidation of primary alcohols or aldehydes.**

$$RCH_2OH \xrightarrow{[O]} RCHO \xrightarrow{[O]} RCO_2H$$ *(Sect. 8.6)*

2. **Oxidation of alkylbenzenes.**

(Sect. 10.4)

3. **Addition of Grignard reagents to CO_2.**

$$RMgX + CO_2 \longrightarrow RCO_2MgX \xrightarrow{H^+} RCO_2H$$ *(Sect. 10.5)*

4. **Hydrolysis of cyano compounds.**

$$RC\equiv N \xrightarrow{H_3O^+} RCO_2H + NH_4^+$$ *(Sect. 10.6)*

10.4   Preparation of Acids by Oxidation

The $-CO_2H$ group is the next stage beyond an aldehyde in the oxidation level of carbon, and carboxylic acids are produced by oxidation of aldehydes or primary alcohols, as discussed in Sections 8.5 and 8.6.

$$RCH_2OH \xrightarrow{H_2CrO_4} R-\overset{O}{\overset{\|}{C}}-H \xrightarrow{H_2CrO_4} R-\overset{O}{\overset{\|}{C}}-OH$$

Vigorous oxidation of alkylbenzenes leads to benzoic acid, with loss of all carbons in the side chain except the one directly adjacent to the ring.

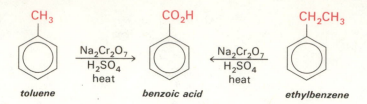

toluene   *benzoic acid*   *ethylbenzene*

Since benzoic acid gives the *meta* isomer on electrophilic substitution, oxidation of the substituted alkylbenzene is the most practical method of obtaining a compound such as *o*- or *p*-chlorobenzoic acid.

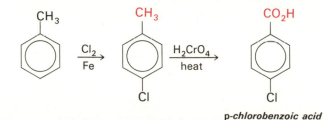

p-chlorobenzoic acid

The oxidation of an aromatic side chain is very facile because the ring can stabilize either carbocation or radical intermediates on the benzyl carbon. An important industrial reaction is the oxidation of *p*-dimethylbenzene to the dicarboxylic acid. As in most large-scale operations, the cheapest reagent possible is used. In this process air is the oxidant and a catalyst is used to promote radical formation. Over two billion pounds of the acid are produced per year for use in making polyester fiber (Section 10.21).

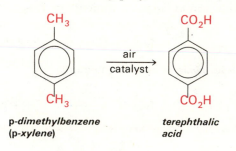

p-dimethylbenzene   *terephthalic*
(p-xylene)   *acid*

Problem 10.3   a) Write the structure and name of an alcohol and an aldehyde that would give β-phenylpropionic acid on oxidation with H_2CrO_4. b) Which of the following compounds would *not* give *p*-nitrobenzoic acid on oxidation: *p*-nitrotoluene, nitrobenzene, *p*-nitroisopropylbenzene?

10.5   Carboxylic Acids from Grignard Reagents

An important method for preparing carboxylic acids is the reaction of a Grignard reagent with carbon dioxide. Addition of the C—Mg bond to CO_2 leads directly to the magnesium salt of the acid, and acidification gives the acid. This sequence can be used with any alkyl or aryl halide that can form a Grignard reagent. Note that the overall process involves adding a carbon to the carbon skeleton of the original halide.

$$R-X \xrightarrow{Mg} \overset{\delta-}{R}\overset{\delta+}{-MgX} \rightarrow \overset{O^{\delta-}}{\underset{O_{\delta-}}{\overset{\parallel}{C^{\delta+}}}} \longrightarrow R\overset{O}{\underset{}{\overset{\parallel}{C}}}-\overset{+2}{O}\overset{-}{MgX} \xrightarrow{H^+} R\overset{O}{\underset{}{\overset{\parallel}{C}}}-OH + Mg^{+2} + X^-$$

$$\underset{\substack{\text{2-bromo-}\\\text{propane}}}{\overset{\overset{Br}{|}}{CH_3\overset{|}{C}HCH_3}} \xrightarrow{Mg} \overset{\overset{MgBr}{|}}{CH_3\overset{|}{C}HCH_3} \xrightarrow[\text{2) } H_3O^+]{\text{1) } CO_2} \underset{\substack{\alpha\text{-methylpropionic acid}}}{\overset{\overset{CO_2H}{|}}{CH_3\overset{|}{C}HCH_3}}$$

10.6 Carboxylic Acids from Cyano Compounds

An alternative method for converting a primary or secondary alkyl halide to a carboxylic acid containing one more carbon involves an alkyl cyanide as an intermediate. Primary and secondary halides can be converted to the corresponding cyanides by S_N2 substitution (Section 6.17), and the cyanide can then be hydrolyzed to the acid. Hydrolysis occurs in two steps to give one mole of carboxylic acid and one mole of ammonia (Section 10.13).

$$RCH_2X \xrightarrow{^-CN} RCH_2C\equiv N \xrightarrow{H_2O} \left[RCH_2\overset{\overset{OH}{|}}{C}=NH \right] \xrightarrow{H_3O^+} RCH_2\overset{O}{\underset{}{\overset{\parallel}{C}}}OH + NH_4^+$$

The cyanide method suffers from the restriction that the intermediate cannot be prepared from a tertiary halide because of competing elimination to give an alkene. However, it does not have the limitations of the Grignard reaction, in that the cyanide route can often be used when hydroxyl or other reactive groups are present. Thus, α-hydroxy acids are obtained by hydrolysis of cyanohydrins (Section 8.11) prepared by addition of cyanide to aldehydes.

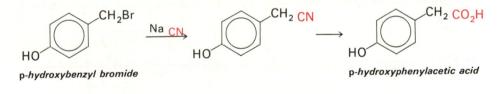

p-hydroxybenzyl bromide   p-hydroxyphenylacetic acid

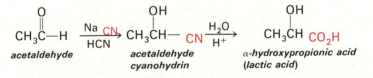

acetaldehyde   acetaldehyde   α-hydroxypropionic acid
   cyanohydrin   (lactic acid)

Problem 10.4 Show reactions that could be used to accomplish the following syntheses:

a) $CH_3C(CH_3)_2CO_2H$ from a four-carbon alcohol of your choice
b) p-chlorophenylacetic acid from p-chlorobenzyl chloride
c) 2-hydroxypentanoic acid from an aldehyde
d) $CH_3CH_2CH_2CO_2H$ from $CH_3CH_2CH_2CH_2OH$

10.7 Acidity of Carboxylic Acids

Although carboxylic acids are weak compared to inorganic acids such as HCl or H_2SO_4, they are much stronger acids than are alcohols. The **strength** of an acid refers to the extent to which its ionization equilibrium lies to the side of the dissociated acid, expressed by the magnitude of the equilibrium constant, K_A.

$$HA + H_2O \rightleftharpoons A^- + H_3O^+ \qquad K_A = \frac{[A^-][H_3O^+]}{[HA]}$$

Since the concentration of water remains essentially constant, $[H_2O]$ does not appear in the equation for K_A.

From the above equation we see that the greater the degree of ionization (that is, the stronger the acid), the larger the K_A. Simple carboxylic acids such as acetic acid have K_A values of 10^{-4} to 10^{-5}. Just as $[H^+]$ is normally expressed as **pH,** it is convenient to use the negative logarithm of K_A or **pK$_A$** in discussing these very small numbers. For acetic acid, with $K_A = 1.8 \times 10^{-5}$, the pK$_A$ value is $-\log(1.8 \times 10^{-5})$ or 4.7. Note that while pK$_A$ values provide convenient numbers for discussion and comparison, two points must be kept in mind when using negative logarithms. (1) The *larger* the pK$_A$, the *smaller* the K_A; and (2) the numbers are exponents and represent powers of 10; an acid with pK$_A = 3$ is *10 times more dissociated* than an acid with pK$_A = 4$.

$$pK_A = -\log K_A \qquad pH = -\log [H^+]$$

If we write the expression for the dissociation constant K_A in terms of pH and pK$_A$ and then rearrange the terms, we obtain a convenient relationship between the dissociation constant of the acid and the pH of a solution of the acid and its conjugate base:

$$K_A = \frac{[A^-][H_3O^+]}{[HA]} \xrightarrow[\text{negative logarithm of both sides}]{\text{rearrange terms and take}} \boxed{pH = pK_A + \log \frac{[A^-]}{[HA]}}$$

This equation tells us that the higher the pK$_A$ value (the *weaker* the acid), the higher the pH needed for a given fraction of the acid HA to exist as the anion A^-.

Since typical carboxylic acids of pK$_A = 4$ to 5 are about 100 times stronger acids than H_2CO_3 (pK$_A = 6.5$), a carboxylic acid is almost completely converted to the carboxylate anion, RCO_2^-, in a solution of sodium bicarbonate. This process has the effect of converting a water-insoluble acid into a water-soluble salt.

$$RCO_2H + Na^+ HCO_3^- \longrightarrow RCO_2^- Na^+ + [H_2CO_3] \longrightarrow H_2O + CO_2$$

water-insoluble   **water-soluble**

Solubility in bicarbonate or hydroxide solution also permits separation of an acid from a mixture that contains non-acidic compounds. To accomplish the separation, a solution of the mixture in ether is extracted with aqueous $NaHCO_3$ solution. Non-acidic compounds remain dissolved in the ether and are removed by separation of the ether layer, and the acid can be recovered by addition of a strong acid to the aqueous solution of the salt.

$$RCO_2^- \ Na^+ + HCl \longrightarrow RCO_2H + NaCl$$

The equivalent weight of an acid, also called the "neutralization equivalent," is the number of grams of RCO_2H corresponding to one equivalent of H^+. This value can be determined by titration of the acid to an end point with standardized base.

10.8   Relative Acid Strengths

Acetic acid, $pK_A = 4.7$, is about 10^{11}-fold (100 billion times) more highly dissociated than ethanol ($pK_A \sim 16$). The high acidity of the OH group in the acid relative to that in ethanol indicates a strong driving force for the ionization of the $-CO_2H$ group. The reason lies in the resonance stabilization of the carboxylate anion, in which the negative charge is shared by the two equivalent oxygen atoms. In contrast, the negative charge in the alkoxide is localized on a single atom.

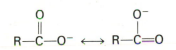

Since acid dissociation is an equilibrium process, the strength of a particular acid should be related to the relative stabilities of RCO_2H and RCO_2^-. Thus, any factors that will stabilize RCO_2^- relative to RCO_2H should shift the equilibrium to the right and create a stronger acid. As we have often seen, charge delocalization is a stabilizing effect. Groups attached to RCO_2^- that withdraw electron density away from the carboxylate anion should therefore lower the energy of the ion relative to the acid. Thus, it is not surprising that introduction of a chlorine atom into acetic acid increases the acid strength 80-fold, and substitution of a second and a third chlorine causes further large increases.

	CH_3-CO_2H	$Cl\leftarrow+CH_2-CO_2H$	$Cl\leftarrow+\overset{\overset{\displaystyle Cl}{\uparrow}}{CH}-CO_2H$	$Cl\leftarrow+\overset{\overset{\displaystyle Cl}{\uparrow}}{\underset{\underset{\displaystyle Cl}{\downarrow}}{C}}-CO_2H$
pK_A	4.7	2.8	1.3	0.7
Relative acid strength	1	80	2800	11,000

Thus, the attachment of an electronegative group on the α-position of a carboxylic acid causes a dramatic effect on the stability of the carboxylate anion.

$$Cl_3CCO_2^- > Cl_2CHCO_2^- > ClCH_2CO_2^- > CH_3CO_2^-$$

stability order of chlorine-substituted acetate ion

The transmission of electronic effects through single bonds, either electron-withdrawing or electron-donating, is called the **inductive effect.** The inductive effect of a substituent falls off sharply as the number of —CH₂— groups between the substituent and the CO₂H increases. This effect of insulation by —CH₂— groups is seen in β-chloropropionic acid, which is only six times stronger than propionic acid.

	$\overset{\beta}{C}H_3—\overset{\alpha}{C}H_2—CO_2H$	$Cl—CH_2—CH_2—CO_2H$	$\overset{Cl}{\underset{}{CH_3—CH—CO_2H}}$
pKₐ	4.9	4.1	2.8
Relative acid strength	1	6	100

Problem 10.5   Arrange the following acids in order of decreasing acid strength: 3-bromobutanoic acid, 2,2-dibromobutanoic acid, 2-bromobutanoic acid, and butanoic acid.

In the dissociation of an acid to a proton and carboxylate ion, a very important factor is the ability of the solvent to form a shell of solvent molecules around each ion. Solvation separates the ions and stabilizes them by dispersal of charge. Water is the ideal solvent for the dissociation of acids because of the efficient solvation provided by hydrogen bonding.

DERIVATIVES OF CARBOXYLIC ACIDS

Several series of compounds with the general formula $R\overset{O}{\overset{\|}{C}}X$ are derived directly or indirectly from carboxylic acids and can be converted to the acids by hydrolysis. The most important are those in which hydrogen in the groups HOR, HNR₂, HCl, or HOCOR is replaced by the group $R\overset{O}{\overset{\|}{C}}—$. These series of compounds, with the derivatives of acetic acid as examples, are illustrated in Table 10.1.

10.9   Names and Structures

The general name of the $R\overset{O}{\overset{\|}{C}}—$ group is **acyl.** For names of specific compounds, the **-ic acid** ending of the acid is changed to **-yl.** The acyl group

TABLE 10.1   DERIVATIVES OF ACETIC ACID

Name of Series	Structure	Name	Boiling Point
acid chloride	$CH_3-\overset{\overset{O}{\|\|}}{C}-Cl$	acetyl chloride	52°
anhydride	$CH_3-\overset{\overset{O}{\|\|}}{C}-O-\overset{\overset{O}{\|\|}}{C}-CH_3$	acetic anhydride	139°
ester	$CH_3-\overset{\overset{O}{\|\|}}{C}-OCH_3$	methyl acetate	59°
amide	$CH_3-\overset{\overset{O}{\|\|}}{C}-NH_2$	acetamide	220°
nitrile	$CH_3-C\equiv N$	acetonitrile	82°

name is used in several ways, including names for ketones with more than one functional group.

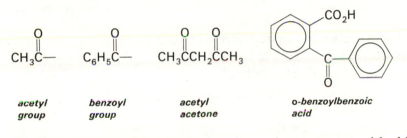

| acetyl group | benzoyl group | acetyl acetone | o-benzoylbenzoic acid |

Acyl halides are named by combining the acyl group name with chloride or bromide.

propionyl chloride   p-chlorobenzoyl chloride

Anhydrides are named by adding the word **anhydride** to the name of the acid from which it was formed; this rule applies also to cyclic anhydrides formed from dicarboxylic acids.

$CH_3CH_2\overset{\overset{O}{\|\|}}{C}-O-\overset{\overset{O}{\|\|}}{C}CH_2CH_3$

propionic anhydride   succinic anhydride

Esters are named as derivatives in which the *hydrogen* of the acid is removed to give RCO_2- and then replaced by an alkyl group. The name of the ester is two words, with the alkyl group first and then the RCO_2— group name, which is formed by changing the **-ic acid** ending of the acid to **-ate**.

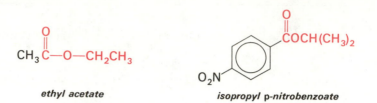

ethyl acetate *isopropyl p-nitrobenzoate*

Amides are named either by replacing the suffix **-oic acid** with **-amide**, or by replacing the **-ic acid** ending of the trivial acid name by **-amide**. When the nitrogen is substituted, this is indicated by starting with the name of the substituent.

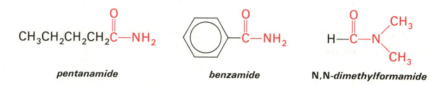

pentanamide *benzamide* *N,N-dimethylformamide*

Nitriles are named as derivatives of acids by replacing the **-ic** ending of the acid's trival name with the ending **-onitrile**. In the IUPAC-derived system they are named as alkanenitriles.

$$CH_3CH_2CN \qquad CH_3\overset{\underset{\displaystyle |}{Cl}}{C}HCH_2CN$$

propiononitrile *β-chlorobutyronitrile*
(or *propanenitrile*) (or *3-chlorobutanenitrile*)

Problem 10.6 Write structures for the following compounds:

a) ethyl formate b) propionitrile
c) *p*-aminobenzamide d) benzoic anhydride

e) *n*-butyl α-chloroacetate f) α-bromoacetyl bromide

10.10   Reactions and Reactivity of Acid Derivatives

Before getting into the behavior of the various acid derivatives, it will be helpful to survey briefly their general reactions and the influence of the X groups. In the RCOX system we have a carbonyl group flanked by an electronegative atom. The characteristic reaction of these compounds is addition of a nucleophile to the carbonyl group to give a tetrahedral intermediate, followed by elimination of X to regenerate the π bond. The overall process is **nucleophilic substitution** of the group X by nucleophile, or it can be considered equally well as replacement of a proton in a nucleophile NuH by an acyl group, *i.e.*, **acylation** of the nucleophile.

$$R-\overset{\overset{\displaystyle O}{\|}}{C}-X + NuH \longrightarrow \left[R-\overset{\overset{\displaystyle OH}{|}}{\underset{\underset{\displaystyle Nu}{|}}{C}}-X \right] \longrightarrow R-\overset{\overset{\displaystyle O}{\|}}{C}-Nu + HX$$

The reactions of acid derivatives with some simple nucleophiles are summarized in the box below. The relative reactivity of acid chlorides,

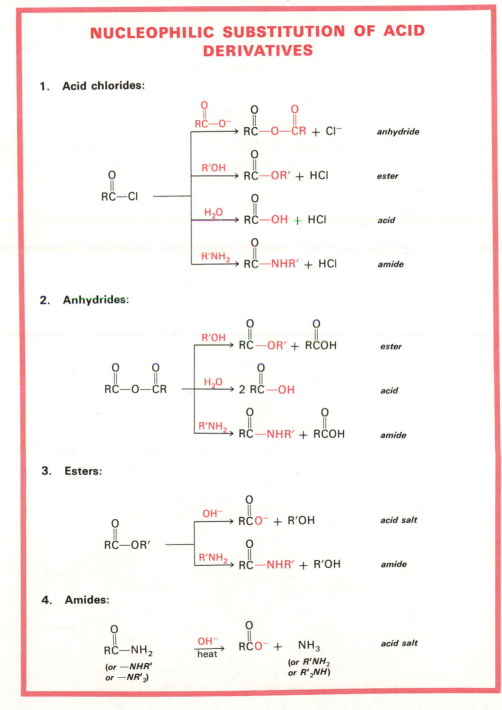

NUCLEOPHILIC SUBSTITUTION OF ACID DERIVATIVES

1. **Acid chlorides:**

2. **Anhydrides:**

3. **Esters:**

4. **Amides:**

anhydrides, esters, and amides decreases in that order, as indicated by the progressively narrower range of reactions as we go down the series. This reactivity sequence is just what we would expect from the decreasing electron-withdrawing effects of the X groups, which can be seen in the acidities of the conjugate acids HX; in other words, the stronger the base (the weaker the conjugate acid), the poorer the leaving group.

$\overset{O}{\overset{\|\|}{RCX}}$	$\overset{O}{\overset{\|\|}{RCCl}}$	$\overset{O\ \ \ \ O}{\overset{\|\|\ \ \ \|\|}{RCOCR}}$	$\overset{O}{\overset{\|\|}{RCOR'}}$	$\overset{O}{\overset{\|\|}{RCNHR'}}$
conjugate acid of X⁻	HCl	RCO_2H	R'OH	$R'NH_2$
pK_A	<0	5	15	25

The high electronegativity of chlorine, which makes HCl such a strong acid, also causes the C=O group in acid chlorides to be strongly electrophilic—more so than the C=O group in most aldehydes and ketones. Anhydrides occupy an intermediate position, with esters next and amides least reactive to nucleophilic attack. A significant factor in the lower reactivity of esters, and especially of amides, is resonance donation of electron density from the unshared electron pair on the —OR or —NHR group, which decreases the electron deficiency of the carbonyl carbon.

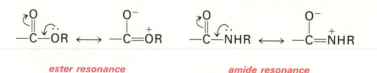

ester resonance   *amide resonance*

10.11   Acid Chlorides

Acid chlorides and acid bromides are obtained from the carboxylic acids with the same reagents that are used to convert alcohols to the alkyl halides—$SOCl_2$, PCl_5, and PBr_3. The most convenient reagent for this purpose is thionyl chloride, since the byproducts SO_2 and HCl are volatile and thus easily removed.

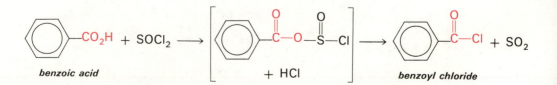

Acid chlorides are highly reactive compounds with sharp, irritating odors. Their main importance is as synthetic intermediates. Acid chlorides react vigorously with water, alcohols, or amines to give the corresponding acid, ester, or amide. In the preparation of an amide, the HCl liberated in the reaction is usually neutralized by using a second mole of amine.

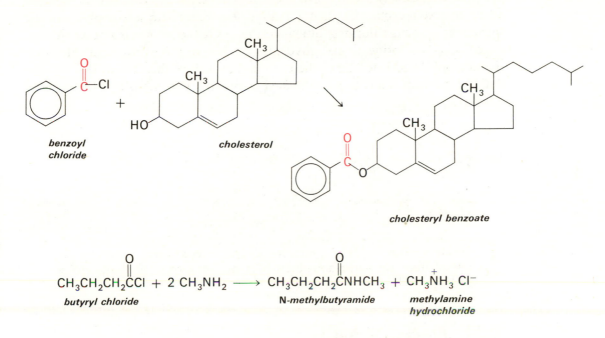

benzoyl chloride

cholesterol

cholesteryl benzoate

$$CH_3CH_2CH_2\overset{O}{\overset{\|}{C}}Cl + 2\ CH_3NH_2 \longrightarrow CH_3CH_2CH_2\overset{O}{\overset{\|}{C}}NHCH_3 + CH_3\overset{+}{N}H_3\ Cl^-$$

butyryl chloride   N-methylbutyramide   methylamine hydrochloride

Acylation of an aromatic ring can be accomplished with an acid chloride in the presence of aluminum chloride. This reaction is completely analogous to that with an alkyl chloride which was seen in Section 4.7. A strongly electrophilic complex is formed between the acid chloride and aluminum chloride, and this complex attacks the ring in a typical aromatic substitution. Loss of a proton from the intermediate gives the aromatic ketone. This **acylation** is not prone to the rearrangements encountered with alkyl chlorides and AlCl$_3$.

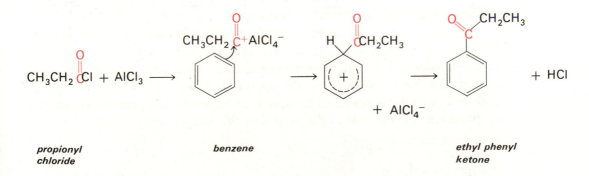

propionyl chloride   benzene   ethyl phenyl ketone

Problem 10.7 Write structures and names for the products that would be obtained from reactions of acetyl chloride with a) benzylamine in the presence of NaOH, b) n-butanol, c) chlorobenzene plus aluminum chloride.

10.12 Acid Anhydrides

Anhydrides are formed by loss of water between two molecules of an acid (or between two ends of a dicarboxylic acid, as discussed in Section 10.20, to give a cyclic anhydride). Anhydrides react with nucleophiles in the same way as acid chlorides, but since the —OCOR group is not as electronegative as a chlorine atom, anhydrides are less reactive than acid chlorides. A nucleophile adds to one of the carbonyl groups of the anhydride, and the —OCOR group is lost to give the acyl derivative plus a mole of the carboxylic acid.

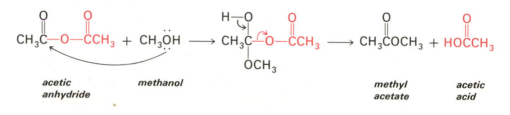

| acetic anhydride | methanol | | methyl acetate | acetic acid |

Problem 10.8 Write structures for the products that would be obtained from propionic anhydride plus a) *n*-propanol, b) benzylamine ($C_6H_5CH_2NH_2$).

10.13 Amides and Nitriles

Simple amides are obtained by acylation of ammonia or an amine with an acid chloride or anhydride, discussed in the preceding sections. The

$$\underset{O}{\overset{\displaystyle O}{R\overset{\|}{C}X}} + R'NH_2 \longrightarrow \underset{O}{\overset{\displaystyle O}{R\overset{\|}{C}NHR'}} + HX$$

formation of amides by linking together complex acids and amines is the key step in the synthesis of proteins (Chapter 14); special methods for this type of amide synthesis are discussed in Section 14.5.

The most important reaction of amides, again in connection with protein chemistry, is hydrolysis to break the CO—NH bond. Vigorous conditions, such as prolonged heating with strong acid or base, are required. In basic solution the final products are the amine and the salt of the acid. Acid-catalyzed hydrolysis proceeds by attack of water on the protonated amide; the products are the carboxylic acid and the amine salt.

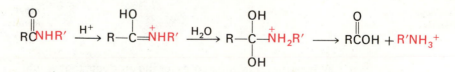

As we noted in Section 10.10, the low reactivity of amides as electrophiles results from the extensive resonance interaction between the unshared

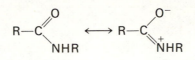

electron pair on nitrogen and the carbonyl group. A further consequence of this resonance is the fact that amides are highly polar compounds. Low molecular weight amides are more soluble in water than in hydrocarbons, and have unusually high boiling points [e.g., $CH_3CONHCH_3$, b.p. 204°; and $CH_3CON(CH_3)_2$, b.p. 165°]. Another manifestation of amide resonance is the much lower basicity of an amide compared to the corresponding amine; if the unshared electron pair on nitrogen is tied up with the carbonyl group, it is less available to an acid. The basicity of amides is quite close to that of alcohols or water, and amides behave as "neutral" compounds with regard to salt formation and solubility in dilute acid.

Problem 10.9 Why does N-methylacetamide, mol. wt. 73, have a higher boiling point than N,N-dimethylacetamide, mol. wt. 87?

The C≡N bond of **nitriles** is similar in properties to the carbonyl group in ketones, and nitriles undergo addition with the same types of nucleophiles. In the presence of acid or base, water adds to the C≡N bond and the resulting amide undergoes hydrolysis to the acid.

$$RC≡N \xrightarrow[H^+ \text{ or } OH^-]{H_2O} RC\overset{OH}{=}NH \longrightarrow RC\overset{O}{-}NH_2 \xrightarrow{H_2O} RCO_2H + NH_3$$

A very useful reaction of nitriles is reduction to primary amines. The C≡N bond can be hydrogenated in the presence of a catalyst, or hydrogen can be added by means of lithium aluminum hydride.

$$CH_3CH_2C≡N \xrightarrow[\substack{\text{or LiAlH}_4, \\ \text{then } H_2O}]{2 H_2, \text{ Ni}} CH_3CH_2CH_2NH_2$$

propionitrile 　　　　　　*n-propylamine*

Problem 10.10 Show how the following syntheses can be carried out using a nitrile as an intermediate:
　　　　a) $C_6H_5CH_2CH_2NH_2$ from benzyl chloride
　　　　b) $CH_2=CHCH_2CH_2NH_2$ from allyl chloride.

ESTERS

Esters are the most important and interesting series of acid derivatives. In contrast to other types of acid derivatives, a wide variety of esters is encountered in biological sources, and esters figure prominently in biochemical processes.

SPECIAL TOPIC: Esters and Odors

Most esters have sweet, pleasant odors, and esters are important constituents of fruit and flower aromas. Natural flavors and fragrances are due to extremely complex mixtures of esters, alcohols, aldehydes, ketones, and hydrocarbons. The analysis of food aromas presents problems of separation and identification that can be approached only with the chromatographic and spectroscopic methods discussed in Chapter 9.

Esters are particularly important in fruit aromas. A vapor phase chromatogram of the volatile essence of ripe bananas is shown in Figure 10.1. All but a

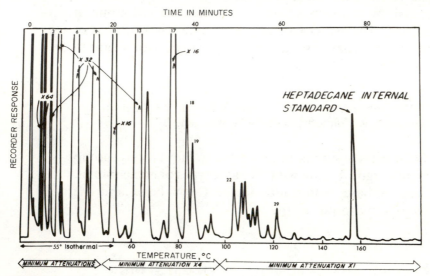

FIGURE 10.1

Vapor phase chromatogram of banana aroma. (Courtesy Dr. Walter Jennings.)

few of the peaks are due to esters, which range from ethyl acetate to hexyl hexanoate. Esters of acetic and butanoic acids are most abundant. Many other constituents, including numerous saturated and unsaturated alcohols, are also present. The amount and complexity of the essence increases markedly during

$$\underset{\substack{\text{CH}_3 \\ | \\ \text{CH}_3\text{CHCH}_2\text{CH}_2\text{OCCH}_3}}{}$$

isopentyl acetate
(used in synthetic banana flavor)

$$\text{CH}_3\text{CH}_2\text{OCH}$$

ethyl formate
(used in artificial rum flavor)

$$\text{CH}_3\text{OCCH}_2\text{CH}_2\text{CH}_3$$

methyl n-butyrate
(apple fragrance)

$$\text{CH}_3(\text{CH}_2)_6\text{CH}_2\text{OCCH}_3$$

n-octyl acetate
(orange flavor)

$$\text{CH}_3\text{CH}_2\text{OCCH}_2\text{CH}_2\text{CH}_3$$

ethyl n-butyrate
(pineapple flavor)

ripening, and the ratio of acetate to butanoate esters has been found to fluctuate in a regular way as ripening of the fruit progresses.

The chemistry of odors is of considerable importance in the food industry. In processed foods, steps such as dehydration may seriously compromise the natural flavor by loss of volatile compounds. The complexity of the aroma is greatly increased in cooking. Over 350 compounds have been identified in the aroma of roasted coffee. The savory smell of broiling meat is due in part to traces of amines and sulfur compounds that are formed by the breakdown of proteins. The pure compounds, undiluted with other aromas, have overpoweringly obnoxious odors.

Odor is the major factor in flavor. Taste buds in different locations of the tongue detect salt, sweet, sour, and bitter, but olfactory receptors in the nose are capable of distinguishing hundreds of individual odors. Occasionally one compound may have an odor associated with a certain flavor or aroma; for example, butyl and pentyl acetate in high dilution are quite reminiscent of bananas. However, the true bouquet of a natural essence, such as a ripe fruit, cannot be fully duplicated with one or two compounds. Artificial aromas and flavors can approach the "real thing" only as they approach the complex composition of the natural mixture.

10.14 Preparation and Reactions of Esters

As discussed in Sections 10.11 and 10.12, esters can be obtained from the reaction of an acid chloride or anhydride with an alcohol. A more important preparation of simple esters is the direct **esterification** of carboxylic acids with alcohols. This reaction is catalyzed by a strong acid, such as H_2SO_4, and is very similar in mechanism to the formation of an acetal (Section 8.13). Protonation of the carboxyl group yields a very stable carbocation which is attacked by the alcohol, acting as a nucleophile. The resulting intermediate loses water to form the ester. The entire reaction is an equilibrium process and is carried to completion by removal of the water as it is formed by distillation.

Overall esterification reaction:

Mechanism:

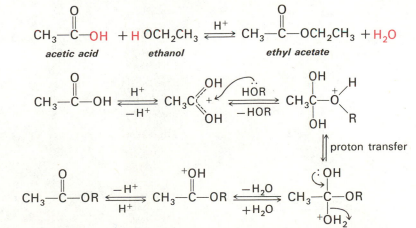

Esters are readily attacked by strong nucleophiles such as hydroxide ion. The intermediate resulting from attack of hydroxide on the carbonyl group loses alkoxide ion, ⁻OR, and gives the carboxylic acid. The acid is immediately neutralized to give the sodium salt and ethanol as the final products.

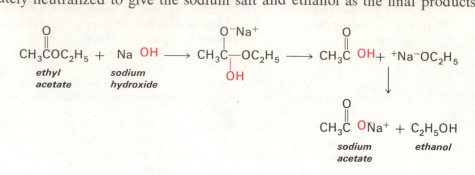

Problem 10.11 Write the products that would be formed by the reaction of potassium hydroxide with a) isobutyl formate and b) diethyl

malonate $\left(C_2H_5O - \overset{\overset{O}{\|}}{C}CH_2\overset{\overset{O}{\|}}{C} - OC_2H_5 \right)$.

Grignard reagents add to the carbonyl group of carboxylic esters in a multistep sequence. In the first step the R group from the Grignard reagent adds as a nucleophile to give an intermediate like that in the reaction with OH⁻. Loss of alkoxide then gives a ketone. However, it is virtually impossible to stop the reaction at this stage because the ketone is more reactive toward Grignard reagent than was the original ester. The ketone therefore undergoes addition of a second molecule of RMgX; the final product, after hydrolysis, is the tertiary alcohol containing *two* R groups from the Grignard reagent.

Overall Grignard reaction:

$$CH_3 - \overset{\overset{O}{\|}}{C} - OC_2H_5 + 2\ C_6H_5MgBr \xrightarrow{H_3O^+} CH_3 - \underset{\underset{C_6H_5}{|}}{\overset{\overset{OH}{|}}{C}} - C_6H_5 + HOC_2H_5$$

1,1-diphenyl ethanol

Mechanism:

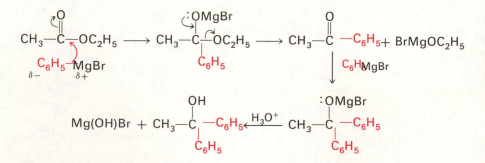

Lithium aluminum hydride, LiAlH₄, donates hydride ion to an ester in a fashion exactly analogous to the Grignard reaction. In this case the addition of H: and loss of :OR gives an intermediate aldehyde. The aldehyde cannot be isolated, since it rapidly undergoes further attack by hydride to give the alkoxide of the primary alcohol.

Overall LiAlH$_4$ reaction:

$$2 \ CH_3(CH_2)_7\overset{\displaystyle O}{\overset{\|}{C}}\!\!-\!\!OCH_3 + LiAlH_4 \longrightarrow \xrightarrow{H_3O^+} 2 \ CH_3(CH_2)_7CH_2OH + 2 \ CH_3OH$$

methyl nonanoate   *1-nonanol*   *methanol*

Problem 10.12 Write the details of the reaction of methyl nonanoate with LiAlH$_4$, showing the structure of the intermediate aldehyde.

10.15 Waxes

Waxes make up the protective water-repellent coating found on the fruits and leaves of plants and secreted also by insects and marine birds. These waxes consist mainly of simple esters in which both acid and alcohol have long aliphatic chains, ranging from C_{16} to C_{34}. Plant waxes contain some higher *n*-alkanes, also. The hardness and melting point of these long-chain esters increase with the molecular weight, and are more or less independent of the position of the ester group in the chain. The natural ester waxes are mixtures of various combinations of acids and alcohols within a range of carbon chain lengths. The major esters in beeswax and carnauba palm wax contain the C_{30} alcohol.

$$C_{15}H_{31}\!\!-\!\!\overset{\displaystyle O}{\overset{\|}{C}}\!\!-\!\!O(CH_2)_{29}CH_3 \qquad\qquad C_{25}H_{51}\!\!-\!\!\overset{\displaystyle O}{\overset{\|}{C}}\!\!-\!\!O(CH_2)_{29}CH_3$$

beeswax   *carnauba wax*

A unique wax is present in the plumage of birds, and is particularly important in waterfowl. The wax is secreted in the uropygial, or preen, gland at the base of the tail. This wax is an extraordinarily complex mixture of branched chain esters, which have been analyzed by special vapor phase chromatographic methods. In contrast to the hard, crystalline plant waxes, preen wax is a viscous fluid. This property, which is necessary for its function in the plumage, is due to the branched-chain structure and the presence of many different compounds in the wax.

The mixtures of esters are characteristic of the genus of bird. In ducks the acid components are mainly 2-methyl- and 4-methylhexanoic, with about 10 higher acids having methyl groups in the 2, 4, or 6 position. The major alcohol in these esters is 1-octadecanol (35 per cent). The remaining alcohols include every possible isomer of methylhexadecanol, methylheptadecanol, and methyloctadecanol with the methyl group on an even-numbered carbon—in all, at least 26 compounds, each constituting 0.5 to 5 per cent of the total alcohol content.

An important practical consideration in these studies of preen wax is the treatment of waterfowl that are stricken by oil spills from marine accidents. Removal of the highly toxic oil also removes the protective wax, and this must be replaced by a substitute that resembles the natural wax as closely as possible. Synthetic octadecyl 2-methylhexanoate has been used with some success for this purpose.

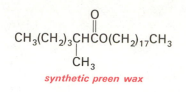

synthetic preen wax

10.16   Fats

Fats are triesters of glycerol, $HOCH_2CHOHCH_2OH$, with long-chain (fatty) acids. These **triglycerides** are found in all living organisms and are the most efficient form of energy storage in plant or animal tissue. Fats usually contain several different acids, ranging from 14 to 18 carbons, always with an *even number* of carbons. Double bonds are often present in the center of the chain (Table 10.2).

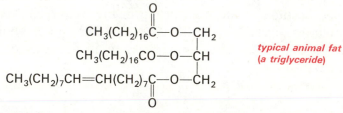

typical animal fat (a triglyceride)

Most animal fats are solids, while the triglycerides in plant seeds are oils at room temperature. The difference in properties of animal and plant triglycerides lies in the structures of the fatty acids. Vegetable oils contain mainly unsaturated acids with one, two, or three double bonds. The triglycerides in animal fats contain relatively large amounts of saturated acids, which have much higher melting points (Table 10.2).

Hydrogenation of vegetable oils with a nickel catalyst is used to reduce the degree of unsaturation. The main purpose of this treatment is to convert linoleic acid units with the $—CH=CH—CH_2—CH=CH—$ system to monosaturated chains. Hydrogenation raises the melting point of the fat and provides a mixture that remains semi-solid over a fairly wide temperature range. After refining to remove color and odor, these "hardened" oils are used as shortening in baking.

Oils that contain a large amount of unsaturated acids are very susceptible to air oxidation. The CH_2 groups flanked by two double bonds form highly stable radicals by hydrogen atom loss. These radicals can polymerize or form hydroperoxides that ultimately break down to oxygenated products. In edible fats, these reactions lead to a rancid taste. In a thin layer, the polymerization of these oils gives a tough, flexible film, and use is made of this property in paints and varnishes that contain linseed oil.

TABLE 10.2   **FATTY ACIDS IN TYPICAL FATS AND OILS**

Acids

No. Carbon / No. Double Bond	Formula	Name	mp, °C
12/0	$CH_3(CH_2)_{10}CO_2H$	Lauric	44
14/0	$CH_3(CH_2)_{12}CO_2H$	Myristic	54
16/0	$CH_3(CH_2)_{14}CO_2H$	Palmitic	63
18/0	$CH_3(CH_2)_{16}CO_2H$	Stearic	70
18/1	$CH_3(CH_2)_7CH\overset{cis}{=\!=}CH(CH_2)_7CO_2H$	Oleic	13
18/2	$CH_3(CH_2)_4(CH\overset{cis}{=\!=}CHCH_2)_2(CH_2)_6CO_2H$	Linoleic	−5
18/3	$CH_3CH_2(CH\overset{cis}{=\!=}CHCH_2)_3(CH_2)_6CO_2H$	α-Linolenic	−10

Fats and Oils

Typical Composition

Source	Lauric	Myristic	Palmitic	Stearic	Oleic	Linoleic	α-Linolenic
Beef tallow	1	3	30	24	40		
Butter*	4	12	28	10	26		
Lard		1	28	16	43	9	
Coconut*	46	18	10	4	6		
Corn			10	2	25	62	
Cottonseed		1	20	1	20	55	
Linseed			10	4	19	24	43
Olive			9	3	78	10	
Peanut			10	3	48	34	
Soybean			9	4	43	40	4

*Also contains lower acids (C_6 to C_{10}).

A specialized group of twenty-carbon acids called **prostaglandins** play an important role in the body as regulators of smooth muscle contraction and gastric secretion. The carbon chain in these molecules is looped and closed in the middle to form a five-membered ring. The prostaglandins are derived in cells from acids containing from three to five double bonds by attack of oxygen and cyclization of a free radical.

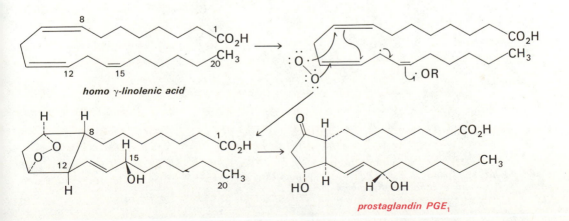

homo γ-linolenic acid

prostaglandin PGE₁

10.17   Soaps

Fats are the starting material for one of the oldest chemical manufacturing processes—the production of **soap** by alkaline hydrolysis. Soaps are the sodium or potassium salts of fatty acids, and crude soap was prepared from fat and wood ash in the early Middle Ages. The term **saponification,** meaning conversion to soap, is still used to denote the reaction of an ester with hydroxide to give the salt of the acid. In the manufacture of soap, the fat and aqueous alkali are heated and stirred together, and the soaps are then precipitated by adding salt. The glycerol is recovered by evaporation of the aqueous solution and the soap is pressed into bars.

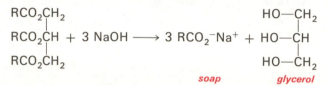

The detergent action of soap is discussed in Chapter 4. For special purposes, bactericidal agents, scents, and emollients can be added to the soap before packaging. Despite the advent of a wide variety of ionic and non-ionic surfactant agents, soap is still one of the most satisfactory and versatile cleaning agents.

10.18   Ester Condensation and Fatty Acid Metabolism

Hydrogens on the α-carbon of an ester, like those in an aldehyde or ketone, are acidic, and in the presence of strong base such as a sodium alkoxide, the resonance-stabilized enolate anion is formed. This anion can react as a nucleophile with another molecule of ester. The result is acylation of the carbanion by the ester to give a **β-keto ester.** The steps in this condensation are reversible, and the equilibrium favors the original ester. However, the position of the equilibrium can be shifted to the side of the keto ester by taking advantage of the fact that this product, with a **C—H** α to *two* carbonyl groups, is a much stronger acid than the original ester. Thus, if one equiva-

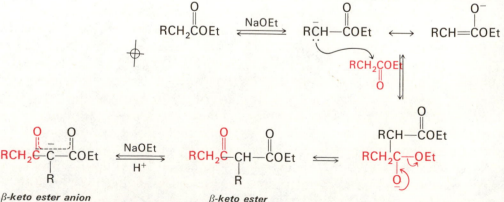

β-keto ester anion
(**Et is an abbreviation for ethyl group**)

β-keto ester

lent of alkoxide is used in the condensation, the stabilized anion is produced, and the β-keto ester can then be obtained by acidification.

If the β-keto ester is treated with dilute aqueous hydroxide, the ester group is saponified to give the β-keto acid. This compound is rather unstable and decomposes on gentle heating to give a ketone plus CO_2. The reaction is called **decarboxylation**, and occurs very readily when there is a carbonyl group in the β-position of the acid to assist in the loss of CO_2.

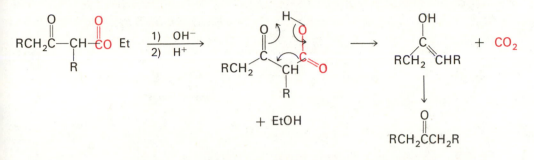

The combination of ester condensation and decarboxylation is thus an important method for the preparation of ketones. This sequence is illustrated by the preparation of diethylketone.

$$2\ CH_3CH_2\overset{O}{\overset{\|}{C}}OC_2H_5 \xrightarrow{NaOC_2H_5} CH_3CH_2\overset{O}{\overset{\|}{C}}\underset{\underset{CH_3}{|}}{CH}\overset{O}{\overset{\|}{C}}OC_2H_5 \xrightarrow{OH^-} CH_3CH_2\overset{O}{\overset{\|}{C}}CH_2CH_3$$

Problem 10.13   Show how $CH_3COCH_2CO_2C_2H_5$ would be prepared by ester condensation; write structures for all intermediates.

Ester condensation occurs in living systems, and is the key step by which fatty acids are built up from acetic acid. Since the reaction in this case must take place in a nearly neutral aqueous environment, the cell must employ a modified form of the reaction. In the biochemical version of ester condensation, the acid is carried through the process as a **thioester** with a thiol group attached to a protein.

To overcome the problem of unfavorable equilibrium in the condensation, the thioester which serves as the nucleophile is first converted to the **malonyl** compound by reaction with CO_2. In the condensation the CO_2 is lost, making this step irreversible. Subsequent steps convert the β-keto thioester to the four-carbon butyryl thioester, and the cycle is then repeated to build up the long fatty acid chain. Throughout the process the growing acyl chain remains attached to a cluster of enzymes that make up the carrier protein and that catalyze the various reactions. Just as important as the construction of the carbon chain is the storage of chemical energy in $—CH_2—$ groups that accompanies the synthesis. This energy is supplied by the coupling of synthetic reactions with energy-releasing processes such as the hydrolysis of pyrophosphate bonds (Section 15.2).

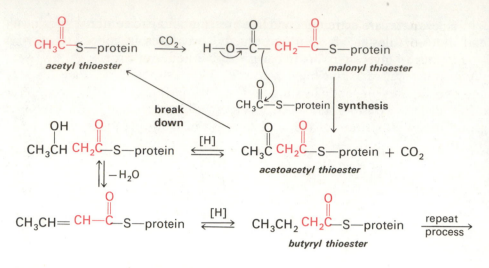

10.19   Ureas and Isocyanates

Carbonic acid, $HO-\overset{O}{\underset{||}{C}}-OH$, and **carbamic acid,** $H_2N-\overset{O}{\underset{||}{C}}-OH$, are compounds at the final oxidation stage of carbon, with four bonds to electronegative atoms. Both substances are unstable, decomposing to carbon dioxide plus water or ammonia, respectively, but several derivatives of these acids are extremely important and useful compounds.

The amide of carbamic acid, called **urea,** is the form in which most of the excess nitrogen in the diet is excreted in urine. An adult person normally excretes about 30 grams of urea in a 24 hour period. Urea is manufactured from ammonia and carbon dioxide on a huge scale for use as a high nitrogen fertilizer and in livestock supplements, plastics, and barbiturate sedatives (Chapter 13).

$$2\,NH_3 \; + \; CO_2 \; \longrightarrow \; H_2N-\overset{O}{\underset{||}{C}}-O^-\,{}^+NH_4 \; \xrightarrow[\text{pressure}]{\text{heat}} \; H_2N-\overset{O}{\underset{||}{C}}-NH_2 \; + \; H_2O$$

ammonium carbamate   urea

Isocyanates, $RN{=}C{=}O$, are derivatives of N-substituted carbamic acids. To prepare an isocyanate, an amine is treated with **phosgene,** the highly toxic dichloride of carbonic acid, and two molecules of HCl are eliminated.

$$C_6H_5NH_2 \; + \; Cl-\overset{O}{\underset{||}{C}}-Cl \; \xrightarrow{-HCl} \; C_6H_5NH-\overset{O}{\underset{||}{C}}-Cl \; \xrightarrow[-HCl]{\text{heat}} \; C_6H_5N{=}C{=}O$$

phosgene   phenyl isocyanate

Isocyanates are extremely reactive electrophiles and combine vigorously with nucleophiles to give carbamic acid derivatives. Alcohols react with isocyanates to give esters of carbamic acid called **urethanes.**

$$RN{=}C{=}O \ + \ R'OH \ \longrightarrow \ \underset{\textit{urethane}}{RNH{-}\overset{\overset{\displaystyle O}{\|}}{C}{-}OR'}$$

10.20 Cyclization of Difunctional Compounds

In this and previous chapters, most of the reactions of functional groups have been illustrated with simple compounds containing a single group. When we turn to compounds containing two reactive groups, the possibility arises of *intramolecular* reactions between two groups in the same molecule to give a cyclic product. When the functional groups are situated so that a five- or six-membered ring can be formed, cyclization will nearly always occur. This behavior was seen in haloalcohols (Section 6.11), and it is also the case with difunctional acids. A 4-hydroxy acid, for example, is generally not a stable compound since the cyclic ester, called a **lactone,** is formed spontaneously in any reaction that would normally give the hydroxy acid. Similarly, a 1,4-dicarboxylic acid, on heating, gives a cyclic anhydride.

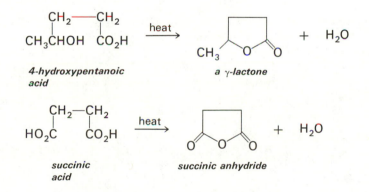

The reason that these reactions occur so readily is very simple. The functional groups are held together by the chain, and in the proper conformation they come within bonding distance. There is no requirement for two separate molecules to collide, and the probability of reaction in a given time is therefore greatly increased.

10.21 Condensation Polymers

If the groups in a difunctional compound are not situated so that ring closure is favorable, *intermolecular* reaction can take place. As this occurs, new difunctional compounds are formed, and if the process is continued, a **condensation polymer** results. This method of polymerization is used to prepare the two most important textile fibers—the **polyester** Dacron and the **polyamide** nylon 66.

Dacron is obtained from ethylene glycol and terephthalic acid (Section 10.4). The monomeric diester is first prepared and then heated to split out one molecule of glycol per unit of diester and build up the chain.

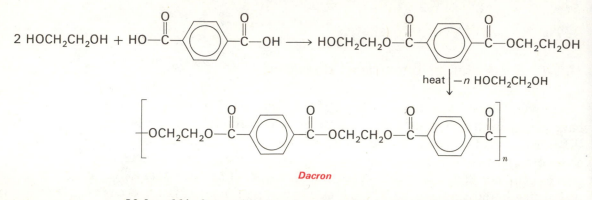

Dacron

Nylon 66 is formed by heating the salt of a six-carbon diacid, adipic acid, and a six-carbon diamine, hexamethylene diamine, to produce a polyamide. The name nylon 66 designates the number of carbons in the acid and amine monomers.

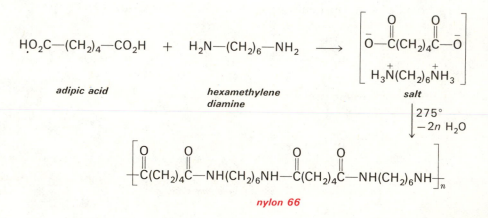

nylon 66

The molten polymers are extruded through a very fine hole to produce a filament. This filament is then **drawn** by stretching in order to orient the polymer chains and produce a highly crystalline chain. The foremost properties of polyester and polyamide fibers are high tensile strength and resistance to abrasion. Variations in the spinning and drawing operations make it possible to tailor the properties of the fiber for different uses, ranging from stockings to tire cord. Polyester is particularly versatile, and can be blended with cotton to give mixed fibers that are more durable than cotton and more comfortable to wear than pure polyester.

Polyamides with a head-to-tail repeating unit can be obtained by a *ring-opening* process. The cyclic amide of an amino acid is treated with a catalyst that attacks the carbonyl group. The reactive end of the open chain intermediate then reacts with a second molecule of the cyclic amide. The reaction continues in a process resembling addition polymerization (Section 3.15). The major polymer produced is nylon 6.

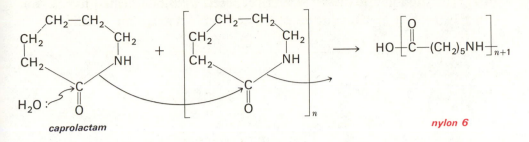

caprolactam

nylon 6

Polyurethanes. Urethane polymers are formed by reaction of a diol and a diisocyanate. The diol is usually a low molecular weight polyether or polyester with OH end groups. Polyurethanes are amorphous, but they have the property of forming stable, flexible foams when small gas bubbles are entrapped during the polymerization. These foams are more durable than foam rubber, and are widely used in upholstery and padding. Gas bubbles to provide the foaming action are formed by adding a small amount of water to the polymerization mixture. Water reacts with —NCO groups to form —NHCO$_2$H groups, which then liberate CO$_2$.

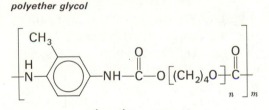

toluene-2,5-diisocyanate   polyether glycol

polyurethane segment

SUMMARY AND HIGHLIGHTS

1. **Carboxylic acids, RĊOH,** include alkanoic acids (RCO$_2$H), benzoic acids (ArCO$_2$H), and dicarboxylic acids (HOCO(CH$_2$)$_n$CO$_2$H).

2. **Acids are prepared** by oxidation of alcohols or aldehydes, reaction of Grignard reagents with CO$_2$, or hydrolysis of nitriles.

$$RCH_2OH \text{ or } RCHO \xrightarrow{[O]} RCO_2H$$
$$RMgX + CO_2 \longrightarrow RCO_2H$$
$$RCN + H_2O \longrightarrow RCO_2H$$

3. The **CO₂H** group is **acidic** (pK$_A$ about 5) because of resonance stabilization of RCO$_2^-$ relative to RCO$_2$H. Electron withdrawing groups such as Cl increase acid strength (lower pK$_A$ value).

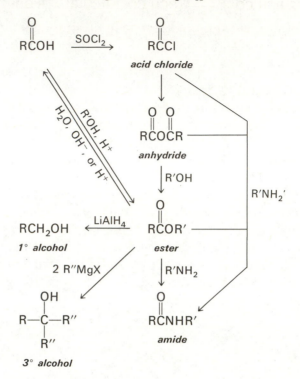

4. **Derivatives of carboxylic acids.**

5. **Acid chlorides** and **anhydrides** are very reactive acylating agents, giving **esters** from alcohols and **amides** from amines; esters are less reactive and amides are least reactive.

6. **Amides** are highly polar owing to resonance:

$$RCNHR \longleftrightarrow R\overset{+}{C}=NHR.$$

7. **Nitriles** undergo addition reactions; hydrogenation gives primary amines:

$$RC\equiv N \xrightarrow{2 H_2} RCH_2NH_2$$

8. **Esters** react with powerful nucleophiles:

$$RCO_2R' + OH^- \longrightarrow RCO_2^- + R'OH \text{ (saponification)}$$
$$RCO_2R' + R''MgX \longrightarrow RC(OH)R''_2$$
$$RCO_2R' + LiAlH_4 \longrightarrow RCH_2OH$$

9. **Plant waxes** are esters with C$_{16}$ to C$_{30}$ acid chains and C$_{24}$ to C$_{30}$ alcohols.

10. **Fats** are triesters of C$_{14}$ to C$_{18}$ acids with glycerol; in vegetable oils the acid chains contain one or two double bonds, which leads to easy

oxidation by air. **Prostaglandins** are C_{20} acids with the chain cyclized in the center to form a five-membered ring.

11.   **Condensation** of **esters** to give β-keto esters occurs in strong base:

$$2\ RCH_2CO_2R' \longrightarrow RCH_2COCHRCO_2R' + R'OH;$$

a modification of this reaction is a key step in the biosynthesis of fatty acids.

12.   **Ureas (RNHCNH$_2$), isocyanates (RN=C=O),** and **urethanes (RNHCO$_2$R')** are useful derivatives of carbonic acid.

13.   **Bifunctional acids** can undergo internal condensation to give cyclic esters (lactones) or anhydrides, or they can give **condensation polymers:** polyesters (Dacron) and polyamides (nylon) are important polymers for fibers; polyurethanes are foams.

ADDITIONAL PROBLEMS

10.14   Write the structures of the following compounds:

a)   2-bromobutanoic acid   b)   chloroacetyl chloride

c)   ethyl benzoate   d)   acetic anhydride

e)   N-n-propyl acetamide   f)   calcium propionate

g)   urea   h)   dimethyl oxalate

i)   N-methylurea   j)   p-chlorobenzonitrile

k)   sodium stearate

10.15   Write names for the following compounds:

a) $C_6H_5CH_2CH{=}CHCOH$ (with =O above C)   b) $BrCH_2COC_2H_5$ (with =O above C)

c) $C_6H_5CO_2\ {}^-Na^+$ d) $C_6H_5COCH_2C_6H_5$ (with =O above C)

e) $(C_6H_5CH_2C)_2O$ (with =O above C)   f) $HCN(CH_3)_2$ (with =O above C)

10.16   Write the structures of the organic products that would be formed by the following reactions:

a)   phenylmagnesium bromide plus CO_2 followed by aqueous HCl
b)   3-chloropropanoic acid plus $SOCl_2$
c)   1,2-dichloroethane plus excess NaCN
d)   benzoic acid plus ethanol, H_2SO_4
e)   benzene plus stearoyl chloride $(C_{17}H_{35}C({=}O)Cl)$ with $AlCl_3$
f)   glycerol (propanetriol) plus large excess of acetic anhydride
g)   phosgene plus excess ethyl alcohol
h)   phenyl isocyanate plus 2-phenylethanol
i)   methyl heptanoate plus excess n-hexylmagnesium bromide
j)   ethyl butyrate plus $LiAlH_4$ followed by aqueous H_2SO_4

10.17   Show all steps and intermediates in the hydrolysis of methyl propionate to propionic acid plus methanol by heating in aqueous acid.

10.18   Match the four pK_A values 1.26, 2.86, 4.05, and 4.82 with the four acids butyric acid, α-chlorobutyric acid, β-chlorobutyric acid, and dichloroacetic acid.

10.19   Show how the following compounds could be prepared from starting materials that contain no more than four carbon atoms:

$$CH_3\overset{\overset{\displaystyle O}{\|}}{C}N(CH_2CH_3)_2$$

a)   (structure above)
b)   3-ethyl-3-pentanol
c)   3-methylbutanoic acid
d)   methyl α-methyl-β-ketopentanoate
e) N-n-pentylacetamide
f)   1-pentanol

10.20   The reaction of ethyl propionate with sodium ethoxide gives a product **A**, having the formula $C_8H_{14}O_3$; its ir spectrum shows two carbonyl groups. Heating **A** with aqueous base followed by acidification and heating gives compound **B**. The mass spectrum of **B** shows a molecular ion at mass 86. The nmr spectrum of **B** shows only a triplet and a quartet corresponding to an ethyl group. Write the structures of **A** and **B**.

10.21   Suggest procedures for separating the following mixtures: (a) benzoic acid and ethylbenzene, and (b) benzoic acid, triethylamine, and ethylbenzene. (Ethylbenzene is a liquid, insoluble in water.)

10.22   Esters can react with alkoxide ion in the same way that they do with hydroxide ion; the reaction is called "ester exchange." In a typical example, a solution of methyl benzoate in ethanol containing sodium ethoxide is heated and then neutralized, and the alcohol is removed by distillation. The product obtained has the formula $C_9H_{10}O_2$. Write the structure and show how the product is formed.

10.23   Ethyl formate, $HCO_2C_2H_5$, is more reactive toward nucleophilic attack than most other esters because the carbonyl group is attached to hydrogen rather than an α-carbon (the difference is the same as that between an aldehyde and a ketone). With this in mind, predict the major product that would be obtained by treatment of a mixture of ethyl formate and ethyl acetate with sodium ethoxide.

10.24   Suggest structures for the compounds indicated by letters:
a)   Compound **A**, $C_4H_6Cl_2O_2$, was a strong acid, with $pK_A = 1.3$.
b)   Compound **B** had an odor similar to bananas. The molecular ion peak in the mass spectrum appeared at mass 144. Refluxing with aqueous NaOH gave a salt and an alcohol identified as 1-butanol. Reaction of **B** with LiAlH$_4$ gave 1-butanol as the only organic product.
c)   Compound **C** had the molecular formula $C_{10}H_{10}O_4$. On heating, a gas was evolved and β-phenylpropionic acid was obtained.
d)   On heating with NaOH solution, compound **D**, C_7H_5N, gave ammonia and the salt of an acid.

11

CARBOHYDRATES

The term **carbohydrate** refers to a large and very important group of naturally occurring organic compounds ranging from simple sugars to the polymers starch and cellulose. Carbohydrates are the compounds that are

$$n\,CO_2 \;+\; n\,H_2O \;\xrightarrow[\text{chlorophyll}]{\text{light}}\; (CH_2O)_n \;+\; n\,O_2$$

formed, together with oxygen, in photosynthesis. The name is derived from the general formula $C_m(H_2O)_n$, *i.e.*, a "hydrate" of carbon.

The basic units of carbohydrates, called **monosaccharides,** include **pentoses** ($C_5H_{10}O_5$) and **hexoses** ($C_6H_{12}O_6$). **Di-** and **trisaccharides** are made up of two and three monosaccharide units, respectively; **polysaccharides** are polymers containing hundreds or thousands of monosaccharide units. The term "sugar" does not have a precise chemical meaning. It is used to designate mono- or disaccharides in a general sense, as well as being the common name for the specific disaccharide **sucrose** (table sugar).

11.1   Glucose

Glucose is the central compound in carbohydrate chemistry. In its various forms, glucose is the most abundant organic compound in nature, and it plays a major role in metabolism and energy storage in all organisms. Glucose is the most readily assimilated food in the diet, and is one of the few nutrients that can be introduced directly into the circulatory system by intravenous feeding. We will begin our survey of sugars with a preliminary look at the structure and properties of this key compound.

Glucose is a hexose, with the formula $C_6H_{12}O_6$. Like most sugars, it is a colorless solid, extremely soluble in water and very sparingly soluble in organic solvents. In early work it was shown by a few simple reactions that glucose is an aldehyde, contains five OH groups, and has an unbranched carbon chain. With only this limited information we have a structural formula for glucose, namely pentahydroxyhexanal. However, some other evidence that we will see later shows that glucose and other sugars actually exist mainly in the form of a **cyclic hemiacetal,** in which a hydroxy group in the chain has added to the carbonyl group. The open chain aldehyde and cyclic hemiacetal structures are in equilibrium, and glucose can *react* as an aldehyde even though it is mainly in the cyclic form. Monosaccharides like glucose that contain a cyclized aldehyde group are called **aldoses** or, more specifically, **aldohexoses** (six-carbon), **aldopentoses** (five-carbon), and so forth. If the carbonyl group is not at the end of the chain, the monosaccharide is called a **ketohexose, ketopentose,** and so on.

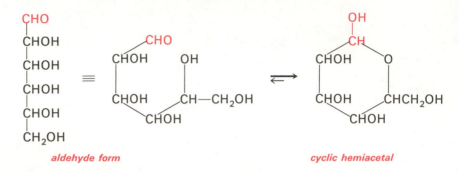

aldehyde form   cyclic hemiacetal

The fact that glucose has an aldehyde group available for reactions is shown by its **reducing** properties. As we saw in Section 8.6, aldehydes are oxidized by alkaline solutions of Ag^+ and Cu^{+2}. Benedict's reagent, containing Cu^{+2}, is the basis of most analytical methods for the determination of glucose in various body fluids. For detection of glucose in urine, a tablet containing Cu^{+2} is added; the appearance of a yellow or orange color due to the reduced form Cu_2^{+2} is a positive test for the presence of reducing sugar. Quantitative methods for blood glucose depend on a colorimetric assay of reduced copper.

11.2 Stereoisomerism of Monosaccharides

When we take a closer look at the pentahydroxyaldehyde structure of glucose, we see that our description of the compound is incomplete. Each of the four central CHOH groups is an asymmetric center, and the structure represents 2^4 or 16 possible stereoisomers. All of these stereoisomers are called **aldohexoses,** and naturally occurring glucose is one of these isomers.

As we saw in Chapter 5, there are two kinds of stereoisomers: enantiomers, which are mirror images, and diastereomers, which are not mirror images. To discuss the stereoisomerism of the aldohexoses, we will arrange the 16 isomers in two series of eight. All of the compounds in each series are diastereomers, and each compound in one series has an enantiomer in the other series. In other words, one series is the mirror image of the other.

The two series of aldohexoses can, in principle, be built up stepwise from the two enantiomers, **R** and **S,** of glyceraldehyde. In sugar chemistry, the enantiomers are called D and L, and the two series of sugars that are built up from the enantiomers are referred to as D-sugars and L-sugars, respectively. To represent the stereoisomerism of the sugars in a convenient way, we use projections of the three-dimensional structures. The chain is held vertically with H and OH on each asymmetric carbon in front, and the carbons above and below in the chain to the rear. The three-dimensional structure is then flattened into the paper.

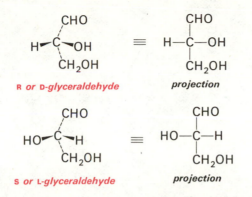

To construct the aldohexose isomers from glyceraldehyde, the CHO group at the top of the chain is converted to —CHOHCHO, and we thus proceed stepwise from the three-carbon sugar, which can be called a triose, to a tetrose, to a pentose, and finally to a hexose. By this process we create a family of sugars in which every compound has the same configuration at the bottom CHOH group as the enantiomer of glyceraldehyde that we began with. At each stage, the new —CHOH group is an additional asymmetric center, and two diastereomers are produced. Thus, the family includes one triose, two tetroses, four pentoses, and eight hexoses.

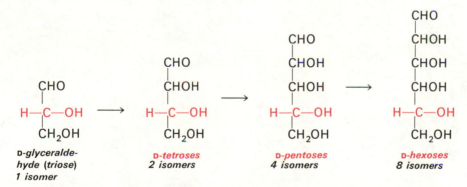

The complete family of D-sugars is shown in Figure 11.1. An enantiomeric set of tetroses, pentoses, and hexoses, the L-sugars, begins with L-glyceraldehyde. Each isomer in Figure 11.1 has an enantiomer in the L series in which the configuration of *every* asymmetric center is reversed.

The eight aldohexoses indicated in Figure 11.1 are diastereomers, each with distinctive properties. Three aldohexoses occur naturally—D-glucose, D-galactose, and D-mannose; the other five D-aldohexoses and many of the L-sugars have been prepared by synthesis. The configurations of the OH groups indicated in Figure 11.1 have been established for each compound by experiments in which various sugars were interconverted, and the configurations of the asymmetric centers in one sugar were correlated with those in others.

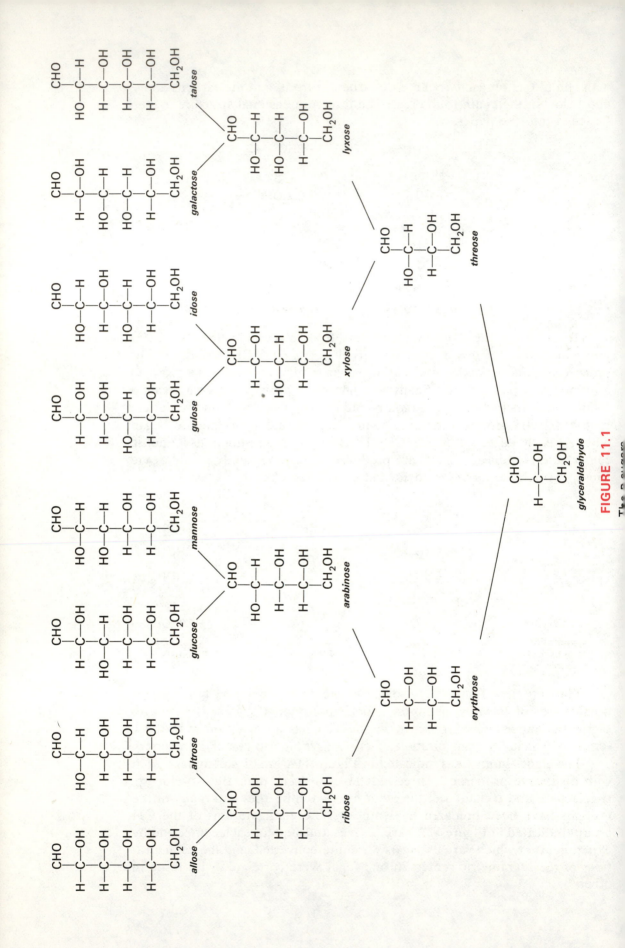

FIGURE 11.1

The D-sugars

REACTIONS OF MONOSACCHARIDES

Glucose and other sugars undergo the usual reactions of —OH and —CHO groups that we have previously seen, as well as some reactions that have been developed for special purposes in sugar chemistry. Although glucose exists mainly as a cyclic hemiacetal, some reactions are more conveniently represented with the open-chain aldehyde structure. As long as we keep in mind that the two forms are rapidly interconverted, it is unimportant which one we write.

11.3   Interconversion of Pentoses and Hexoses

Chain Lengthening.   In Section 11.2 it was mentioned that the sugar chain can be built up, carbon-by-carbon, from the aldehyde end. The first step in this process is addition of HCN to the carbonyl group to give the cyanohydrin. The C≡N group is then reduced to CH=NH, and this group is hydrolyzed by water to CHO.

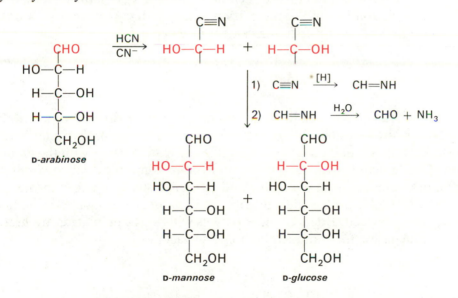

Since the CHO group in the original sugar becomes a new asymmetric center, we obtain two products by this sequence: thus, D-arabinose gives D-mannose plus D-glucose.

Problem 11.1   Write the structures of the hexoses that would be obtained by the cyanohydrin chain lengthening process starting with D-ribose.

Problem 11.2   Would you expect D-mannose and D-glucose to be formed in equal amounts from D-arabinose? Explain. (See Section 5.7.)

Chain Shortening.   The opposite process, in which the chain is shortened by removing the CHO group to give the next lower sugar, can be accomplished by taking advantage of the fact that cyanohydrin formation is a reversible reaction (Section 8.11). Using glucose as an example, reaction of the sugar with hydroxylamine gives the oxime, and dehydration of the oxime gives a C≡N group. This product is the cyanohydrin of the next lower sugar, and treatment with base causes loss of CN^- and formation of arabinose.

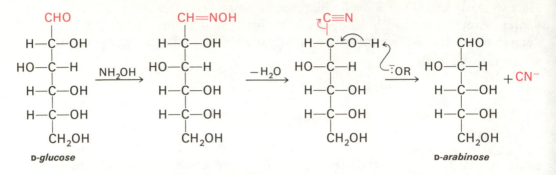

In these interconversions, the configurations of the three asymmetric carbons in the pentose are unaffected. By carrying out the interconversion of glucose and arabinose in either direction, therefore, it is established that the relative configurations of the three asymmetric centers in arabinose are the same as those of the three lower centers in glucose.

11.4 Reduction and Oxidation

Sugars can be reduced to the corresponding hexahydroxy compounds by catalytic hydrogenation or by treatment with sodium borohydride. These sugar alcohols are named by indicating the sugar stem name with the ending **-itol.** Thus glucose is reduced to glucitol (more commonly called **sorbitol**). Sorbitol is present in many fruits, and is used instead of sugar in dietetic foods. Mannitol is quite widely distributed in nature; it is obtained commercially from seaweed. Esters of these polyols with long-chain acids are useful as emulsifiers for food and as non-ionic surfactants.

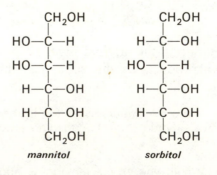

Aldohexoses can give rise to three series of acids, in which either the CHO, the CH_2OH, or both groups are oxidized to CO_2H. Oxidation of the CHO group to CO_2H can be carried out with bromine in water or by enzyme-catalyzed air oxidation. The acid derived from glucose, called **gluconic acid,** is an effective complexing agent for metal ions, and is used to prevent precipitation of calcium ion in beverages and other solutions. Gluconic acid and the corresponding acids from other hexoses are in equilibrium with the lactones which are formed by spontaneous esterification of the acid group with the OH group at C-5.

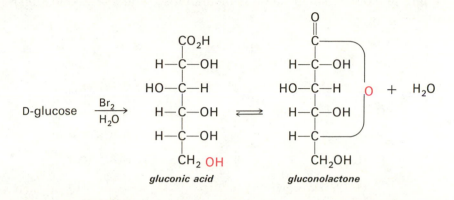

Acids with only the CH_2OH converted to CO_2H cannot be made by direct chemical oxidation of aldohexoses. However, some of these compounds, called **-uronic acids,** are formed from the hexose in living cells. As we will see in Section 11.7, glucuronic acid plays a central role in the metabolism and excretion of drugs.

Oxidation of an aldohexose with dilute nitric acid gives the **saccharic acid** in which both the CHO and the CH_2OH groups are converted to CO_2H. These dicarboxylic acids were of great importance in determining the steric configurations of the sugars. If the two ends of the chain are the same, the diacids from some sugars are *meso* structures and have zero optical rotation. Thus, two of the four pentose sugars give *meso*-trihydroxy acids, with a plane of symmetry passing through the central carbon.

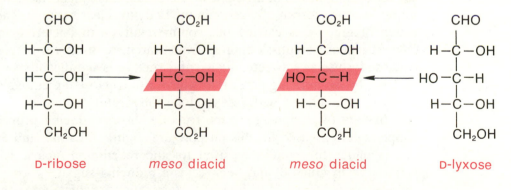

Problem 11.3   Which of the hexoses in Figure 11.1 would give *meso*-saccharic acids?

11.5   Enolization—Aldose and Ketose Sugars

An aldohexose, like any carbonyl compound with α-hydrogens, undergoes enolization. In the case of a sugar, the enolic structure is an **enediol,** since the α-carbon bears an OH group. When the carbonyl group is formed again by protonation at C-2, both the original aldohexose and the isomer with the opposite configuration of the OH group can be obtained. Moreover, a third carbonyl compound can arise by protonation of the enediol at C-1; in this case, the carbonyl group is at C-2, and the product is a **2-ketohexose.** Thus, treatment of glucose with dilute base gives an equilibrium mixture containing glucose, a small amount of mannose, and the 2-keto sugar D-fructose.

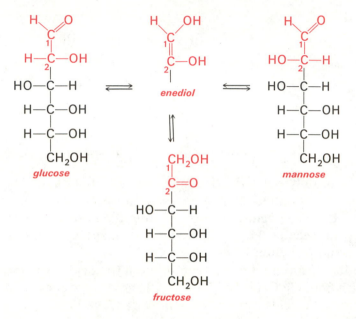

Fructose is present in a number of natural sources; it is the principal sugar in honey. Fructose and glucose are very closely related biochemically. In carbohydrate metabolism (Section 15.5), glucose and fructose are inter-converted in the form of their phosphate esters; this transformation is the first step in the breakdown of glucose in the body to release energy. The isomerization of glucose is carried out commercially with bacterial enzymes to convert glucose syrups, obtained from starch, to mixtures of glucose and fructose for use as a sweetening agent. Fructose is significantly sweeter than glucose; by isomerizing part of the glucose to fructose, an equivalent sweet-ness can be achieved with less total carbohydrate.

In tests for reducing sugars, fructose shows reducing properties. The copper solutions used for this purpose are slightly alkaline, and enolization occurs under the conditions used for the reaction. Fructose is therefore isomerized to glucose, and behaves as a reducing sugar.

11.6   Cyclic Structures

As noted earlier, despite the fact that glucose has reducing properties and behaves as an aldehyde in many reactions, glucose and other sugars are actually **cyclic hemiacetals.** A hemiacetal group is normally not stable; i.e., the equilibrium between an alcohol and an aldehyde usually favors the

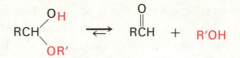

separate compounds. In a sugar, however, an OH group is present in the same molecule at the right distance for addition to form either a five- or a six-membered ring. With glucose and most aldohexoses the six-membered ring is favored; this is called the **pyranose** form. Fructose and pentose sugars have five-membered rings, referred to as the **furanose** form.

In writing the cyclic structure of glucose, the chain is tipped over horizontally to the right. Hydroxyl groups that are on the right in the vertical projection are thus below the ring, and those to the left are above the ring.

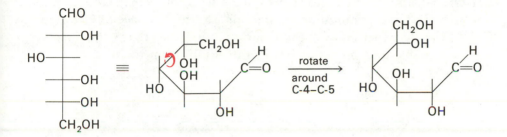

The chain is then rotated between C-4 and C-5 to bring the C-5 OH group into position for addition; this puts the —CH₂OH group above the ring.

When cyclization occurs, the C-5 OH group can attack the carbonyl double bond either from above or from below. A new asymmetric center is formed at C-1, and the cyclic sugar can therefore exist as two diastereomers, with the new OH group either below the ring (α form) or above the ring (β form). The two forms of glucose can be obtained by crystallization of the sugar under different conditions.

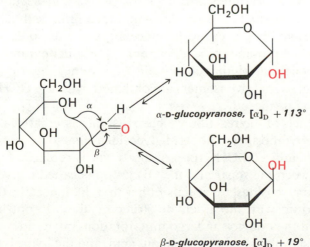

α-D-*glucopyranose*, $[\alpha]_D$ +113°

β-D-*glucopyranose*, $[\alpha]_D$ +19°

As expected for diastereomers, the α and β forms have different properties; and when they are dissolved in water, the initial values of the optical rotations of the two isomers differ considerably. As the solutions stand, however, the optical rotations change until the same value is observed for both. These changes in rotation, called **mutarotation,** are due to the fact that cyclization to the hemiacetal structures is reversible, and an equilibrium is set up between the two cyclic isomers by way of a very small concentration of the aldehyde form.

Problem 11.4   Write pyranose structures of α-D-galactose and β-D-mannose.

The six-membered ring in the pyranose sugars has a chair conformation like that of cyclohexane. In the stable conformation of β-glucopyranose, the CH_2OH substituent and all four OH groups are in the more stable equatorial positions. In the α-pyranose form, the bond to the hemiacetal OH group at C-1 is axial. As expected from this conformational difference, the β-isomer is the more stable and predominates in the equilibrium mixture.

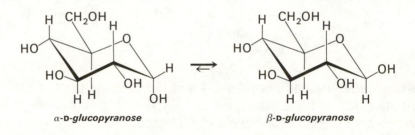

α-D-*glucopyranose*   β-D-*glucopyranose*

11.7   Glycosides

When glucose is warmed with methanol and an acid catalyst, the hemiacetal group is transformed to an acetal, with the C-1 OH replaced by OCH_3. This reaction occurs by the same steps that we saw in Section 8.13 for the conversion of an aldehyde to an acetal, with the difference that one OR group is already present in the cyclic hemiacetal, and thus only one OCH_3 is introduced. In the presence of acid, either α- or β-glucopyranose gives rise to the same carbocation. Methanol can add from either the top or the bottom side of the ring, so that two isomers are obtained, with the OCH_3 down (β) in one and up (α) in the other.

The products of the reaction we have just discussed are called methyl α- and β-D-glucopyranosides. The term **glycoside** refers to the general structure in which the hemiacetal OH group of any sugar is replaced by OR or OAr. The bond between the sugar and the OR group is called a glycoside linkage or bond. Since glycosides are acetals rather than hemiacetals, they are not in rapid equilibrium with the aldehyde structure, they do not have reducing properties, and they do not undergo mutarotation. Glycosides can be hydrolyzed to the sugar plus alcohol in aqueous acid, and the hydrolysis can also be

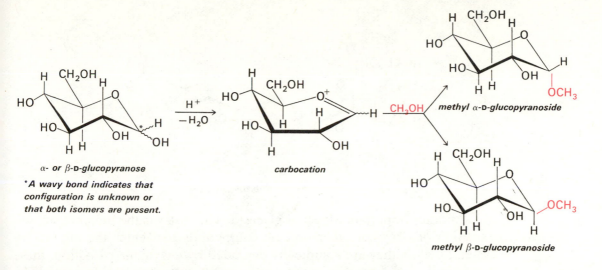

α- or β-D-glucopyranose

*A wavy bond indicates that configuration is unknown or that both isomers are present.

carbocation

methyl α-D-glucopyranoside

methyl β-D-glucopyranoside

catalyzed by enzymes. As in all enzymatic reactions, glycoside hydrolysis is stereospecific, particularly with respect to the configuration of the glycoside bond. Thus, certain enzymes catalyze the hydrolysis only for α-glycosides and others are selective for β-glycosides, regardless of the R group or the configuration of the other hydroxyl groups in the sugar.

Glycosides are very widely distributed in nature, and many naturally occurring hydroxy compounds such as sterols, terpenes, and plant pigments are present in plants in the form of glycosides. These compounds are, of course, much more soluble in water, and thus more readily transported in tissues, than are their parent hydroxy compounds. Examples of plant glycosides are peonin, which is the violet pigment of peonies, and amygdalin, present in bitter almonds. The latter contains two glucose units linked together through a 1→6 bond; the non-sugar unit is the cyanohydrin of benzaldehyde. Enzymatic hydrolysis of this glycoside releases glucose, benzaldehyde, and cyanide ion. It is the benzaldehyde that imparts most of the "almond flavor" to oil of bitter almonds.

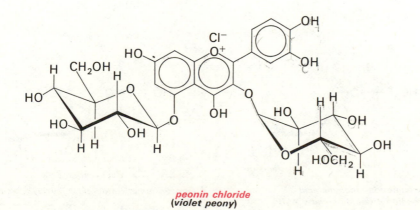

peonin chloride
(violet peony)

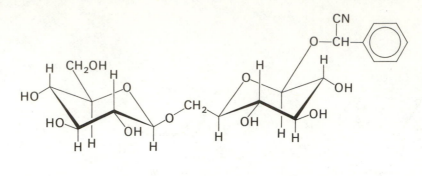

amygdalin (*bitter almond*)

Another important aspect of glycoside chemistry is the metabolism and excretion of drugs or other organic compounds that enter the circulatory system. Unless they are completely degraded by enzymatic oxidation, these compounds must be eliminated in the urine in water-soluble form. The mechanism by which this is usually accomplished is formation of a glycoside with **glucuronic acid.** Thus, compounds containing an aromatic ring, such as the pain-relieving drug phenacetin (page 313), are converted to a hydroxy compound which is then excreted as the glucuronide.

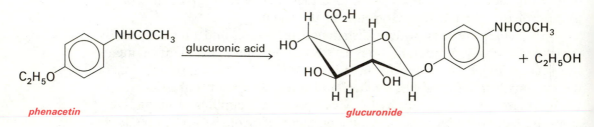

phenacetin   *glucuronide*

11.8   Nucleosides and Nucleotides

Glycoside bonds can be formed with NH as well as OH groups. *N*-Glycosides containing purine and pyrimidine bases attached to D-ribose or

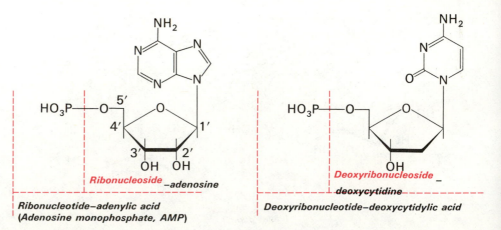

Ribonucleoside—adenosine

Ribonucleotide–adenylic acid
(Adenosine monophosphate, AMP)

Deoxyribonucleoside — deoxycytidine

Deoxyribonucleotide–deoxycytidylic acid

2-deoxy-D-ribose* are called **ribonucleosides** and **deoxyribonucleosides,** respectively. The ribose units are in the furanose form, with the *N*-substituent in the *β*-configuration. The 5'-phosphate esters of these glycosides are called **nucleotides,** and are the monomer units of nucleic acids (Chapter 16).

11.9   Disaccharides

Disaccharides contain two monosaccharide units combined as a glycoside, with a hydroxyl group of one hexose or pentose linked to the hemiacetal carbon of the other. Disaccharides are intermediates in the build-up and breakdown of polysaccharides, and a few are important sugars in their own right. The glycoside bond between the sugar units is formed in cells by displacement of a 1-phosphate group in one sugar by an OH group of another mono- or polysaccharide.

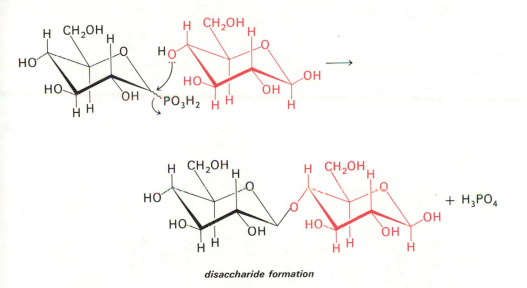

disaccharide formation

Lactose, the principal carbohydrate in milk, is a typical disaccharide. The structure of lactose and the reactions on which the structure is based are shown in Figure 11.2; let us look at how the structure can be deduced from this evidence.

a)   Hydrolysis of lactose with an enzyme that is known to catalyze the cleavage of *β*-glycosides liberates equal amounts of glucose and galactose. This result establishes the two sugar units in the disaccharide and shows that the glycoside bond has the *β*-configuration.

b)   Lactose reduces the Cu^{+2} reagent, and reacts with bromine water to give an acid. These facts show that one of the two sugar units has a reducing group, *i.e.*, aldehyde $\rightleftharpoons$ hemiacetal.

---

*The prefix **de-** indicates *absence* of a group, in this case, the 2-hydroxy group of ribose.

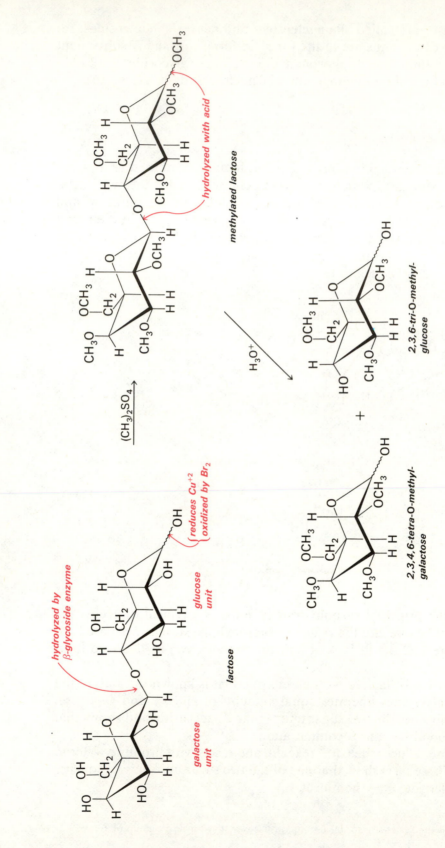

FIGURE 11.2

Determination of structure of lactose.

270

c)   When the acid from bromine oxidation is hydrolyzed, the products are galactose and gluconic acid. From this experiment we know that the galactose unit is the one in which the reducing carbon is "tied up" as a glycoside, and glucose is the sugar with the reducing group which was oxidized.

d)   The final question concerns which position in the glucose unit is attached through the glycoside bond to galactose. To determine this, lactose is treated with excess dimethyl sulfate ($CH_3OSO_2OCH_3$) in the presence of base. Under these conditions, all the OH groups present in the disaccharide are converted to OCH_3 groups. The methylated disaccharide is then hydrolyzed. The point of this experiment is that the OCH_3 groups which are simple ethers are not affected by hydrolysis, and remain intact when the glycoside bond is hydrolyzed. The products obtained from these steps are 2,3,4,6-tetra-*O*-methylgalactose and 2,3,6-tri-*O*-methylglucose. (The *O* in these names indicates that the methyl groups are attached through the oxygen atoms.) The fact that the OH group at C-4 of the glucose has no CH_3 group means that this is the position which is connected to galactose in the disaccharide.

With the information just outlined, we have defined the structure of lactose and in the process learned quite a bit about its chemistry.

Problem 11.5   In the hydrolysis of methylated lactose, one of the eight OCH_3 groups is hydrolyzed. Explain why this one group, and not the others, is removed.

Maltose is a disaccharide containing two glucose units, linked from C-1 of one glucose to the 4-OH of the other. Since C-1 of the second glucose is a hemiacetal, maltose has reducing properties. Maltose is obtained by partial hydrolysis of starch, and has an α-glycoside bond.

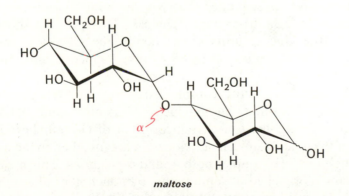

maltose

Sucrose is common sugar, isolated from sugar cane or beets. Sugar cane juice contains about 15% sucrose. After evaporating most of the water, crude sucrose is crystallized directly; the mother liquor from the crystallization is molasses.

Sucrose is a disaccharide which, on hydrolysis with acid or the enzyme invertase, gives glucose and fructose. Sucrose does not reduce copper solution and therefore does not contain a hemiacetal $\rightleftharpoons$ aldehyde group. The simple fact that sucrose is a non-reducing sugar tells us a great deal about the structure, since it means that the monosaccharide units can be linked together in only one way: The carbonyl groups of both glucose and fructose must be "tied up" in the glycoside bond. In other words, each sugar is a glycoside of the other. The glycoside bond has the α configuration in the glucose unit and is β in the fructose unit, which is in the furanose form.

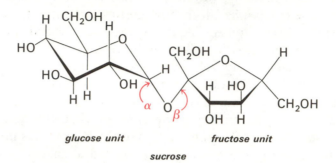

glucose unit   fructose unit

sucrose

11.10   Polysaccharides

Polysaccharides are present in the cells of all organisms and serve both as food reserves and as structural material in the cell wall. Quite a variety of polysaccharides have been isolated from plants. Straw, corncobs, and similar agricultural wastes are largely polysaccharides of pentose sugars. Gums and mucilages from seed pods, seaweed, and other plants contain polysaccharides made up of galacturonic acid. By far the most important are the two polymers of glucose—**starch** and **cellulose.**

Starch consists of two fractions, called **amylose** and **amylopectin.** Both are hydrolyzed by α-glycoside enzymes to give D-glucose. By applying the methylation-hydrolysis sequence that we saw with lactose, amylose is degraded mainly to 2,3,6-tri-O-methylglucose. This product shows that the 4-position of the glucose units is tied up in the glycoside linkages, and indicates a chain made up of $1\alpha \rightarrow 4$ linked glucose units in a regular sequence.

Amylopectin has a more complex structure. Methylation followed by hydrolysis gives 2,3,6-tri-O-methylglucose, but also some 2,3,4,6-tetra-O-methylglucose and 2,3-di-O-methylglucose. The di-O-methyl product reveals that some of the glucose units in amylopectin are attached to the amylopectin chain through the OH group at both 4- and 6-positions. This means that the $1\alpha \rightarrow 4$ chain is branched at some points by attachment of another glucose (through its 1α position) to the 6-position of a glucose unit in the chain. A new chain can then extend from the 4-position of the branching unit. Eventually each branch must have an **end group** with a "free" OH at the 4-position, and this group will give 2,3,4,6-tetra-O-methylglucose on degradation. The amount of the tetramethyl compound relative to trimethyl is a measure of the number of end groups, and thus the degree of branching. In amylopectins, a branch occurs at about every 30 to 35 glucose units.

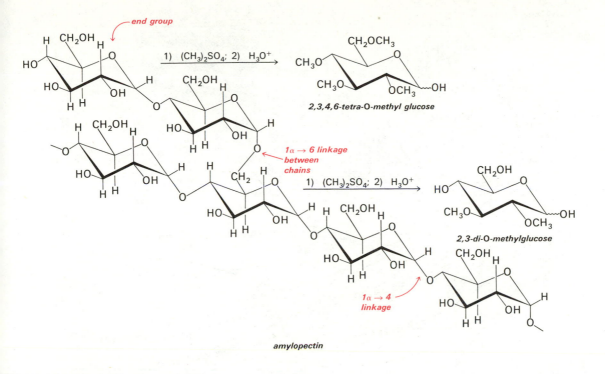

2,3,4,6-tetra-O-methyl glucose

2,3-di-O-methylglucose

amylopectin

Glycogen, or animal starch, is a reserve polysaccharide, stored in the liver and muscle tissue of mammals. Glycogen is similar in structure to amylopectin, but the degree of branching is higher, with a side chain at about every 8 to 10 glucose units.

Cellulose, like starch, contains only glucose units linked through the 1- and 4-positions. However, cellulose is totally unlike starch in its properties. Cellulose is unaffected by the mammalian digestive enzymes that degrade starch, and it can be digested as food only by a cud-chewing animal, such as a cow or a goat. These animals are equipped with a rumen containing bacteria that have the necessary enzymes. Cellulose in the form of recovered newspapers mixed with molasses can, in fact, be used as cattle feed, but traces of lead and other heavy metals from the printing ink are likely to be carried over into the milk. In human diets, cellulose provides only bulk and roughage.

The difference between cellulose and starch lies solely in the configuration of the glycoside bond. In cellulose, the 4-OH group of one glucose unit is linked as a β-glucoside to another unit. This configuration not only presents a different stereoisomeric situation to an enzyme surface, but also leads to a different macromolecular structure. Since a β-substituent is *equatorial,* the cellulose chains are flat ribbons that can pack neatly together as crystalline bundles in the thickened cell wall of plant tissue. These bundles twine together in hollow cylinders called **fibrils.** The cellulose fibrils in wood are cemented together with lignin to give a strong, tough structure (Figure 11.3).

Cellulose has a very high molecular weight (estimated as high as 10^6), and is much more resistant than starch to chemical hydrolysis. Cellulose can

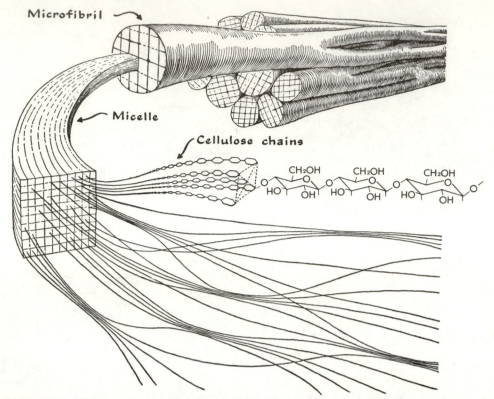

FIGURE 11.3

Cellulose molecule in the cell wall. Cell walls are composed of microfibrils, which are composed of micellar strands. (From *Principles of Plant Physiology* by James Bonner and Arthur W. Galston. W. H. Freeman and Company. Copyright © 1952.)

be converted to ester and ether derivatives with partial degradation of the chain. Acetylation of cellulose gives an inexpensive textile fiber ("acetate") with a silky texture. Several derivatives of cellulose, such as cellophane, were formerly used as plastics and films; these materials are now being displaced by synthetic polymers.

SUMMARY AND HIGHLIGHTS

1. **Carbohydrates**
$$\begin{bmatrix} \text{Monosaccharides} & \begin{bmatrix} \text{Pentose—}C_5H_{10}O_5 \\ \text{Hexoses—}C_6H_{12}O_6 \end{bmatrix} \\ \text{(simple sugars)} \\ \text{Disaccharides—two hexose units} \\ \text{Polysaccharides—polymers of pentose or} \\ \quad \text{hexose units} \end{bmatrix}$$

2. **Glucose** is an aldohexose—a pentahydroxyhexanal that is mainly in the hemiacetal form. Analysis of glucose in blood is based on the reduction of Cu^{+2} by the aldehyde group.

3. **Aldohexoses** in the open-chain form contain four asymmetric centers, giving a total of 16 stereoisomers which are grouped in two series, D and L.

4. **Chain lengthening** or **shortening** can be accomplished *via* cyanohydrins.

5. **Oxidation** of aldose sugars gives acids with CHO, CH_2OH, or both groups converted to CO_2H. **Reduction** gives polyalcohols.

6. **Enolization** of glucose gives an enediol that is in equilibrium with the isomeric aldohexose mannose and the ketohexose fructose.

7. **Cyclic** forms of sugars are hemiacetals with an additional asymmetric center, giving rise to two diastereoisomers, α and β.

8. **Glycosides** are acetals derived from the cyclic hemiacetal form of sugars. Alcohols are present in plants in the form of *O*-glycosides, and drugs are excreted as glycosides.

9. **Disaccharides** such as maltose, lactose, and sucrose contain two sugar units with a hydroxyl group of one unit linked as a glycoside to the other.

10. **Starch** and **cellulose** are polymers of glucose units with a 1,4-glycoside structure. The glycoside bonds in cellulose are β (equatorial) and lead to a flat, linear structure; in starch the glycoside bonds are α (axial), causing a spiral structure.

ADDITIONAL PROBLEMS

11.6   Identify the pentose or hexose unit in each of the following structures (refer to Figure 11.1), and name the compound, specifying α or β configuration at C-1 and ring size.

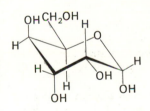
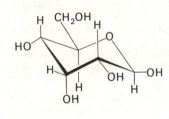

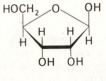
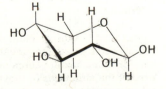

11.7   Which of the following would reduce Benedict's reagent (Cu^{+2})?

    a) galactose   b) ribose

    c) sorbitol   d) methyl β-glucopyranoside

    e) sucrose   f) maltose

11.8 Draw projection formulas (open chain) for the starting compounds and products indicated by letter in the following reactions.

a) L-arabinose $\xrightarrow{NH_2OH}$ oxime **(A)** $\xrightarrow{-H_2O}$ **B** $\xrightarrow{OR^-}$ **C**

b) D-mannose $\xrightarrow[H_2O]{Br_2}$ mannonic acid, **D**

c) D-galactose $\xrightarrow{NaBH_4}$ **E**

d) D-glucose $\xrightarrow[H^+]{CH_3OH}$ **F**

11.9 One of the early observations of the chemistry of glucose was its reaction with excess acetic anhydride to give a product that contained *five* acetate groups. This reaction was the first evidence that glucose contains five hydroxyl groups.
a) Show this reaction and write the structure of the product, using the open-chain aldehyde structure of glucose.
b) Further work revealed that the penta-acetate product was actually a mixture of two compounds, neither of which showed the properties of a carbonyl group (reaction with $ArNHNH_2$, etc.). Suggest structures for the two penta-acetates.

11.10 The reaction of glucose and other sugars with excess phenylhydrazine gives **osazones,** in which the 2-OH group is oxidized and then forms a second hydrazone group: How many different osazones can be formed from the D-aldohexoses?

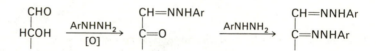

11.11 The following statement is *not* true; explain why not. "In the formation of a mixture of methyl D-glucopyranosides from D-glucose, α-D-glucose gives methyl α-D-gluco-pyranoside and β-D-glucose gives methyl β-D-glucopyranoside."

11.12 Two compounds, both with the formula $C_4H_8O_4$, are produced by the chain-lengthening sequence (HCN; reduce; hydrolyze) from D-glyceraldehyde. Write the structures and give the names of these products. After the two compounds are separated, how could it be determined which product has which structure?

11.13 A key point in establishing the structures of the aldohexoses was the observation that a certain pentose and *both* of the aldohexoses that were obtained from it by chain extension all gave optically active dicarboxylic acids on oxidation with nitric acid. Referring to Figure 11.1, give the names of the hexoses that can be eliminated on the basis of this information.

11.14 In the reaction of glucose with aqueous base, the equilibrium with fructose is rapidly established. On longer standing, a number of other products begin to appear, and a very complex mixture containing very little fructose is eventually obtained. Without trying to predict the structures of these further products, suggest the first step in their formation; in other words, how and why does further reaction occur?

11.15 Write the structure of the compound that would be obtained on complete methylation of cellulose with $(CH_3)_2SO_4$, followed by acid hydrolysis.

11.16 A glycoside containing two sugar units was hydrolyzed with an α-glycosidase enzyme to give one mole each of glucose, mannose, and 1-butanol. After methylation of hydroxyl groups, acid hydrolysis gave 2,3,4,6-tetra-O-methylglucose, 2,3,6-tri-O-methyl-mannose, and butanol. Write the structure of the glycoside.

11.17 In the hydrolysis of amylopectin by malt enzymes, a series of progressively smaller polysaccharides can be detected, giving the disaccharide maltose and finally glucose. By careful separation work, a small amount of another disaccharide can be isolated. This compound, called isomaltose, has reducing properties and gives two moles of glucose on hydrolysis. Examine the structure of amylopectin and suggest a possible structure for isomaltose.

11.18 The disaccharide trehalose, isolated from yeast, gives only D-glucose on hydrolysis. Trehalose does not reduce Benedict's solution; what information does this fact provide about the structure of trehalose? Is the structure of the disaccharide completely known from the information given? If so, write the structure; if not, what questions must be settled? How can they be answered?

11.19 Write the steps in the mechanism of hydrolysis of lactose by acid to give glucose and galactose; show where the acid attacks, and which bond breaks. If the reaction were carried out in the presence of $H_2^{18}O$, what positions in the products would contain ^{18}O?

12
PHENOLS AND AROMATIC AMINES

When a hydroxyl or amino group is attached directly to an aromatic ring, the properties of the groups are significantly different from those in aliphatic alcohols and amines. Moreover, the reactivity of the benzene ring toward electrophilic substitution is greatly increased compared to that of the aromatic compounds discussed in Chapter 4. For these reasons, ArOH and $ArNH_2$ compounds are best treated in a separate chapter.

12.1 Names and Properties of Phenols

Compounds with an OH group bonded directly to a benzene ring are called **phenols.** The parent substance, C_6H_5OH, is phenol, and compounds with additional substituents are named as derivatives of phenol, with the carbon carrying the OH group numbered 1. Some of these compounds have common names; *e.g.*, methylphenols are called **cresols,** and hydroxy derivatives of naphthalene are **naphthols.**

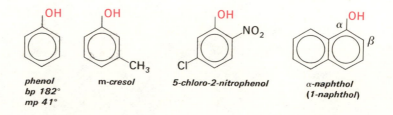

phenol
bp 182°
mp 41°

m-cresol

5-chloro-2-nitrophenol

α-naphthol
(1-naphthol)

Phenols are low-melting solids or oily liquids with a sharp odor reminiscent of Lysol or creosote. These products and various germicidal soaps and cleansers contain phenol and cresols as the active bactericidal agents. **Ger-**

micides and **antiseptics** are substances applied to live tissue to prevent the growth of microorganisms. **Disinfectants** are chemical agents applied to floors and other surfaces to destroy pathogenic microorganisms but not bacterial spores. A wide variety of chemicals, including halogens, mercury salts, hydrogen peroxide, and alcohols, have been used for these purposes. All these compounds, as well as phenol, are irritating to tissue.

The bactericidal action of phenols depends in part on the relative solubilities of the compounds in fat and in water. Alkyl substituents usually increase the germicidal potency. **Hexylresorcinol,** with two hydroxyl groups, is one of the most active phenolic antiseptics. Certain other phenols have higher potency for specific organisms, but a good *general* antiseptic must have a broad antibacterial spectrum.

A compound that has been widely used in germicidal skin cleansers is **hexachlorophene,** which is effective in controlling the bacterial flora that cause minor skin eruptions. Hexachlorophene can lead to serious toxic reactions when it is absorbed into the body. Skin dusting powder containing grossly excessive concentrations of hexachlorophene has caused fatalities in infants, and restrictions have been placed on its use.

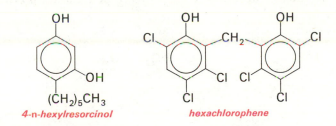

4-n-hexylresorcinol   *hexachlorophene*

12.2   Acidity of Phenols

The pK_A of phenol is 9.8, which may be compared to a value around 16 for a typical alcohol. Thus, the acidity of phenol is about one-million-fold greater than that of an alcohol. As a result of this acidity, phenol is more than 99 per cent dissociated in dilute sodium hydroxide solution (pH 12–13); unlike alcohols, phenols are soluble in dilute NaOH as the phenoxide salts. On the other hand, the acidity of phenol is about 10^5-fold less than that of a carboxylic acid. Since phenol is also a considerably weaker acid than carbonic acid, it does not dissociate or dissolve, as a carboxylic acid does, in aqueous sodium bicarbonate.

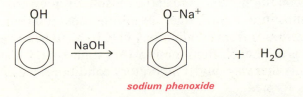

sodium phenoxide

The high acidity of the OH group of phenols relative to alcohols is due to resonance stabilization of the phenoxide anion. Unlike an alkoxide anion, RO^-, the phenoxide anion is stabilized by resonance delocalization of the negative charge, which occurs by donation of an electron pair on oxygen into the aromatic ring.

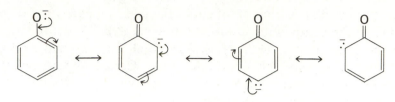

resonance stabilized phenoxide anion

Electron-withdrawing groups on the ring that further delocalize the negative charge and stabilize the phenoxide anion greatly increase the acidity of phenols. For example, nitro groups in the *ortho* and *para* positions have a large effect. With two nitro groups, a phenol has an acidity roughly equal to that of a carboxylic acid, while with nitro groups occupying all *o-* and *p*-positions, the phenol becomes a very strong acid called **picric acid.**

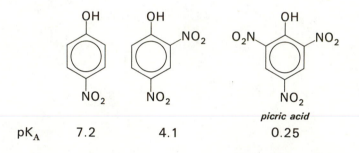

pK$_A$	7.2	4.1	0.25

Problem 12.1 Write five resonance structures showing delocalization of the negative charge in the anion of *p*-nitrophenol.

12.3   Synthesis and Occurrence of Phenols

Phenol is a large volume industrial chemical, and several processes are used in its manufacture. Two of these methods bring out points that deserve mention. Production of phenol from chlorobenzene and NaOH is carried out under drastic conditions: 260° and 260 atmospheres pressure. The reaction is not a direct substitution but rather an elimination-addition sequence. The intermediate resulting from elimination of HCl, called **benzyne,** reacts rapidly with water to give phenol. Benzyne is a very high-energy species which can be generated in other ways under laboratory conditions; it cannot be isolated, but is used as formed.

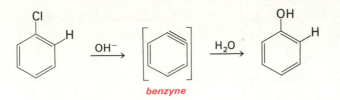

Another major industrial process for preparing phenol is the oxidation of isopropylbenzene (cumene). The mechanism of this process is complex, but the process is a good example of the economic considerations involved in an industrial process. The reagents involved—benzene, propene, oxygen, and sulfuric acid—are all relatively inexpensive. However, this process probably would not be economically feasible except for the fact that *two* important commercial products are produced, phenol *and* acetone.

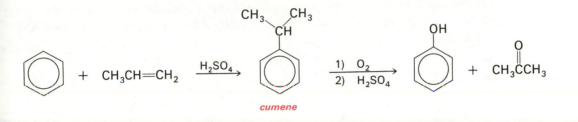

A number of phenolic compounds occur in nature. For example, the active constituent of the allergenic oil of poison ivy, *urushiol,* is a mixture of several phenolic compounds in which the side chain varies in the number of double bonds.

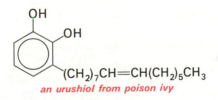

an urushiol from poison ivy

Several phenolic compounds are obtained in quantity as byproducts in paper manufacturing. Wood is a mixture of cellulose (Chapter 11) and a polymer called **lignin.** To obtain cellulose fibers for paper making, the lignin is broken down by cooking wood chips with alkaline sodium bisulfite. The black liquor which remains after removing the cellulose is terribly damaging if discharged into streams or lakes. The effluent contains phenolic hydrolysis products derived from lignin and large amounts of inorganic salts.

The monomer units of lignin, called **lignans,** are variations of a general structure consisting of a 1-phenylpropane skeleton with hydroxy or methoxy groups at the 3 and 4 positions. A typical lignan is coniferyl alcohol. One of the useful products that can be obtained from lignans is **vanillin,** which is the principal flavor constituent of the vanilla bean. Much of the vanilla flavoring used today comes from pine trees rather than vanilla plants, but the bouquet of vanilla extract is not matched by vanillin alone.

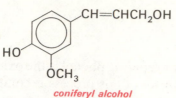

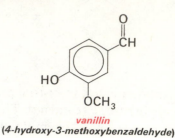

coniferyl alcohol
[3-(4-hydroxy-3-methoxyphenyl)-2-propen-1-ol]

vanillin
(4-hydroxy-3-methoxybenzaldehyde)

12.4 Aromatic Amines

The parent aromatic amine, $C_6H_5NH_2$, is called **aniline;** like phenol, the name is used as a basis for naming substituted members of this series. If another functional group takes precedence, the prefix **amino-** is used to denote the $-NH_2$ group.

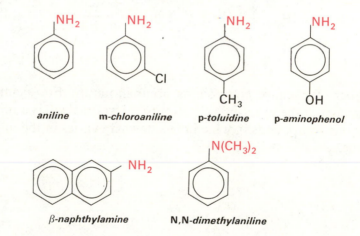

aniline   m-chloroaniline   p-toluidine   p-aminophenol

β-naphthylamine   N,N-dimethylaniline

Aniline is an oily liquid without the fishy smell characteristic of aliphatic amines. Although colorless when pure, liquid aromatic amines are usually yellow or brown because of the presence of air oxidation products, which are readily formed and are difficult to remove. Aromatic amines are highly toxic substances, and a number of them are **carcinogenic** (cancer-initiating). β-Naphthylamine, which was at one time an important dye intermediate, was one of the first compounds established as a causative agent for tumors. Its use for any manufacturing purpose is now banned in the United States.

12.5 Basicity of Amines

The aromatic ring causes aniline to be about 10^6-fold *less* basic than a typical alkylamine. Resonance donation of the unshared pair of electrons on nitrogen to the ring stabilizes aniline by delocalization. In the anilinium ion formed by protonation of aniline, this electron pair is tied up as a σ bond and is unavailable for resonance delocalization. Alkylamines do not have the opportunity for resonance stabilization of either the neutral or the protonated form. The overall result is that the energy gap between the neutral and protonated forms is increased for the aromatic amine relative to the aliphatic amine.

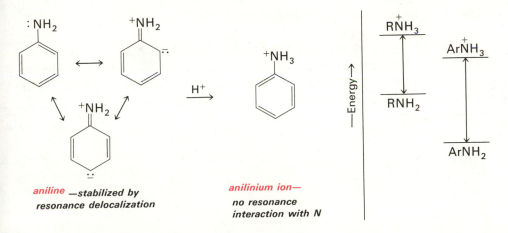

aniline —stabilized by resonance delocalization

anilinium ion—
no resonance interaction with N

Another way of looking at this problem is that the electron pair on nitrogen in an aromatic amine is less accessible for bonding with an acid because of the resonance interaction with the ring. Either way, the important thing to realize is that salt formation with an aromatic amine requires a loss in resonance energy.

The relative basicity of two amines such as RNH_2 and $ArNH_2$ is expressed in terms of their equilibria in water with the conjugate acids RNH_3^+ and $ArNH_3^+$. The stronger the base, the greater the tendency for the species to be protonated and for the equilibrium to lie to the left. In other words, since RNH_2 is a stronger base than $ArNH_2$, RNH_3^+ is a weaker acid, with a higher pK_A value, than $ArNH_3^+$.

$$RNH_3^+ + H_2O \rightleftharpoons RNH_2 + H_3O^+$$
$$ArNH_3^+ + H_2O \rightleftharpoons ArNH_2 + H_3O^+$$

The pK_A value for a typical alkylammonium ion is about 10, which means that an alkylamine exists almost entirely as the ammonium salt even in weakly basic solution. Anilinium ion has a pK_A of 4.6, and is therefore protonated and dissolves in aqueous acid. Just as in the case of the acidity of phenols, a nitro group in the ortho- or para-position greatly increases the acidity of $ArNH_3^+$ and lowers the basicity of the nitroaniline to the point that they are not soluble in dilute aqueous acid.

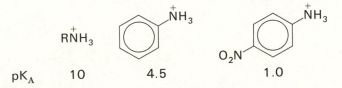

	RNH_3^+		
pK_A	10	4.5	1.0

Problem 12.2   The pK$_A$ of *m*-nitroanilinium ion is 2.5. Is *m*-nitroaniline a stronger or weaker base than *p*-nitroaniline? Suggest a reason for the difference in basicity of the two isomers in terms of resonance structures of the two amines.

12.6   Preparation of Aromatic Amines

The only general method for preparation of aromatic amines is the reduction of nitro compounds. A nitro group can be introduced by electrophilic substitution, and reduction to the amine can be accomplished by catalytic hydrogenation or by metals such as zinc, tin, or iron in the presence of mineral acid.

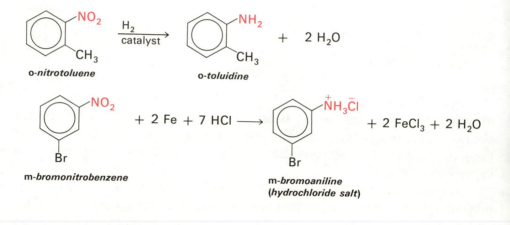

o-nitrotoluene   *o-toluidine*

m-bromonitrobenzene

m-bromoaniline
(hydrochloride salt)

12.7   Electrophilic Substitution in Phenols and Anilines

As discussed in Chapter 4, an unshared electron pair on an atom attached directly to the aromatic ring greatly stabilizes the intermediates formed by *o*- or *p*-attack of an electrophilic reagent. Electrophilic substitution in phenols and anilines is therefore rapid and extensive. Chlorine and bromine, for example, bring about substitution of *all ortho* and *para* hydrogens, even in the absence of a catalyst.

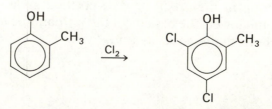

2,4-dichloro-6-methylphenol

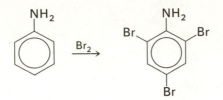

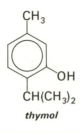

2,4,6-tribromoaniline

Problem 12.3   Thymol, obtained from thyme oil, is an effective disinfectant and is used in many antiseptic preparations. Suggest how thymol could be synthesized from *m*-cresol and propene.

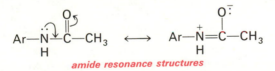

thymol

Problem 12.4   Nitration of phenol with dilute aqueous HNO_3 gives two products with the formula $C_6H_5NO_3$. What are the probable structures?

Amides of aniline, such as acetanilide, are less reactive to electrophilic substitution than are the corresponding amines. This is because the adjacent carbonyl group "drains off" some of the lone pair electron density from the nitrogen, thus decreasing the ability of the system to stabilize the cationic intermediates generated by electrophilic attack. This decrease in the reactivity

Ar—N—C—CH₃ ⟷ Ar—N=C—CH₃

amide resonance structures

of the ring is an advantage in obtaining a mono-substituted aniline, since aniline itself is often too reactive toward further substitution. Acetanilide, however, usually undergoes electrophilic substitution selectively to give the *para*-substituted derivative in good yield. A common procedure, therefore, is to carry out *para*-substitution with acetanilide and then regenerate the amine by hydrolysis of the amide. This approach is illustrated by the preparation of the important drug sulfanilamide (see Section 13.7). Sulfonation of acetanilide is carried out with excess chlorosulfonic acid to give the sulfonyl chloride. After conversion of the sulfonyl chloride to the sulfonamide with ammonia, the acetyl group is removed by hydrolysis.

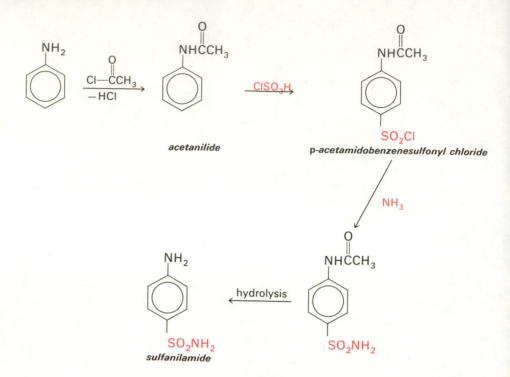

A *meta*-substituted aniline can be prepared by introducing the substituent at the stage of the nitro aromatic. Reduction of the resulting *meta*-substituted nitro compound then gives the desired amine.

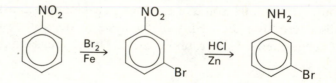

Problem 12.5   Write reaction sequences with structural formulas for the following syntheses. Show all steps from the starting material indicated.

    a) *m*-chloroaniline from benzene
    b) *p*-chloroacetanilide from benzene
    c) 2-methyl-4-nitroaniline from *o*-nitrotoluene

Several useful reactions of phenols occur under alkaline conditions in which the phenoxide anion is the actual reactant. Since the *ortho* and *para* positions share the negative charge, sodium phenoxide can react as a *carbanion*. At elevated temperatures, condensation at the *ortho* position with CO_2 gives sodium salicylate, which is the key starting material in the preparation of aspirin.

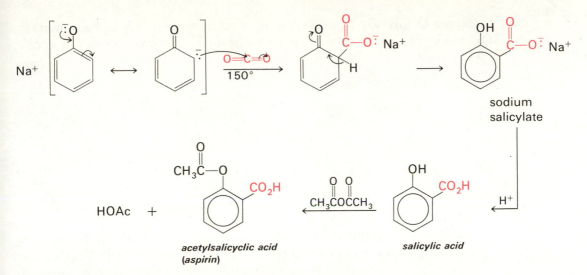

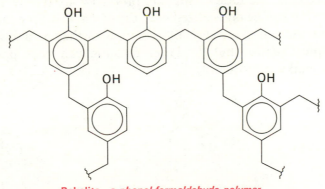

acetylsalicyclic acid
(aspirin)

salicylic acid

The reaction of phenol with formaldehyde in the presence of dilute alkali gives a mixture of *o*- and *p*-hydroxybenzyl alcohols. This reaction is

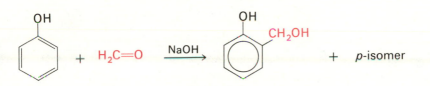

difficult to control because of further condensations that lead to a polymeric product. Under the proper conditions the final product is a dark, brittle, cross-linked resin called **Bakelite,** one of the first synthetic polymers.

Bakelite—a phenol-formaldehyde polymer

12.8   Oxidation of Phenols and Anilines

The high electron density of the phenol and aniline rings makes these compounds very susceptible to oxidation. With certain reagents, oxidation can lead to highly colored condensation products; the dark pigment **melanin** is formed in this way by oxidation of phenolic compounds in hair and skin. Oxidation of phenol or aniline with chromic acid gives *p*-benzoquinone, often called simply **quinone.**

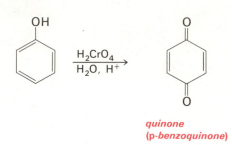

quinone
(p-benzoquinone)

Substituted quinones are highly colored compounds and are often found as pigments in plants and molds. Quinones are strongly electrophilic and are readily reduced to the corresponding dihydroxybenzenes. Interconversion of the quinone and the dihydro compound is a rapid, reversible redox reaction. Hydroquinone is the reducing agent used in photographic development to convert unexposed silver ion to metallic silver.

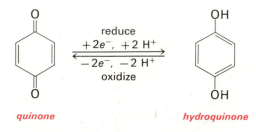

quinone   hydroquinone

Phenols are also oxidized by odd-electron reagents, such as peroxides or molecular oxygen. The radicals that are produced are relatively stable, since the odd electron can be delocalized over the oxygen and the three carbons in the ring. The radicals can eventually undergo further oxidation to quinones or can couple together, but the latter reaction can be suppressed by the presence of bulky substituents.

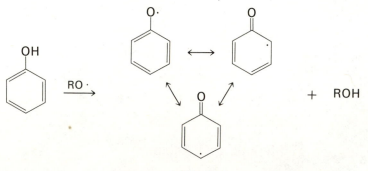

resonance-stabilized phenoxy radical

The ease with which phenols react with oxidizing radicals makes them useful as **antioxidants.** Air oxidation of unsaturated compounds occurs by radical chain reactions (Section 3.13), and a phenol can interrupt this process by reacting with one of the radicals. An important application of phenolic antioxidants is in foods, particularly those containing unsaturated fats. Two compounds approved for use in food by U.S. Food and Drug regulations are "butylated hydroxyanisole (BHA)" and "butylated hydroxytoluene (BHT)"; these phenols may be added to a total concentration of 0.02 per cent of the total fat content.

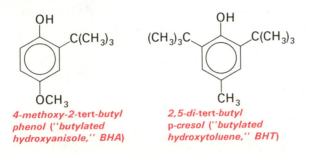

4-methoxy-2-tert-butyl phenol ("butylated hydroxyanisole," BHA)

2,5-di-tert-butyl p-cresol ("butylated hydroxytoluene," BHT)

Tocopherols (vitamin E complex) are naturally occurring antioxidants present in animal and plant cells. Vegetable oils are a particularly rich source. These compounds are hydroquinone derivatives and are readily oxidized to the quinone, thereby functioning to protect tissues against oxidation or deterioration by free radicals, both of which are involved in the complex process of aging.

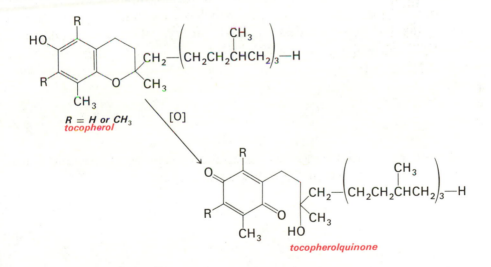

R = H or CH₃
tocopherol

tocopherolquinone

12.9 Aromatic Diazonium Compounds

One of the most important properties of aromatic amines is their ability to form **diazonium salts,** $ArN_2^+X^-$. These compounds are obtained by treating the amine with nitrous acid, which is prepared in solution from sodium nitrite and a mineral acid such as HCl or H_2SO_4. In the acid solution, nitrous acid is protonated and then dissociates to form the electrophile NO^+.

$$NaNO_2 + HCl \longrightarrow HO{-}N{=}O \xrightarrow{H^+} H{-}\overset{\underset{|}{H}}{\underset{+}{O}}{-}N{=}O \longrightarrow {}^+N{=}O + H_2O$$

Although NO^+ is a relatively weak electrophile, it reacts readily with amines. After a series of proton shifts and loss of water, the diazonium ion is formed. The overall process is called **diazotization** of the amine. Aromatic diazonium ions are stable in cold solution and are used in this form in reactions. Solid diazonium salts are dangerously explosive and are rarely isolated. Some diazonium salts are now known to be highly carcinogenic.

$$Ar\overset{..}{N}H_2 + {}^+NO \longrightarrow Ar\overset{\underset{|}{H}}{\underset{|}{N}}{}^+{-}NO \xrightarrow{-H^+} Ar\overset{\underset{|}{H}}{N}{-}N{=}O$$

$$Ar\overset{+}{N}{\equiv}\overset{..}{N} \longleftrightarrow Ar\overset{..}{N}{=}\overset{+}{N} \xleftarrow{-H_2O} ArN{=}N{-}\overset{+}{O}H_2 \xleftarrow{H^+} ArN{=}N{-}OH$$

aromatic diazonium ion

12.10   Substitution Reactions of Diazonium Salts

The N_2^+ group in diazonium salts is a potential nitrogen molecule and can function as an excellent leaving group in nucleophilic substitutions. The simplest reaction is replacement by OH, which occurs on heating the aqueous solution. The diazonium ion decomposes to an aryl cation, and this immediately reacts with water to give a phenol.

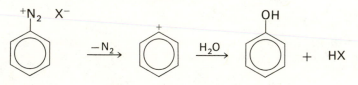

A number of other important substitution reactions of aromatic diazonium salts are illustrated in the following examples.

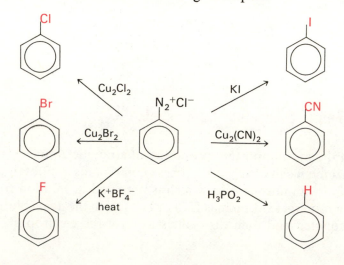

These substitution reactions are particularly useful in obtaining substitution patterns different from those available by direct electrophilic substitution. For example, it is impossible to prepare a *meta*-dihalobenzene by direct halogenation because the first halogen on the ring will direct the second one to an *ortho* or *para* position.

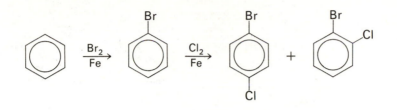

However, the *meta*-isomer can easily be obtained by using a diazonium salt.

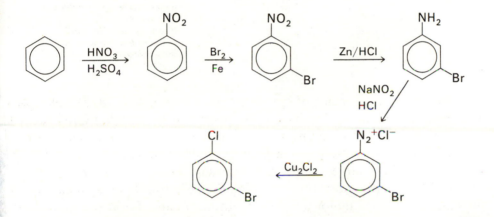

Another illustration is the preparation of 1,3,5-tribromobenzene. This arrangement of bromine atoms is available only by bromination of a compound with a very strong *o-*, *p*-activating group, such as phenol or aniline. The synthesis therefore involves bromination of aniline, followed by removal of the amino group through diazotization and treatment with hypophosphorous acid.

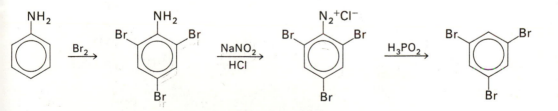

Problem 12.6 Show how the following syntheses could be accomplished using a diazonium reaction. Write structural formulas for all steps from the starting compound indicated.

a) *p*-chloroiodobenzene from chlorobenzene
b) *m*-bromophenol from nitrobenzene
c) 2,6-dibromotoluene from *p*-nitrotoluene.

12.11 Azo Coupling

The diazonium group can also behave as an electrophile toward strongly activated aromatic rings. This electrophilic substitution reaction yields highly colored **azo compounds,** Ar—N=N—Ar.

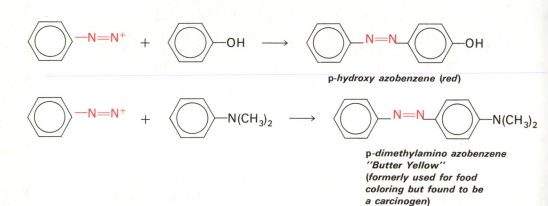

p-*hydroxy azobenzene* (red)

p-*dimethylamino azobenzene*
"Butter Yellow"
(formerly used for food
coloring but found to be
a carcinogen)

The reaction of a diazonium salt with an activated ring is called **azo coupling.** The highly colored azo compounds are formed rapidly at low temperature simply by mixing solutions of a phenol or an aromatic amine and the diazonium salt. This coupling reaction is the basic process for the preparation of the largest and most important group of synthetic dyes.

The azo coupling reaction is used in the detection of marijuana, *Cannabis sativa.* The three major constituents of marijuana, tetrahydrocannabinol (THC), cannabinol (CBN), and cannabidiol (CBD), are phenolic. The marijuana plant material reacts with *p*-nitrobenzenediazonium chloride to give a red-brown color which is presumably due to azo dyes formed by *o*- and *p*-coupling. This is a useful forensic test, since tobacco does not give a color reaction with this reagent.

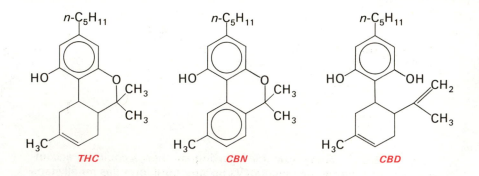

Problem 12.7   Write the structure of the compound responsible for the red color in a positive test for tetrahydrocannabinol.

CHEMISTRY OF COLOR AND DYES

The dyeing of cloth and the fermentation of sugar to produce alcohol are two specialized areas of organic chemistry that date back to antiquity. Fabrics dyed with indigo and madder have been found in tombs of predynastic Egypt. These, and a few coloring matters extracted from insects and tropical woods, were the dyes available until the preparation in 1856 of the first **synthetic dye,** which was obtained by the oxidation of impure aniline. A few years later, the structures of the natural dyes indigo and alizarin were determined, and these compounds were then prepared by synthesis. However, the major development in the chemistry of synthetic dyes was the discovery of diazotization and azo coupling. From this point, the synthetic dye industry in Germany provided fertile ground for the growth of benzene chemistry in the late nineteenth century.

12.12   Color and Constitution

As noted in Chapter 9, color in an organic compound is due to *absorption* of light of a certain wavelength. When this band is removed from the transmitted or reflected light, the complementary color is perceived. Absorption in the blue region (400 to 450 nm) imparts a yellow color; absorption at the red end of the visible spectrum (600 to 700 nm) results in a blue or green color.

The structures required for strong absorption in the visible region are highly conjugated systems, with extensive delocalization of π electrons. In order to obtain stable compounds with this feature, chromophoric groups such as —C=C—, —C≡N, —N=N—, and —N=O are combined with aromatic rings. Electron delocalization is increased, and absorption is shifted to a longer wavelength (toward red), by groups such as —NH$_2$ and —OH, which can donate an unshared electron pair to extend the conjugated system. This effect is seen in a comparison of fuchsone imine, which is colorless, and the purple dye called Doebner's violet. The chromophoric system of the dye can be represented as a resonance hybrid, in which one substituted ring has a benzenoid structure and the other has a quinone structure.

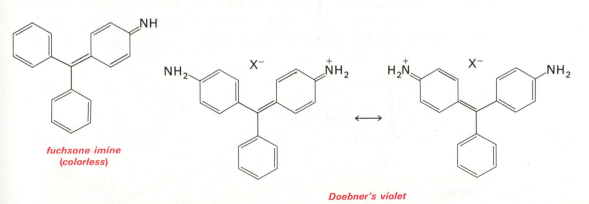

fuchsone imine
(colorless)

Doebner's violet

The extended chromophoric system resulting from the interaction of benzenoid and quinoid rings is responsible for the color of many dyes and indicators. **Indicators** are compounds that undergo a sharp color change during a reversible reaction, such as protonation-deprotonation (acid-base indicators) or oxidation-reduction. One of the most familiar indicators is **phenolphthalein,** which is readily prepared by condensation of phthalic anhydride and phenol. When the pH is raised to between 8.5 and 9.0, a dianion is formed by loss of two protons and opening of the cyclic ester. Delocalization of the negative charge in the benzenoid-quinoid system gives rise to the intense color of the anion.

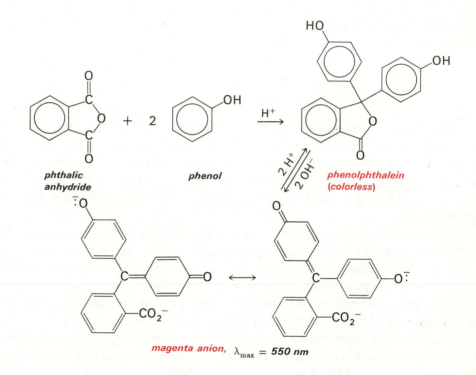

phthalic anhydride   *phenol*   *phenolphthalein (colorless)*

magenta anion, $\lambda_{max} = 550$ *nm*

Another type of benzenoid-quinoid structure is present in the redox indicator **methylene blue.** Because of its reversible color change on mild reduction, the compound is used in biochemical studies of oxidation-reduction processes and also as a histological stain.

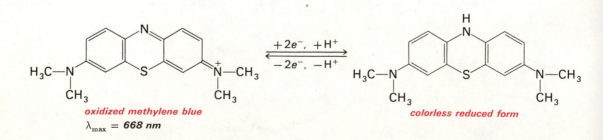

oxidized methylene blue
$\lambda_{max} = 668$ *nm*

colorless reduced form

12.13 Dyes

A **dye** is a substance that imparts permanent color to a fabric. The general features of colored compounds were summarized above, but correlation of the color and structure of dyes is quite complex. Bright primary colors, such as red or blue, are obtained with compounds having narrow absorption bands. Dyes for brown or black have relatively broad absorption throughout most or all of the visible spectrum. In general, dyes in the blue end of the spectrum, absorbing in the longer wavelength (lower energy) red region, have larger molecules and more extended chromophoric systems than dyes for red and yellow shades. The same chromophoric groups in different positions can produce quite different colors, as illustrated by the brown and red azo dyes obtained from α- and β-naphthol. These dyes were developed over 100 years ago and are still in wide use.

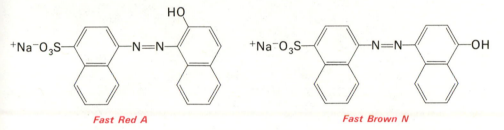

Fast Red A   Fast Brown N

The color of a dyed fabric depends not only on the structure of the dye but also on the method of application to the fiber. In addition to the shade, the depth or intensity of the color, the degree of brilliance, and—very importantly—the fastness (permanence) and reproducibility are properties that must be controlled in the dyeing process. To arrive at the best balance of these factors, a large variety of dyes for every color and type of fiber are required.

The three major types of dye structures are **azo compounds, triphenylmethane** derivatives, and **anthraquinones.** Dyes of most colors can be obtained with all of these structures, and groups such as $-SO_3H$ or $-CO_2H$ can be introduced to adapt the dyes to different fibers. Typical triphenylmethane and anthraquinone dyes for blue shades are Patent Blue A and Celliton Fast Blue B.

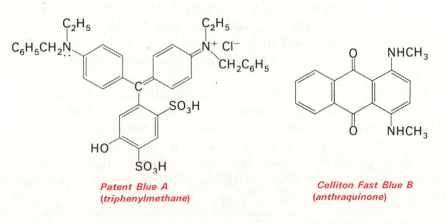

Patent Blue A
(triphenylmethane)

Celliton Fast Blue B
(anthraquinone)

Azo dyes are the largest and most important group. A huge number of combinations, giving a broad range of colors, is available by coupling diazonium salts with phenols, naphthols, and naphthylamines. Compounds with two azo groups ("disazo" dyes) can be obtained from a diaminobiphenyl and two moles of phenol, naphthol, or amine. By controlling the pH of the solution, a disazo dye with two different end groups can be prepared. Coupling to the active position of an amine occurs in acid solution; coupling to a phenol or naphthol is carried out at higher pH. A typical disazo dye is Diamine Fast Red F.

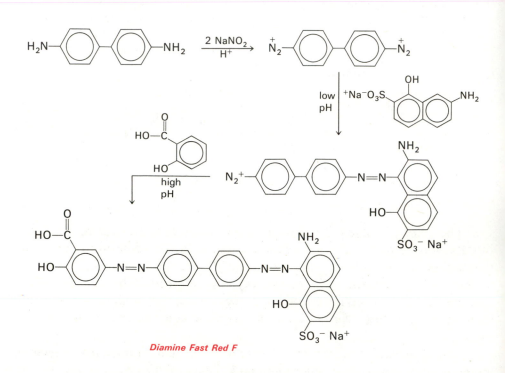

Diamine Fast Red F

To obtain permanently dyed fabric, some kind of interaction between the dye molecule and the fiber is required. The chemical nature of textile fibers ranges from a highly polar protein in wool, which has many basic and acidic groups, to cotton, which contains hydroxyl groups, to polyester and polypropylene, which have no polar or reactive groups. Triphenylmethane dyes, which are cations, are strongly bound to wool, and beautifully dyed fabrics can be obtained simply by immersing the cloth in an aqueous bath of the dye.

In dyeing cotton, a widely used process is formation of an azo dye by carrying out the coupling reaction directly on the fiber. The cloth is first soaked in the phenol or amine; after drying, it is passed through a bath of the diazonium salt. This process is also used in textile printing. Azo dyes with very reactive groups have been developed for use in high quality cottons. With these **fiber reactive** dyes, a covalent bond is formed with a hydroxyl group in the cotton.

Synthetic polyamide and polyester fibers require somewhat different dyeing methods, and the polymer structure can be modified to improve dyeability. Nylon (Section 10.21) can be prepared with either basic $-NH_2$ or acidic $-CO_2H$ end groups in excess; $-SO_3^-Na^+$ groups can be incorporated into a polyester molecule. Nylon with excess $-NH_2$ groups can be dyed with **"acid"** dyes that contain $-SO_3^-Na^+$ groups. Dye molecules are strongly absorbed by formation of ionic $-SO_3^-NH_3^+$ bonds, and the depth of the shade can be controlled by varying the number of amino groups in the polymer. Fibers with acidic groups are dyed with **cationic** dyes similar in structure to the indicator methylene blue.

Synthetic fibers can also be dyed by **disperse** dyes, which are insoluble in water. The dye bath contains a fine suspension of the dye, which becomes dissolved in the polymer during the process. Since there is no strong attractive force between dye and fiber, these dyes are less fast with respect to washing than are acid or cationic dyes. When three or four types of fibers with differing dye characteristics are combined in a fabric, vividly contrasting color effects can be achieved by using several dyes together in a single dye bath; this process is used in dyeing nylon carpet (Color Plate 7).

SUMMARY AND HIGHLIGHTS

1. **Phenols,** ArOH, are much stronger acids than are alcohols because of resonance stabilization of the phenoxide anion, ArO^-.
2. Many phenols act as bactericides.
3. Phenol is commercially produced from chlorobenzene plus NaOH *via* the benzyne intermediate, or by the oxidation of cumene.
4. **Anilines,** $ArNH_2$, are much weaker bases than are alkylamines because resonance interaction between the ring and N is *lost* in the anilinium ion, $ArNH_3^+$.
5. Anilines are prepared by the reduction of nitrobenzenes.
6. Both phenols and anilines are highly reactive toward electrophilic substitution; reactions with aniline are often carried out with the less reactive acetanilide, $C_6H_5NHCOCH_3$, as an intermediate.
7. The phenoxide anion can react as a carbanion and can react with carbonyl compounds at the *o-* and *p-*positions (*e.g.*, in the syntheses of aspirin and Bakelite).
8. Phenols are important **antioxidants** because of their ability to form relatively stable free radicals.
9. Complete oxidation of phenol gives **quinone,** which can be reversibly reduced to **hydroquinone.**
10. **Diazonium salts,** $ArN{=}N^+$, obtained from anilines and HNO_2, are useful in substitution reactions in which nitrogen is replaced: $ArN_2^+ + X^- \longrightarrow ArX + N_2$.
11. **Azo compounds** are obtained by the coupling of diazonium salts with phenols or anilines: $ArN_2^+ + C_6H_5OH \longrightarrow ArN{=}NC_6H_4OH$.

12. Azo compounds and triarylcarbocations with benzenoid-quinoid systems are dyes and indicators.

ADDITIONAL PROBLEMS

12.8 Draw the structures of the following compounds.

a) *p*-chlorophenol
b) *β*-naphthol
c) *o*-nitroacetanilide
d) 2,3-dichlorobenzoquinone
e) picric acid
f) *N*-ethylaniline
g) methyl salicylate
    (oil of wintergreen)
h) azobenzene
i) *m*-toluidine

12.9 Write the predominant form (anion, cation, or uncharged) of the following compounds at the pH indicated.

a) phenol at pH 12
b) phenol at pH 8
c) *p*-nitrophenol at pH 8
d) 2,4-dinitrophenol at pH 3
e) aniline at pH 8
f) benzylamine at pH 8
g) aniline at pH 3
h) *p*-nitroaniline at pH 3

12.10 Write the structures of the products that would be obtained in the following reactions.

a) phenol plus acetic anhydride
b) *p*-cresol plus one mole of bromine
c) 2,5-dihydroxytoluene plus Tollens' reagent (see Section 8.6)
d) acetanilide plus chlorine, $FeCl_3$; then hot H_3O^+
e) *p*-chloroaniline plus $NaNO_2$ and H_2SO_4; then heat

12.11 Although phenoxide anions can readily undergo ring substitution, they behave like alkoxides in most reactions with alkyl halides or sulfates, and give alkyl aryl ethers. This reaction is involved in the synthesis of two important agricultural chemicals, the herbicide 2,4-D and methoxychlor, which is a nonpersistent analog of DDT. Outline the steps in the preparation of these compounds from phenol and other starting materials.

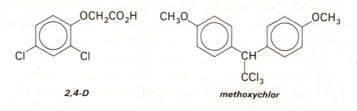

2,4-D   *methoxychlor*

12.12 Show how the following could be prepared from the compound indicated and other starting materials. More than one step will be needed; show all steps and reagents.

a) *p*-bromoaniline from benzene (two routes are possible; give both and evaluate their relative merits)
b) *o*-chlorobenzonitrile from *o*-chloronitrobenzene
c) 3-bromo-4-methylacetanilide from toluene
d) 2-bromo-4-methylacetanilide from toluene
e) *o*-bromotoluene from 3-bromo-4-methylaniline

12.13 *o*-Aminobenzoic acid is treated with a source of NO^+ ion in the absence of a proton-containing solvent, and in the presence of cyclopentadiene; nitrogen, CO_2, and the reaction product shown below are obtained. Suggest an intermediate in this process, and show how it would be formed and how it could react.

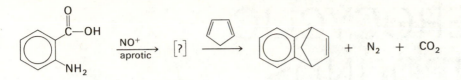

12.14 Write out the steps in detail for the preparation of phenolphthalein from phenol and phthalic anhydride. The first step, after protonation of the anhydride, is electrophilic attack at the *para* position of a molecule of phenol.

12.15 Write structures of the diazonium ions and coupling components that would be combined to give the following dyes:

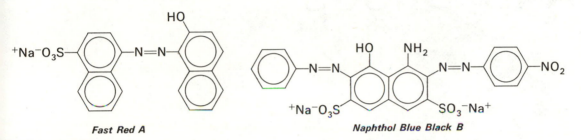

Fast Red A　　　　　　　　　　　　　　　**Naphthol Blue Black B**

12.16 Anthraquinone dyes can be used in the vat dyeing process, in which the quinone is reduced in alkaline solution to the dihydro compound. The latter is applied to the cloth and then reoxidized in air to the quinone. What is the structure of the dihydro derivative that would be obtained by adding two electrons to anthraquinone? (The dihydro compound is present in the solution as the dianion.)

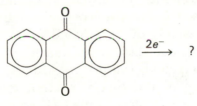

anthraquinone

13
HETEROCYCLIC COMPOUNDS, ALKALOIDS, AND DRUGS

HETEROCYCLIC COMPOUNDS

The term **heterocycle** means a cyclic structure with one or more atoms other than carbon in the ring. Heterocyclic compounds include simple cyclic ethers (Section 6.11) and amines, and also *aromatic rings* containing nitrogen, oxygen, or sulfur atoms. The chemistry of the first group differs little from that of open chain compounds, but the second group, called **heteroaromatic compounds,** forms a large and distinctive branch of organic chemistry.

Heteroaromatic compounds can be grouped into two major classes. In one group the structure of the ring is that of cyclopentadiene with a heteroatom such as nitrogen in place of the CH_2 group. An unshared pair of

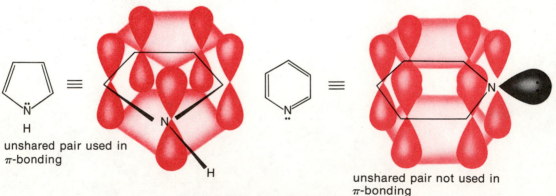

unshared pair used in
π-bonding

aromatic π electron systems
in heterocyclic compounds

unshared pair not used in
π-bonding

electrons from the heteroatom makes up the "aromatic sextet" of π electrons. In the second type of heteroaromatic ring, a nitrogen atom replaces a —CH= unit in a benzene ring. The unshared electron pair is not required to complete the aromatic π electron system, and these electrons can be used for bonding.

13.1 Five-Membered Rings

The three parent compounds are **pyrrole, furan,** and **thiophene.** These aromatic rings contain six π electrons (one from each carbon and two from

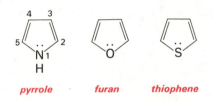

pyrrole   *furan*   *thiophene*

the heteroatom) distributed over five atoms. As in benzene, the π bonding is delocalized; but, as seen in the resonance structures for pyrrole, dipolar forms contribute to the hybrid. Since the nitrogen electron pair is delocalized into the ring, pyrrole is a much weaker base than a typical secondary amine

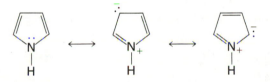

would be. On the other hand, the ring is electron-rich and extremely reactive toward electrophilic attack at carbon.

Problem 13.1   Write the structures of a) 1,3-dimethylpyrrole, b) 2-furoic acid (analogous to benzoic acid), and c) 3-aminothiophene.

Problem 13.2   How many isomers are there of a) dichlorothiophene, b) methylnitrofuran?

Pyrroles.   Owing to the high electron density in the ring, pyrrole undergoes electrophilic substitution with mild reagents. A characteristic reaction is the condensation of pyrrole and a formic acid derivative. The product is a stable cation in which the conjugated system extends over both rings.

Problem 13.3   Another useful reaction of pyrrole is condensation with formaldehyde to give a product with the molecular formula $C_9H_{10}N_2$. Oxidation of this substance gives the same compound obtained in the condensation of pyrrole with a formic acid derivative. Suggest the structure of the formaldehyde product.

A condensation product of this type involving four pyrrole rings is the basis for the important pigments **heme** and **chlorophyll,** which play the central roles in respiration and plant photosynthesis, respectively. The ring system of heme is called a **porphyrin.** The flat, conjugated double bond system holds the four nitrogens in the ideal geometry for complexing the metal atom. Oxygen transport in respiration occurs by coordination of O_2 at the iron atom in heme. In carbon monoxide poisoning, a more stable coordination complex with the iron is formed by CO molecules, and oxygen transport is blocked.

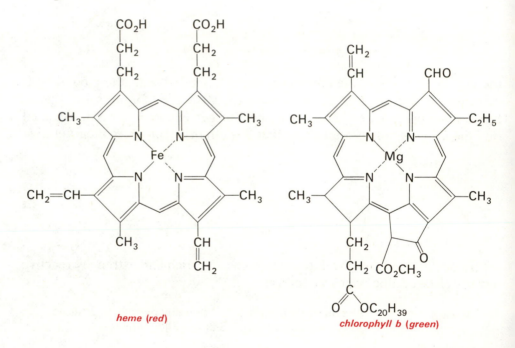

heme (*red*)   *chlorophyll b* (*green*)

Furans.   Like pyrrole, furan is very reactive toward acids and other electrophiles. The furan ring is present in a number of naturally occurring compounds and is commercially obtained as the 2-aldehyde, furfural, from corn cobs and oat hulls. The acid hydrolysis of the polysaccharides in these agricultural wastes yields large amounts of pentoses, which in turn are cyclized and dehydrated to furfural. Large quantities of furfural are used in

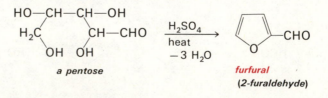

a pentose   *furfural*
          (*2-furaldehyde*)

the manufacture of furfural-phenol plastics (Durite), varnishes, insecticides, fungicides, and germicides.

Thiophenes.   Thiophene is less reactive toward electrophiles than pyrrole or furan, and closely resembles benzene in properties. Thiophene rings are present among the sulfur compounds of crude oil residual fractions, and they present one of the main difficulties in desulfurization of these fractions. Because the boiling point of thiophene (87°) is quite close to that of benzene (80°), thiophene is always an impurity in coal-tar–derived benzene.

> **Problem 13.4**   In order to remove thiophene, crude benzene can be stirred with cold sulfuric acid; after separation of the sulfuric acid layer, the benzene is washed with water, dried, and distilled. Explain the rationale for this method, *i.e.*, why it works, and how the operation removes the thiophene.

Five-membered Rings with Two or More Heteroatoms.   About fifteen aromatic ring systems with combinations of two to four nitrogen, oxygen, and sulfur atoms are known. Several of these rings are incorporated in drugs; imidazole and thiazole are found in naturally occurring amino acids and vitamins. These ring systems can be written by replacing one or more of the —CH= groups in pyrrole, furan, or thiophene by nitrogen. The names are derived by designating oxygen as **ox-**, sulfur as **thi-**, and nitrogen as **az-**, with the ending **-ole.**

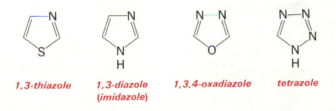

1,3-thiazole	*1,3-diazole*	*1,3,4-oxadiazole*	*tetrazole*
	(imidazole)		

All these rings are stable aromatic systems. Oxazoles and diazoles are less reactive toward electrophiles than are the parent furan and pyrrole, because the additional electronegative nitrogen reduces the electron availability. The main interest in these ring systems is the preparation and utilization of derivatives for drugs and agricultural chemicals.

The synthesis of rings containing several heteroatoms is a straightforward extension of the chemistry seen in earlier chapters. When the proper combination of functional groups is assembled, formation of the aromatic ring provides a driving force for cyclization. A simple example is the preparation of the herbicide 3-amino-1,2,4-triazole (amitrole).

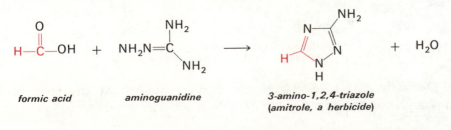

formic acid	*aminoguanidine*	*3-amino-1,2,4-triazole*
		(amitrole, a herbicide)

Problem 13.5   Write structures and give names for four different five-membered aromatic rings containing one sulfur and two nitrogen atoms in the ring.

13.2   Six-membered Rings

The most important six-membered heterocyclic ring is **pyridine.** Pyridine is an example of an aromatic heterocycle in which the lone pair of electrons on the heteroatom is *not* involved in the π bonding. Thus pyridine, unlike pyrrole, is basic and forms salts with acids and alkylating agents without destroying the aromatic character of the ring.

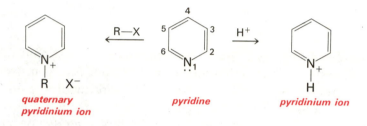

quaternary
pyridinium ion   *pyridine*   *pyridinium ion*

The electronegative heteroatom in pyridine strongly *deactivates* the ring toward electrophilic substitution. However, the pyridinium ion is quite susceptible to attack by *nucleophiles* at the 2- and 4-positions. Two very important biochemical processes (see Chapter 15) depend upon the ability of pyridinium compounds to react in this way.

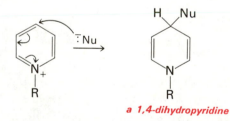

a 1,4-dihydropyridine

Three **diazines** (the ending **-ine** denotes the six-membered nitrogen-containing ring) are possible. Of these, **pyrimidine** (1,3-diazine) is by far the most important. Pyrimidines are an integral part of the nucleotides that make up the "genetic material" of cells (Chapter 16), and the formation of pyrimidine rings was a key event in the beginning of life on this planet.

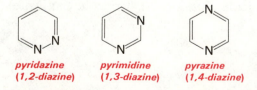

pyridazine   *pyrimidine*   *pyrazine*
(1,2-diazine)   *(1,3-diazine)*   *(1,4-diazine)*

Derivatives of **1,3,5-triazine** are readily obtained by trimerization of compounds containing a C=N or C≡N bond. The trimerization of cyanogen chloride gives 2,4,6-trichlorotriazine, in which the chlorine atoms are susceptible to nucleophilic attack. Displacement of two chlorines by amino groups occurs very readily to give 2-chloro-4,6-diaminotriazines, which are important herbicides.

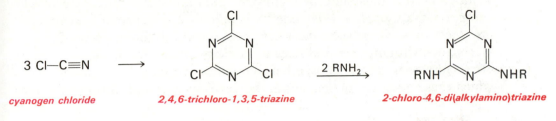

3 Cl—C≡N ⟶

cyanogen chloride *2,4,6-trichloro-1,3,5-triazine* *2-chloro-4,6-di(alkylamino)triazine*

13.3 Condensed Ring Heterocycles

Heterocyclic units can be fused together, or fused to benzene, to give a wide variety of ring systems, a number of which occur in nature. **Indole** and **quinoline** illustrate fusion to a benzene ring. The indole ring system is present in one of the amino acids in proteins and in some important neurohormones. One of the most famous (or infamous) indoles is a complex molecule called lysergic acid diethyl amide or LSD.

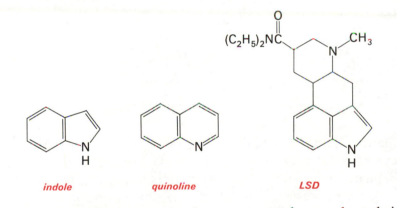

indole *quinoline* *LSD*

Purine and **pteridine** are somewhat more complex condensed ring systems. Compounds with the purine ring are complementary to simple pyrimidines in nucleic acids; pteridines are found in such diverse materials as coenzymes and wing-pigments of butterflies.

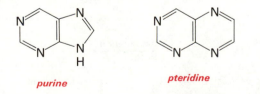

purine *pteridine*

Problem 13.6   Indole and quinoline have the same π electron systems as naphthalene (Section 4.6). Predict the relative reactivities of these three compounds in electrophilic substitution and explain the reason for your prediction.

13.4 Herbicides and Plant Growth-Regulating Agents

A major contribution of organic chemistry in the past 20 years has been the development of compounds that control or inhibit plant growth. A number of the most effective growth-regulating agents are heterocyclic compounds, although a wide variety of aliphatic and benzenoid compounds also have uses. These substances can exert several different types of action. Herbicides, with more-or-less selective toxicity for certain plants, are used to eliminate weeds in field crops; other compounds are growth-regulating agents which cause a variety of changes in the development and fruiting of plants.

Herbicides. Weed control is a major factor in increasing crop yields and reducing labor in food production. An effective herbicide must have a selective action; in addition, it must have low toxicity to animals and a limited lifetime so that residues do not accumulate. Chlorinated phenoxyacetic acids, *e.g.*, **2,4-D,** kill nearly all broad-leaved plants and are the most widely used weed control agents. Another type of herbicide has a dipyridinium structure such as **paraquat.** The quaternary pyridinium compounds are toxic to most plants, but since they are ionic, they are rapidly absorbed and then deactivated on clay soils. They can therefore be used as "pre-emergence" herbicides to kill weed plants which appear before crop seedlings.

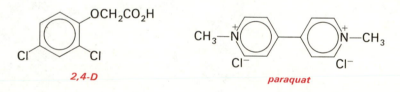

2,4-D   *paraquat*

Highly selective herbicidal action is illustrated by chlorodiaminotriazines such as **atrazine.** These compounds, applied at a level of 3 lb per acre, eliminate weeds by blocking a key reaction in photosynthesis. Corn and sorghum are quite resistant to the effect of the chlorotriazine, because the compound is transformed in the root system of these species to the nontoxic diaminotriazinone. The effect of one application of atrazine can be seen in Plate 8*A* and *B*, which shows a plot of corn treated with herbicide and an untreated control plot.

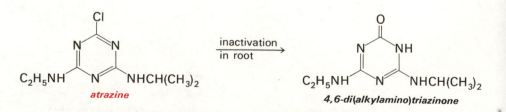

atrazine   *4,6-di(alkylamino)triazinone*

Growth-Regulating Substances. Vascular plants have elaborate chemical mechanisms which control the development and maturation of various tissues and reproductive organs. Four types of compounds are recognized as natural plant hormones or growth regulators. Three of these, **auxins, gibberellins,** and **cytokinins,** have growth-promoting properties; the fourth group are growth inhibitors. Each group contains several closely related compounds; one representative of each type is shown here:

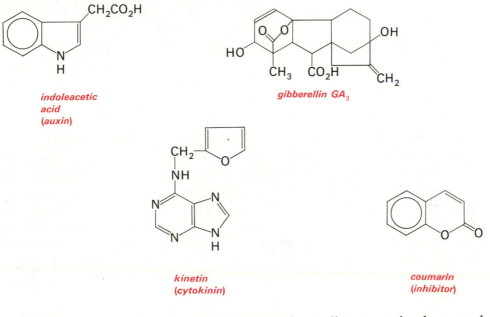

indoleacetic
acid
(auxin)

gibberellin GA₃

kinetin
(cytokinin)

coumarin
(inhibitor)

These compounds are present in extremely small amounts in plants, and the concentrations vary in different parts of the plant and at different stages of development. The effects of the different types of growth regulators can be seen by treating plants with the natural compounds or with synthetic compounds that have analogous structures and activity. Auxins promote root and bud growth, but at higher dosage they can cause disorganized growth and eventually kill the plant. Application of gibberellins causes enormously elongated cells, and the normal height of a plant can be doubled. Cytokinins are required in cell division and differentiation.

The mechanisms by which these compounds affect the plant interact and overlap, so that various overall responses can result from their application to crop plants and orchards. The number of blossoms that set fruit can be increased or reduced, and the time of dropping of fruit can be controlled for optimum harvesting. Bushier and stronger plants are produced by compounds that inhibit stem growth; higher yields of grain can be achieved by compounds that promote germination.

The relationship of chemical structure and plant growth activity is not well defined or easily predictable. The discovery of useful growth-regulating compounds is largely an empirical, trial-and-error process in which many compounds are synthesized and screened. Auxin activity is found in many aromatic acids; the herbicidal activity of 2,4-D is due to its effect as an auxin. Naphthaleneacetic acid is another synthetic auxin; it is used to control the time of fruit drop. A growth inhibitor that finds wide use is the heterocyclic compound maleic hydrazide. This compound can be used as a "chemical lawnmower" to control the height of grasses without harming the plant, as shown in Plate 8C. Another compound with similar inhibitory activity is morphactin, whose structure bears a vague resemblance to the gibberellins.

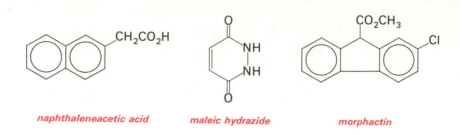

naphthaleneacetic acid   maleic hydrazide   morphactin

A further aspect of food production is the maturing and ripening of crops, particularly if the growing season is limited or if produce must be shipped long distances. The compound that is primarily responsible for ripening is the simple alkene ethylene, which is generated by plants during the ripening process. The rate of ripening of green bananas can be closely controlled by regulating the concentration of ethylene in a warehouse or the hold of a ship. The role of ethylene is more complicated than terminal ripening, since compounds such as 2-chloroethanephosphonic acid, which slowly liberates ethylene, can bring about significant increases in the yield of certain crops.

$$ClCH_2CH_2PO_3H_2 \longrightarrow CH_2{=}CH_2 + HCl + HPO_3$$

2-chloroethane-
phosphonic acid

ALKALOIDS

Alkaloids are basic heterocylic compounds present in many higher plants. Several alkaloids are among the longest known and most famous organic compounds: morphine, strychnine, and quinine were isolated in pure crystalline form in the early 1800's. The early and long-sustained interest in alkaloids was due to their pharmacological properties and importance as drugs. Cinchona bark, the source of quinine, was brought to Europe from Peru in the 16th century by Spanish explorers, and until 1940, quinine was the only compound available for treatment of malaria. **Morphine** and **cocaine** were the major analgesic (pain-relieving) drugs of the 19th century. **Strychnine, nicotine,** and the **curare** bases, used as arrow poisons, are extremely toxic compounds.

Over a thousand alkaloids have been isolated from plants in all parts of the world. Plant extracts frequently contain up to a dozen closely related alkaloids, and extensive fractionation by chromatography is required to obtain individual compounds. Determination of the structures of alkaloids has been a challenge to organic chemists since the isolation of morphine and strychnine in 1818. These compounds contain only 15 and 19 carbons, respectively, but they are two of the most complex organic structures known, and more than 100 years were required to unravel the molecular architecture. As the structures of a number of alkaloids were established, structural patterns emerged which led to the suggestion that alkaloids are derived in the plant from certain amino acids found in proteins (Chapter 14).

The origin of most alkaloids from a few amino acids has been verified experimentally, and this is the most useful basis for classifying alkaloids. To study the biogenetic origin of alkaloids or other plant substances, a compound suspected of being a precursor in the biosynthesis is prepared with an **isotopically labeled atom** at a certain position. The most widely used label is ^{14}C, a radioisotope of carbon with a long half-life. A solution of the labeled precursor is then "fed" to the plant by a wick drawn through the stem, or by growing a seedling plant hydroponically in the solution. After the feeding period, the alkaloid is isolated; if it is radioactive, the molecule is degraded, atom by atom, to locate the position of the isotopic carbon.

13.5   Alkaloids Derived from Tyrosine

Several important groups of alkaloids are derived from the phenolic amino acids tyrosine and dihydroxyphenylalanine. A typical incorporation experiment is illustrated in Figure 13.1. After feeding ^{14}C-tyrosine to seedlings of the **opium poppy,** *Papaver somniferum,* the alkaloid reticuline can be isolated and is found to contain ^{14}C at two positions. The reaction leading to the alkaloid can be represented as condensation of one molecule of an amine and one of an aldehyde. Both of these intermediates are formed from the amino acid by well-known enzymatic reactions (Chapter 14). Introduction of O- or N-methyl groups is a very common further step in alkaloid biosynthesis.

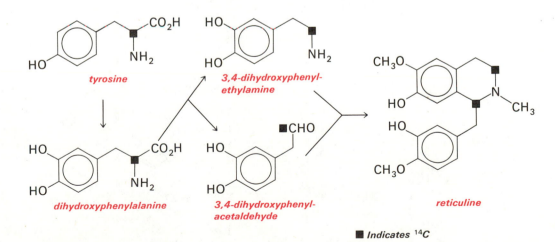

Indicates ^{14}C

FIGURE 13.1

Alkaloid biosynthesis in *Papaver somniferum.*

Condensation of the aldehyde and amine, as shown in Figure 13.1, is an example of a useful general reaction, called the Mannich condensation, which is seen in the pathways leading to a number of alkaloids. The reaction involves the stepwise attack of an amine and a carbon nucleophile on an aldehyde carbonyl group. Primary or secondary amines can be used; the nucleophile can be a phenolic ring or any C—H group adjacent to a carbonyl (carbanion, enol, enamine, and so on).

$$R'NH_2 \ + \ R''\overset{\overset{\displaystyle O}{\|}}{C}H \ \xrightarrow{H^+} \ R'NH{=}CH \ \xrightarrow{CH_3\overset{\overset{\displaystyle O}{\|}}{C}R'''} \ R'N\overset{\overset{\displaystyle R''}{|}}{H}CHCH_2\overset{\overset{\displaystyle O}{\|}}{C}R'''$$

Reticuline (Figure 13.1) is one of the first alkaloids in the biosynthetic pathway leading from tyrosine. A further step that occurs in the plant with this and other benzylisoquinolines is a free radical coupling reaction in which the two benzene rings are linked at positions *ortho* or *para* to a free phenolic OH group. Reactions of this type can be carried out *in vitro** with oxidizing agents that produce phenolic radicals. The alkaloids formed in certain plants by coupling at the *ortho* position are called **aporphines.**

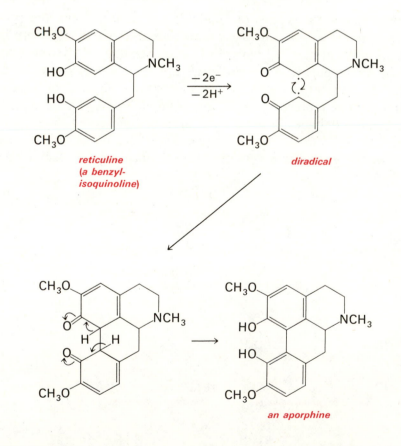

reticuline
(a benzyl-
isoquinoline)

diradical

an aporphine

In vitro means in glass apparatus, as opposed to *in vivo,* which refers to a reaction in a living cell.

In *P. somniferum*, the oxidative coupling of reticuline occurs at the *para* position of one ring; the enzyme apparently holds the precursor in a conformation with the benzyl group bent over the isoquinoline ring. The result of this coupling is a molecule in which one phenolic ring becomes fixed as a nonaromatic dienone. Addition of the remaining phenolic OH group to the dienone and subsequent steps lead to the alkaloids codeine and morphine (Figure 13.2). The latter makes up about 10 per cent of the weight of crude opium.

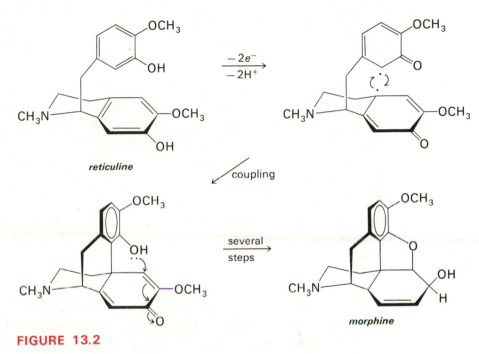

FIGURE 13.2

Biosynthesis of morphine.

13.6   Alkaloids Derived from Other Amino Acids

Tryptophane.   About 600 complex alkaloids arise from the indole amino acid tryptophane. Condensation of the amino acid with a 10-carbon aldehyde is followed by a series of cyclization steps that lead to structures with three or four additional rings. Among the indole alkaloids are **reserpine,** one of the first tranquilizing drugs, the very toxic compound **strychnine,** and the **curare** alkaloids.

The **curare alkaloids** are extremely complex mixtures of quaternary ammonium compounds, in which a number of strychnos bases are condensed into symmetrical or unsymmetrical dimers. These mixtures are prepared by tribes in the upper Amazon and Orinoco jungles of South America by grinding roots and bark with resin. Applied to darts or arrow tips, the alkaloids cause rapid, fatal paralysis. Clay pots, gourds, and bamboo tubes packed with crude alkaloids were brought to Europe by anthropologists, and

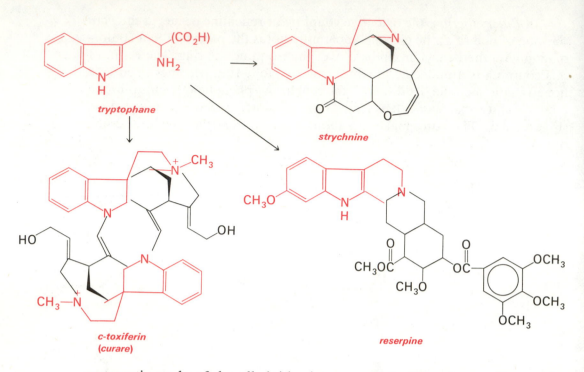

tryptophane

strychnine

*c-toxiferin
(curare)*

reserpine

systematic study of the alkaloid mixtures, which differed according to the container, led to the modern curarizing agents, which are a valuable adjunct in surgery.

Aliphatic Amino Acids. Some important nonaromatic alkaloids are derived from the diamino acid ornithine, as illustrated below. The diamine putrescine (so named because it is one of the compounds responsible for the odor of decaying meat) is the immediate precursor. One of the amino groups is oxidized to an aldehyde which then undergoes cyclization. Condensations with another molecule of diamine or with other components of the plant cell lead to alkaloids containing the five-membered ring in a variety of forms.

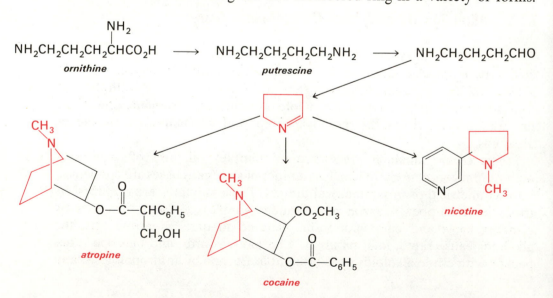

$$NH_2CH_2CH_2CH_2CHCO_2H \longrightarrow NH_2CH_2CH_2CH_2CH_2NH_2 \longrightarrow NH_2CH_2CH_2CH_2CHO$$

ornithine

putrescine

atropine

cocaine

nicotine

Cocaine and **atropine,** despite their similar structures, occur in different plant families and have very different effects as drugs. Cocaine was the first effective local anesthetic known, and it is also an addicting drug. Atropine is one of the belladonna alkaloids, and was one of the more popular poisons in earlier centuries; it is still a source of occasional poisoning from ingestion of berries of the deadly nightshade (*Atropa belladonna*) or jimsonweed (*Datura stramonium*). Atropine is a valuable drug for pupil dilation in ophthalmology (the name belladonna stems from the use of the alkaloid by women in the seventeenth century to dilate their pupils for cosmetic effect).

DRUGS

The use of powdered herbs and roots as medicinal agents was described in Chinese writings of 2700 B.C., which record the stimulatory effects of the plant *ma huang* that are now known to be due to the presence of ephedrine. With the exception of a few inorganic compounds, such as mercury and antimony salts, alkaloid-containing plants were the sole source of effective drugs until the introduction of salicylic acid and acetanilide as mild pain- and fever-relieving agents in the 1880's. From this time, **pharmacology,** the study of drugs in the body, and **medicinal chemistry,** the design and synthesis of organic compounds as drugs, have become major and closely intertwined disciplines.

Classification of drugs is complicated since some compounds have several pharmacological actions, or have different actions at different dose levels. On the other hand, certain clinical effects, such as relief of pain, may be accomplished by compounds of unlike structure acting by different mechanisms. Finally, compounds of similar overall structure may have contrasting pharmacological properties because one passes more readily through a cell membrane, or is more rapidly excreted from the body.

These factors are strikingly illustrated by the properties of a few common drugs. **Aspirin,** or acetylsalicylic acid, is consumed at a rate of about 40 million pounds per year. Aspirin is an **antipyretic,** lowering body temperature in fever; but in large overdoses due to accidental ingestion by children, the body temperature becomes dangerously elevated. Aspirin is a mild analgesic, raising the pain threshold, and is also very valuable in relieving **inflammation** in rheumatic fever. These effects are not brought about by the same mechanism, however, and are displayed to different extents by various derivatives of salicylic acid.

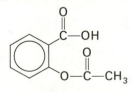

acetylsalicylic acid
(aspirin)

acetaminophen (R = H)
phenacetin (R = C₂H₅)

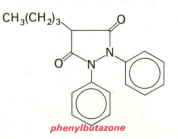

phenylbutazone

Another simple aromatic compound, *p*-hydroxyacetanilide (acetaminophen), and the ethyl ether (phenacetin) have antipyretic and analgesic properties similar to those of aspirin, and are also widely used drugs. Both compounds are transformed in the body to a sugar derivative of a phenolic compound.

Neither acetaminophen nor phenacetin has anti-inflammatory properties, but another chemically related compound, phenylbutazone, whose structure can be thought of as two acetanilide units fused together, is a very potent anti-inflammatory substance. Phenylbutazone is useful in severe arthritis, but it has only weak activity for general pain relief.

There are several dozen well-defined types of pharmacological action and drugs to bring them about. These include drugs to combat specific diseases and also to regulate or restore normal physiological function. A few drugs are natural hormones or alkaloids from plants, but by far the bulk of them are synthetic heterocyclic compounds which have either been designed, or accidentally found, to have useful properties as drugs. A great many drugs fall into one of two broad categories—chemotherapeutic agents and compounds acting on the central nervous system.

13.7   Chemotherapy

A **chemotherapeutic agent** is a substance that inhibits or destroys an infectious organism, such as pathogenic bacteria or parasites, or a cancerous growth, which has invaded a host. To accomplish this action, a drug must reach the site of the organism and cause selective damage, usually by interfering with a specific metabolic process in the cell of the invader. To obtain an adequate level of the drug in blood or plasma, its degradation must not be excessively rapid.

Sulfa Drugs.   Chemotherapy had its beginning in the observation that many bacterial cells have a high affinity for synthetic dyes, and certain azo dyes were shown to be effective drugs in treating parasitic diseases. In 1930 it was found that the dye prontosil, prepared from diazotized sulfanilamide (*p*-aminobenzenesulfonamide), prevented streptococcal infections in mice. It was then shown that the antibacterial activity was about the same with a variety of dyes containing the —SO_2NH_2 group. Finally, **sulfanilamide** itself was found to be the crucial part of the drug.

Over 10,000 compounds containing the sulfanilamide unit have been prepared and tested for antibacterial activity. These efforts were directed

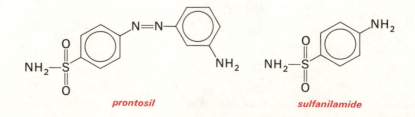

prontosil   *sulfanilamide*

toward finding compounds with optimum duration of action and antibacterial spectrum, and minimum side effects. The most useful derivatives contain heterocyclic substituents, such as pyrimidine and 1,2-oxazole rings.

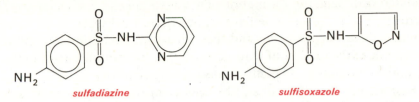

sulfadiazine   *sulfisoxazole*

The antibacterial properties of sulfanilamides are due to their action as **antimetabolites** in the bacterial cell. An antimetabolite is a compound that replaces a natural metabolite in a biochemical reaction, blocking the process at a certain step. Sulfanilamides are antimetabolites of *p*-aminobenzoic acid in the bacterial synthesis of an essential growth factor, folic acid. Sulfa drugs have been eclipsed to a considerable extent by antibiotics, but they remain powerful chemotherapeutic agents for certain organisms, particularly in urinary tract infections.

Antibiotics are a chemically diverse group of compounds that are produced by microorganisms and are toxic to *other* organisms, particularly

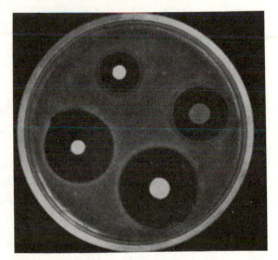

FIGURE 13.3

Agar plate antibiotic assay—disk method. The surface of the agar is inoculated with the organism to be tested. Paper disks of the antibiotic agent to be tested are then placed on the agar surface. The agent diffuses into the agar and prevents growth of the bacteria in a zone around the disk. The width of the zone indicates the sensitivity of the organism to the antibiotic agent. (Courtesy of Dr. R. W. Fairbrother and Dr. A. Rao, Department of Clinical Pathology, Manchester Royal Infirmary, Manchester, England. In J. Clin. Path., Vol. 7.)

bacteria. This antibiotic activity permits the detection and isolation of the compounds. Molds and actinomycetes from soil or other sources are cultured, and samples of the filtrate are then applied to nutrient plates that are inoculated with bacteria. Formation of a zone that is free of bacteria indicates the presence of a substance lethal to that organism (Figure 13.3). Over 1200 antibiotics have been isolated and their structures determined. About 60 of these compounds are manufactured for use in human and veterinary medicine, and in treatment of plant diseases in agriculture.

Historically the first and still by any criterion the most important antibiotic is **penicillin.** Several related penicillins differ in the group R in the following formula. Benzylpenicillin is the major drug, and since it became available in 1945 it has saved thousands of human lives from otherwise fatal bacillary and coccal infections. Only one other antibiotic, **cephalosporin,** bears any chemical resemblance to penicillin. Both compounds contain the strained four-membered cyclic amide (β-lactam) unit, which is quite sensitive to ring opening.

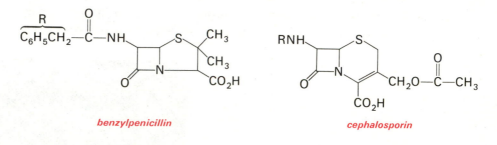

benzylpenicillin   *cephalosporin*

Many of the useful antibiotics belong to one of a few general chemical groups. These include cyclic esters, called macrolides, with rings of 12 to 16 atoms, and tetracyclines, with four fused rings. Several compounds of each type are widely used in medicine. Closely related to the tetracyclines are the anthracyclines such as adriamycin, which are promising drugs against solid tumors.

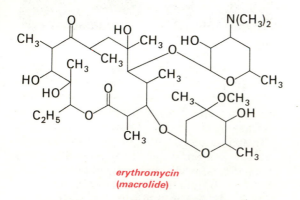

erythromycin
(macrolide)

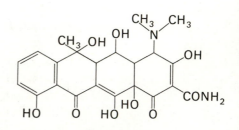

terramycin
(tetracycline)

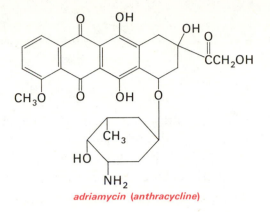

adriamycin (anthracycline)

A few common antibiotics follow no structural pattern. They contain un-usual structural features, such as chloro and nitro groups and triple bonds. Some of these compounds are relatively simple structures from the stand-point of chemical synthesis, and are produced commercially by synthesis rather than fermentation. Chloramphenicol is one of the few antibiotics that is effective against typhoid, typhus, and rickettsial infections. Pyrrolnitrin is an antifungal agent; cellocidin is used in large amounts to control plant pathogens.

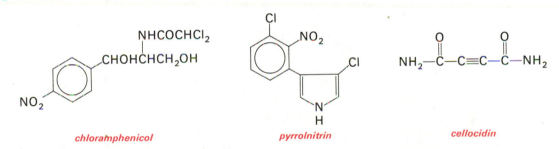

chloramphenicol   *pyrrolnitrin*   *cellocidin*

Cancer Chemotherapy. Great effort has been made to develop effective drugs to combat cancer, but success has been limited. Cancer is the abnormal and uncontrolled multiplication of cells, and the basic rationale of cancer chemotherapy is administration of a drug that is highly toxic to proliferating cells. The two types of drug that have been most widely explored are alkylating agents and antimetabolites.

Alkylating agents, as the name indicates, are highly reactive substrates for nucleophilic substitution at a saturated carbon. The potential of alkyla-ting agents in cancer chemotherapy was first recognized from the effect of mustard gas poisoning in World War I. Exposure to mustard gas, $ClCH_2CH_2SCH_2CH_2Cl$, caused severe damage to bone marrow and lymph nodes, both rapidly growing tissues. Similar effects are observed with the somewhat less toxic **nitrogen mustards,** such as di(2-chloroethyl)-methylamine. These compounds cause relatively greater damage to cancer-ous tissue than to normal tissue and bring about remission of the disease, but a complete cure is rare.

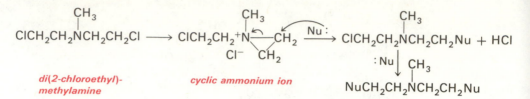

di(2-chloroethyl)-
methylamine   cyclic ammonium ion

Alkylation with nitrogen mustards occurs by formation of the cyclic quaternary ammonium ion. The combination of ring strain and positive charge makes this intermediate highly susceptible to nucleophilic attack by an —OH, >NH, or —SH group. Since reaction occurs with both chloroalkyl groups, alkylation can tie together two nucleotide chains and disrupt the function of the cell.

Two of the antimetabolites that are used in cancer chemotherapy are structural analogs of pyrimidine and purine bases in nucleic acids. Thymine, a pyrimidine in DNA (Chapter 16), is formed in the cell by methylation of uracil. The antimetabolite is 5-fluorouracil. The effective size of fluorine is not much greater than that of hydrogen, and 5-fluorouracil can fit on an enzyme in place of uracil. The fluorine atom blocks methylation to thymine, and the cell is destroyed. 5-Fluorouracil has been used in the form of an ointment for treatment of skin cancer.

uracil   thymine   5-fluorouracil   6-mercaptopurine

Another antimetabolite is 6-mercaptopurine, in which the —SH group replaces—NH$_2$ in the important purine adenine. The usefulness of these antimetabolites in cancer chemotherapy, as with the nitrogen mustards, is severely limited by their *general* toxicity. The ability of these drugs to bring about temporary remissions is encouraging, however, since it represents a small measure of success in the rational design of drugs based on chemical principles.

13.8   Central Nervous System (CNS) Drugs

Drugs acting on the CNS include anesthetics, analgesics, sedatives, antipsychotic agents, antianxiety agents, antidepressants, and stimulants. This lengthy list reflects the vastly complex chemistry of the brain and nervous system. Several of these activities represent overall effects rather than specific pharmacological actions. Many drugs have one primary effect, but can be placed in more than one group.

Anesthetics such as ether or halothane cause depression of consciousness, but they also produce muscle relaxation and analgesia (insensitivity to pain). **Analgesics** specifically prevent the sensation of pain without producing loss of consciousness.

Sedative is a broad term for drugs that produce various types and levels of depression of the CNS. Sedative action ranges from a hypnotic (sleep-producing) effect to mild tranquilization. The most widely used sedatives are the barbiturates. Next to aspirin and phenacetin, barbiturates are the drugs consumed in largest volume, and they are also responsible for the largest number of cases of fatal drug poisoning.

Barbituric acid is pyrimidine-2,4,6-trione. Barbiturate drugs are 5,5-disubstituted derivatives, easily prepared by condensation of a disubstituted malonic ester with urea. The compounds are weak acids (pK_A 8) because of the stabilization of the imide anion; either the free acid or the sodium salt (both called barbiturates) can be administered as a sedative drug.

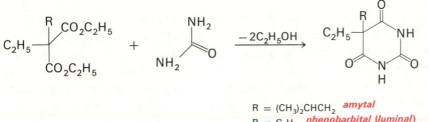

$R = (CH_3)_2CHCH_2$ *amytal*
$R = C_6H_5$ *phenobarbital (luminal)*
$R = CH_3(CH_2)_2CHCH_3$ *pentobarbital (nembutal)*

The potency and rapidity of action of barbiturates increase with increasing length of the alkyl group. A range of pharmacological activity can be achieved by structural variations. Replacement of the 2-oxo group in pentobarbital by sulfur gives the ultra-short acting pentothal, which can be used for anesthesia.

Cyclic imides, such as 3-ethyl-3-phenylpiperidinedione, with one of the CONH groups replaced by —CH_2CH_2—, have sedative properties similar to those of barbiturates, and are useful in cases of sensitivity to the latter. The phthalimido derivative also has satisfactory sedative properties. However, this compound, called **thalidomide,** has an amino substituent instead of two carbon substituents and is therefore broken down in the body in a different way. This fact led to a horrible disaster in Europe and Great Britain, where use of thalidomide by pregnant women caused the birth of hundreds of tragically deformed infants.

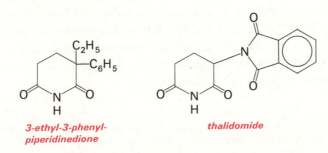

3-ethyl-3-phenyl-
piperidinedione

thalidomide

Tranquilizers and Antianxiety Agents. Drugs in this group have several pharmacological actions. Some are sedatives, but without strong central depressant activity. The tranquilizing properties are of value in a number of anxiety conditions, including those induced by withdrawal from chronic alcoholism. Benzodiazepines, which are fused seven-membered heterocyclic compounds, are among the most effective drugs of this type.

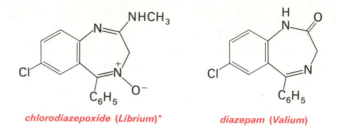

chlorodiazepoxide (Librium)*   diazepam (Valium)

Antipsychotic Drugs. Another group of tranquilizing drugs with a different spectrum of effects in the CNS is useful in treating psychiatric disorders. Foremost among these are **phenothiazine** derivatives. As in the case of barbiturates, a number of phenothiazines have been developed. The drug properties are similar, but small differences in structure make certain compounds more suitable than others for a specific situation. A negative substituent at the 2 position, usually —Cl or —CF$_3$, and a basic side chain at the 10 position are essential for antipsychotic activity. These drugs have been of immense value in relieving psychoses and have permitted many mental patients to return to society.

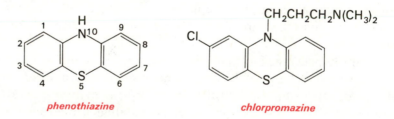

phenothiazine   chlorpromazine

Antidepressants and Stimulants. In an attempt to find structural analogs of phenothiazine with similar properties, compounds with the sulfur atom replaced by a —CH=CH— unit were evaluated. Instead of quieting agitated individuals, these substances, with the dibenzazepine nucleus, were effective in relieving severe depression, and several derivatives have come into use for this purpose. Since —S— and —CH=CH— are similar in size and electronegativity, the contrasting effects of the two series of drugs serve as a reminder that structure-activity relationships are elusive and empirical in an area as complex as psychopharmacology.

---

*The first name given is the **generic** drug name used in official pharmacopeias. The name in parentheses is the proprietary trademark name used by one manufacturer.

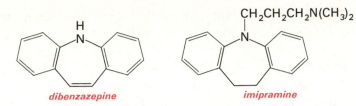

dibenzazepine

imipramine

Relief of depression and stimulation are different pharmacological actions. The principal stimulant drugs are **amphetamines,** a group of compounds that are structurally related to 2-phenylethylamine. This structure is present also in norepinephrine, which transmits impulses in the peripheral nervous system. Some of the pharmacological properties of amphetamines are related to those of norepinephrine, but, in addition, the amphetamines are powerful CNS stimulants. Both enantiomers of amphetamine have comparable peripheral effects, but the **s**-isomer is more active in the central nervous system.

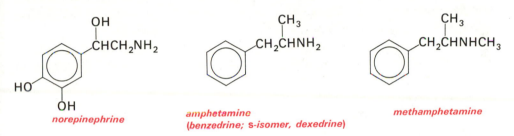

norepinephrine

amphetamine
(benzedrine; s-isomer, dexedrine)

methamphetamine

Amphetamines have several medical uses, including the property of reducing appetite, which is helpful in controlling obesity. The stimulatory effects are valuable in emergency situations to prevent drowsiness and postpone fatigue, but frequent usage can lead to **tolerance,** meaning that dosage must be continually increased.

Narcotic Analgesics.   Opium has been used as a drug since the third century B.C. and has been a medical, social, economic, legal, and political issue since that time. **Morphine,** the major alkaloid, is the most effective analgesic drug known; it not only raises the pain threshold, but increases the ability to tolerate pain that is perceived. Morphine also induces euphoria, or a sense of well-being, and this property leads to the liability of **addiction.** With repeated use, tolerance and **physical dependence** occur, and withdrawal of the drug causes severe physiological reactions.

Drug addiction is an extremely complex problem, since both psychological and physical dependence are involved. Nearly all the morphine derivatives and synthetic analogs that are highly active* analgesics can bring about physical dependence. Certain compounds, in particular **heroin,** the diacetyl derivative of morphine, because of the rapidity of distribution in tissues, have

---

*Activity is measured by mg of drug per kg of body weight required to block pain from a standard stimulus; the more active the drug, the smaller the dose.

greater liability than others to cause the psychological dependence that is part of the phenomenon of addiction. Heroin is the drug responsible for most narcotic abuse and the huge social and economic burdens that result.

Many structural variations of morphine have been examined in an effort to find an active *nonaddicting* analgesic. Two of the rings can be removed to give the benzomorphan derivative phenazocine, which has properties similar to those of morphine. Two compounds, **meperidine** and **methadone,** which resemble morphine only in having a fully substituted carbon with phenyl and $-CH_2CHRNR_2$ groups, are also highly active narcotic analgesics, although they differ somewhat from morphine in side effects and activity by various means of administration. Meperidine (pethidine) has a cumulative *toxicity* on prolonged usage; methadone has low toxicity and is more active than morphine when taken orally. However, all the active analgesics containing the $C_6H_5CR_2CH_2CHRNR_2$ system exhibit **cross-tolerance** and **cross-dependence** with morphine or heroin; that is, one drug can substitute for another in maintaining physical dependence and preventing withdrawal symptoms in an addicted person. This suggests that at the actual receptor site in the central nervous system, the compounds are structurally equivalent.

Although morphine-like activity is retained with major structural changes in the carbon framework, substitution of the $N-CH_3$ group in

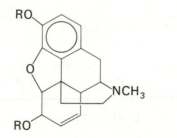

morphine R = H
heroin R = COCH₃

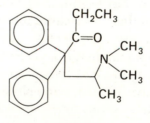

benzomorphan
phenazocine R = CH₂CH₂C₆H₅
cyclazocine R = CH₂–◁

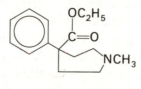

meperidine
(pethidine)

methadone

morphine by $NCH_2CH=CH_2$ causes a drastic change in properties. The N-allyl compound is an **antagonist** of morphine and other narcotic drugs; an even more powerful antagonist is the cyclopropylmethyl derivative cyclazocine in the benzomorphan series. These compounds are also analgesics, but they *block* the effect of morphine; a very small dose administered to a heroin addict causes immediate withdrawal symptoms.

Since antagonists such as cyclazocine block the development of dependence, they can be administered to a person who has been withdrawn from drugs to prevent the recurrence of addiction. Another approach to reduce the ravages of heroin addiction is **substitution therapy** with methadone. An addicted person can be maintained by oral doses of methadone which remove the psychological dependence on heroin and avoid the debilitating effect of heroin or other drugs; however, physical dependence on methadone continues.

SUMMARY AND HIGHLIGHTS

1. **Pyrrole, furan,** and **thiophene** contain five-membered rings with a heteroatom (N, O, or S) supplying an electron pair to the aromatic system. They are *electron-rich* and reactive toward electrophiles. Pyrrole is the basic unit of porphyrins.

2. **Five-membered rings** with two or more heteroatoms (oxazoles, thiazoles, triazoles, and so on) are obtained by condensation reactions and are also present in certain vitamins and natural amino acids.

3. **Pyridine,** with N replacing a $=CH-$ group in benzene, is aromatic but *electron-poor;* pyridinium salts are reactive toward nucleophiles.

4. **Alkaloids** are complex, naturally occurring amines derived biogenetically from amino acids. They include important drugs, such as quinine, morphine, and reserpine.

5. **Chemotherapeutic agent:** substance (usually synthetic) used to combat infectious organisms or cancer.

Antibiotic: anti-infectious compound obtained from microorganisms.
Sedative: drug that depresses the central nervous system, *e.g.*, barbiturates.
Tranquilizer: drug that relieves anxiety and stress symptoms, *e.g.*, benzodiazepines.
Narcotic: compound, usually powerful analgesic, that causes dependence and addiction, *e.g.*, morphine.

ADDITIONAL PROBLEMS

13.7 Write the structures of:

a) 2,4-dimethylpyrrole
b) 5-methylfuran-2-aldehyde
c) 2-phenyl-1,3-oxazole
d) 2-methyl-5-bromothiophene
e) pyridine-3-carboxylic acid
f) 1,3,4-thiadiazole
g) 2-methyl-4,5-diaminopyrimidine
h) indole-3-acetic acid

13.8 Show the steps in the following condensations, using curved arrows to follow the combinations of nucleophilic and electrophilic groups and enolizations.

Example:

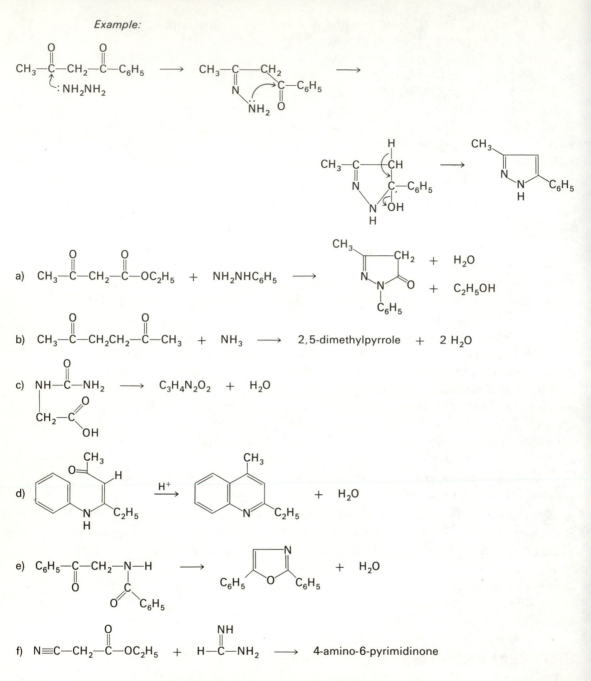

a) CH$_3$—C(=O)—CH$_2$—C(=O)—OC$_2$H$_5$ + NH$_2$NHC$_6$H$_5$ ⟶

b) CH$_3$—C(=O)—CH$_2$CH$_2$—C(=O)—CH$_3$ + NH$_3$ ⟶ 2,5-dimethylpyrrole + 2 H$_2$O

c) ⟶ C$_3$H$_4$N$_2$O$_2$ + H$_2$O

d) $\xrightarrow{\text{H}^+}$ + H$_2$O

e) C$_6$H$_5$—C(=O)—CH$_2$—N—H ⟶ + H$_2$O

f) N≡C—CH$_2$—C(=O)—OC$_2$H$_5$ + H—C(=NH)—NH$_2$ ⟶ 4-amino-6-pyrimidinone

13.9 Draw the structure of morphine (Figure 13.2) and show the locations of the two ^{14}C-labeled atoms that would be incorporated in the *in vivo* synthesis of morphine from the labeled reticuline shown in Figure 13.1.

13.10 Certain strains of pathogenic organisms, particularly *Staphylococcus,* become resistant to penicillin because a penicillinase enzyme is developed by the organism. The enzyme inactivates penicillin by catalyzing hydrolysis of the cyclic amide bond. Write the structure of the product that is formed.

13.11 The pK$_A$ of most barbiturates is about 8. In what form (ionic or neutral) would these compounds exist in the stomach (pH 2)? In the blood (pH 7.4)?

14

AMINO ACIDS, PEPTIDES AND PROTEINS

Proteins are a major component of all living matter, serving both as fibrous structural material and as enzymes in the vital role of catalysts for metabolic reactions. Proteins are condensation polymers of α-amino acids connected as amides from the carboxylic acid group of one amino acid to the amino group of the next. The amide linkage between amino acid units is called a **peptide bond.**

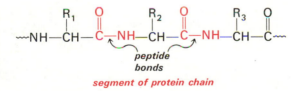

peptide
bonds

segment of protein chain

14.1   Amino Acids

Twenty different amino acids are found in proteins. Their structures and names are given in Table 14.1. Eight of the amino acids, indicated by asterisks in the table, are **essential** for mammals; *i.e.*, they cannot be synthesized in the body, and must be supplied in the diet. In all the amino acids except glycine, the α-carbon is an asymmetric center. Stereoisomerism is simple, since the absolute configuration is **s** in all these amino acids. Using a convention similar to that for the sugars, the protein amino acids are designated L. A few D-amino acids are found in polypeptide antibiotics, and single D-amino acids occur in animal cells, but they are not incorporated into proteins.

TABLE 14.1 AMINO ACIDS PRESENT IN PROTEINS

Name	Abbreviation	Formula
		A. Neutral
Glycine	Gly	CH_2-COOH NH_2
Alanine	Ala	$CH_3-CH-COOH$ NH_2
Phenylalanine*	Phe	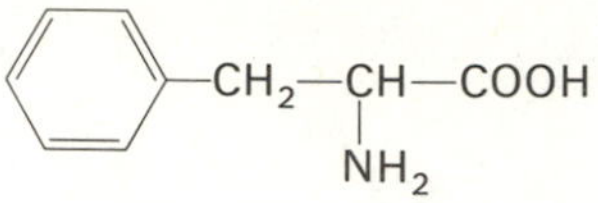 $\langle\bigcirc\rangle-CH_2-CH-COOH$ NH_2
Tyrosine	Tyr	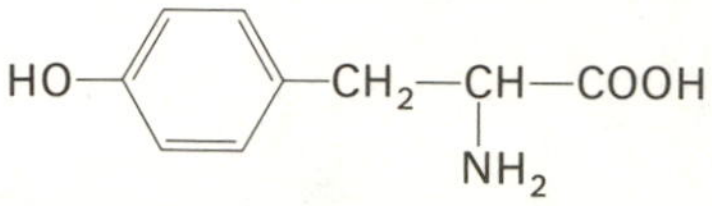 $HO-\langle\bigcirc\rangle-CH_2-CH-COOH$ NH_2
Tryptophan*	Try	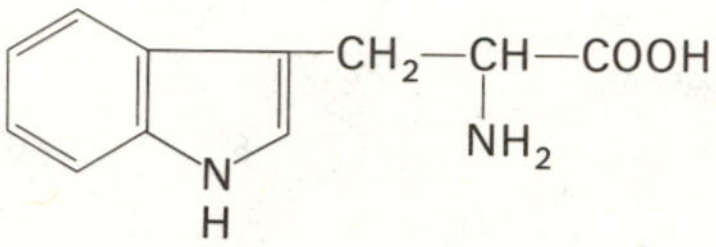 indole$-CH_2-CH-COOH$ NH_2
Proline	Pro	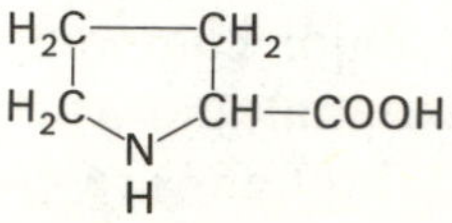 H_2C-CH_2 $H_2C \quad CH-COOH$ N H
Cysteine	Cys	$HS-CH_2-CH-COOH$ NH_2
Serine	Ser	$CH_2-CH-COOH$ $OH \quad NH_2$
Threonine*	Thr	$CH_3-CH-CH-COOH$ $OH \quad NH_2$
Valine*	Val	$CH_3-CH-CH-COOH$ $CH_3 \quad NH_2$
Leucine*	Leu	$CH_3-CH-CH_2-CH-COOH$ $CH_3 \qquad NH_2$

Table continued on opposite page.

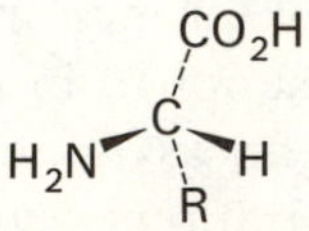

$$CO_2H$$
$$H_2N-C-H$$
$$R$$

L-amino acid

Amino acids are highly polar compounds, with very low solubility in organic solvents and very high melting points. Their solubility in water is

TABLE 14.1   **AMINO ACIDS PRESENT IN PROTEINS**
(*continued*)

Name	*Abbreviation*	*Formula*
		A. Neutral (continued)

Isoleucine*	Ile	$CH_3-CH_2-CH-CH-COOH$
		     $|$ $|$
		     CH_3 NH_2

Methionine*	Met	$CH_3-S-CH_2-CH_2-CH-COOH$
		           $|$
		           NH_2

		B. Basic amino acids

Arginine	Arg	NH_2
		      $|$
		$HN=C-N-CH_2-CH_2-CH_2-CH-COOH$
		    $|$ $|$
		    H NH_2

Lysine*	Lys	$CH_2-CH_2-CH_2-CH_2-CH-COOH$ $|$ $|$ NH_2 NH_2
Histidine	His	$HC=\!\!=C-CH_2-CH-COOH$ $|$ $|$ N NH NH_2 $\backslash CH$
		C. Acidic amino acids

Aspartic acid	Asp	$HOOC-CH_2-CH-COOH$
		       $|$
		       NH_2

Glutamic acid	Glu	$HOOC-CH_2-CH_2-CH-COOH$
		         $|$
		         NH_2

		D. Amides

Asparagine	Asn	O
		  $\|\|$
		$H_2N-C-CH_2-CH-COOH$
		      $|$
		      NH_2

Glutamine	Gln	O
		  $\|\|$
		$H_2N-C-CH_2-CH_2-CH-COOH$
		         $|$
		         NH_2

*Essential amino acid.

somewhat greater than in alcohols, and the compounds are very soluble in aqueous acid or base.

When a simple amino acid such as glycine dissolves in aqueous acid, it is present in the form of the ammonium cation, $^+NH_3CH_2CO_2H$, which has two acidic groups, $^+NH_3$ and CO_2H. Upon titration with base, two equivalents of ^-OH are consumed and the species present is $NH_2CH_2CO_2^-$. The pK_A values determined by titration (Section 10.7) are 2.5 and 9.0. The lower pK_A corresponds to removal of the proton from the CO_2H group. This group is a stronger acid than acetic acid ($pK_A = 4.8$) because the adjacent $^+NH_3$ group exerts a strong inductive effect in the direction $^+NH_3 \leftarrow\!\!\!|\!\!\!- CH_2CO_2H$. The higher pK_A value, at a basic pH, corresponds to removal of a proton from the $^+NH_3$ group. Thus, glycine (and the other α-amino acids in neutral solution) exists as a **zwitterion** or dipolar ion in which the amino and carboxyl groups are mutually neutralized to give an "inner salt."

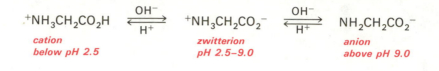

$$^.NH_3CH_2CO_2H \underset{H^+}{\overset{OH^-}{\rightleftharpoons}} \;^+NH_3CH_2CO_2^- \underset{H^+}{\overset{OH^-}{\rightleftharpoons}} NH_2CH_2CO_2^-$$

cation
below pH 2.5

zwitterion
pH 2.5–9.0

anion
above pH 9.0

Problem 14.1   Predict the pK_A values of the ester $^+NH_3CH_2CO_2CH_3$ and the amide $RCONHCH_2CO_2H$.

The dipolar structure of amino acids accounts for the salt-like properties of the compounds. Although the formulas of amino acids may be written as if they were neutral compounds (*e.g.*, $NH_2CH_2CO_2H$), it must be kept in mind that they are in fact dipolar ions. The pH at which the positive and negative charges are exactly balanced to give an electrically neutral molecule is called the **isoelectric point.** For most amino acids this value is in the range of pH from 5.5 to 6.5. If an additional NH_2 group is present, as in lysine, the isoelectric point will be higher (9.5), and if an additional CO_2H group is present, the isoelectric point will be lower.

14.2   Amino Acid Analysis

One of the major steps in elucidating the structure of a peptide or protein is determination of the amino acid composition, *i.e.*, how many of each amino acid are present in the chain. To accomplish this, a sample of the polypeptide is hydrolyzed by acid under vigorous conditions which break all peptide bonds. The amino acid mixture is then separated by passing it over a column of **ion exchange resin** which contains SO_3H groups attached to a matrix of polystyrene. The amino acids are adsorbed as cations on the resin and are then displaced in order of increasing base strength by passing buffer solutions of increasing pH through the column of resin. The amino acids are identified by the sequence of their appearance in the solution that leaves the column. The operation is carefully standardized and calibrated with known mixtures of amino acids so that a given amino acid always appears at the same point. To measure the relative amounts of the amino acids the solution, as it leaves the column, is mixed with a reagent that produces a color, and the absorbance is continuously monitored and recorded (Section 9.3).

The color reaction used in automatic amino acid analysis and in paper chromatography of amino acids is a condensation with the highly reactive central carbonyl group of a 1,2,3-triketone called **ninhydrin.** Decarboxylation, hydrolysis, and further condensation give an intense blue dye.

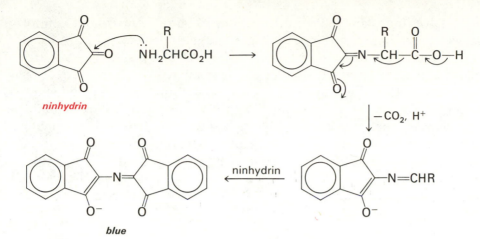

ninhydrin

blue

$-CO_2,\ H^+$

ninhydrin

PEPTIDES

14.3 Structure and Occurrence of Peptides

The term **peptide** is used somewhat loosely to designate any chain of amino acids connected by peptide bonds. Di-, tri-, and tetrapeptides refer to compounds with two, three, and four amino acids, respectively. Polypeptide (or simply peptide) usually means a chain of from 3 or 4 up to 50 or more amino acids; there is no sharp boundary between a "large peptide" and a "small protein." A peptide chain is always written with the terminal amino group on the left end and the terminal carboxyl group on the right, *i.e.*, $NH_2CHRCO(NHCHRCO)NHCHRCO_2H$. The amino acids are designated by the abbreviations given in Table 14.1, as illustrated for the tetrapeptide glycylleucylaspartyltyrosine, written in the usual abbreviation as gly.leu.asp.tyr. Since gly is at the left, it is the unit with the NH_2 group, or *N*-terminal amino acid; tyr is the *C*-terminal unit.

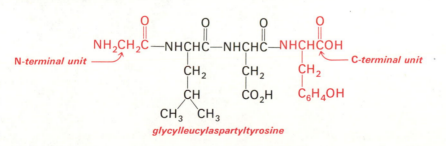

N-terminal unit

C-terminal unit

glycylleucylaspartyltyrosine

Problem 14.2 Refer to Table 14.1 and write the structures of the following peptides: a) ala.gly and b) phe.ser.lys.

Peptides are obtained by partial hydrolysis of proteins and are also isolated from various fluids and tissues in the body. A number of peptides with chains containing as few as three and as many as about forty amino acids have important biological functions ranging from overall control of growth to regulation of blood pressure and gastric secretion.

The pituitary or "master" gland at the base of the skull is the source of many important peptides. One of the best known is **adrenocorticotrophic hormone,** or **ACTH,** which controls the output of the adrenal gland and thus in turn a number of other hormones (p. 000). Human ACTH contains 39 amino acids. The sequence begins with serine at the free NH₂ terminal and ends with phenylalanine. A smaller peptide with only amino acids 1 through 24 has nearly the same biological activity as the full ACTH molecule, as determined by a specific bioassay. Moreover, ACTH molecules from various animals contain this same sequence of amino acids 1 to 24, although there are differences in the rest of the chain.

ser · tyr · ser · met · glu · his · phe · arg · trp · gly · lys · pro · val · gly · lys · lys · arg · arg · pro · val
  1   2   3   4   5   6   7   8   9   10   11   12   13   14   15   16   17   18   19   20

lys · val · tyr · pro · asn · gly · ala · glu · asp · glu · ser · ala · glu · ala · phe · pro · leu · glu · phe
 21   22   23   24   25   26   27   28   29   30   31   32   33   34   35   36   37   38   39

amino acid sequence of human ACTH

Species differences in segments of peptide chains are found in many peptide hormones and enzymes. Another example is **insulin,** the peptide hormone formed by certain cells in the pancreas. The physiologically active form of insulin contains two separate chains linked together at two points by disulfide bonds. The amino acid sequence in these chains (Figure 14.1) varies only slightly from one species to another. The hormone is synthesized in the cell in a form called **proinsulin,** which has a continuous chain containing a

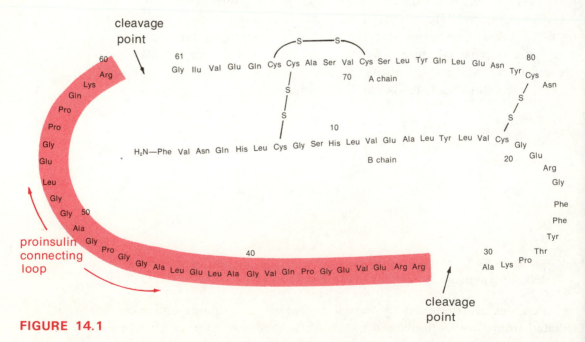

FIGURE 14.1

Bovine insulin.

connecting loop that is lost to produce the active hormone. This central segment, which is not essential for the hormonal function, varies quite widely among species in the number and sequence of amino acids. It thus appears that genetic mutations that cause a change in an amino acid can be tolerated if they affect a non-essential part of the polypeptide, but are lethal and are not preserved if the change occurs in the functional part of the chain.

14.4 Determination of Amino Acid Sequence

The sequence of amino acids in a peptide or protein is the primary structural problem that must be solved before we can understand the biological functions of the molecule. This problem is approached by two types of experiments: 1) identification of the amino acids at the ends of the chains, and 2) cleavage of the chain and isolation of smaller overlapping segments. The amino acid composition and, if needed, the end groups of these fragments are then determined and the overall sequence is deduced by fitting together the ends and pieces.

The general principle can be illustrated with the simple case of a hexapeptide containing six different amino acids, A to F, as determined by amino acid analysis. By methods described below, B is found to be the N-terminal amino acid, and we can write the structure as B-(A, C, D, E, F), with the sequence inside the parentheses unknown. The peptide is partially hydrolyzed to give a mixture of smaller peptides and individual amino acids. Let us assume that from this mixture we can isolate three smaller peptides whose amino acids are: (I) A, D; (II) A, C, E; and (III) B, C, F. These three peptides are fragments of the original sequence, and since they contain all of the six amino acids they must represent the entire peptide. Two amino acids which appear twice, A and C, are points of overlap.

Fragment III contains the N-terminal amino acid B, and the other two amino acids, C and F, that appear together with B in fragment III must occupy the next two positions. Since C occurs in both fragments III and II, it must occupy the overlapping position between these two tripeptides, and we can therefore place fragment II adjacent to and overlapped with III. Similarly, A is in the overlapping position between fragments II and I. We can

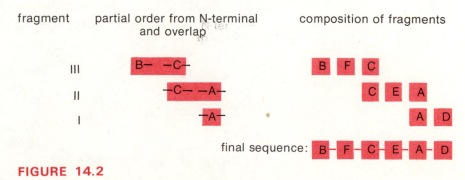

FIGURE 14.2

Amino acid sequence determination.

now arrange the fragments according to these relationships as in Figure 14.2, writing the *N*-terminus on the left and the *C*-terminus on the right. The remaining amino acids D, E, and F can be fitted into place from their occurrence in the three fragments, and our sequence is complete.

It must be emphasized that this example is an idealized and greatly oversimplified case. In practice, the separation and purification of many small peptide fragments is a very difficult and laborious task, and the fragments needed to work out the full sequence may be formed in very low yield. If the same amino acid is present two or three times, as frequently happens, much more information is required to solve the sequence. With larger peptides, each fragment obtained in the initial cleavage is treated as a new peptide and subjected to end group analysis and further cleavage.

Problem 14.3   How many smaller peptides (containing two to five amino acids) could be obtained by hydrolysis of a hexapeptide containing six different amino acids?

Problem 14.4   A tetrapeptide contains four amino acids, A to D; D is found to be the *N*-terminal amino acid. Two dipeptides are isolated after hydrolysis. One of these contains A and D, and the other contains A and B. Deduce the sequence of amino acids in the tetrapeptide.

N-Terminal Amino Acid.   To determine the *N*-terminal amino acid, the peptide is treated with a reagent which reacts selectively with the NH_2 group to "label" it in some way. One of the most effective reagents for this purpose is 5-dimethylaminonaphthalenesulfonyl chloride, or "dansyl chloride," which forms a sulfonamide with the NH_2 group. The peptide chain is then hydrolyzed under conditions which cleave all peptide bonds. The sulfonamide group is much less readily hydrolyzed, and remains intact. The *N*-terminal amino acid is easily isolated and identified as its dansyl derivative by comparison with reference dansyl amino acids.

Problem 14.5   A tripeptide has the amino acid sequence seryl-ala-nyl-proline. a) From the structures in Table 14.1, write the complete structure of the peptide. b) Write the structures of the products that would be obtained by treatment of the tripeptide with dansyl chloride followed by hydrolysis.

Stepwise Removal of N-Terminal Amino Acid—Automated Sequencing.   An enormous advance in peptide sequencing is a procedure, pioneered by Edman, which permits removal and identification of the *N*-terminal amino acid while leaving the remainder of the peptide intact. The process can be repeated many times, permitting the direct determination of sequence in an automated instrument.

The basis for the **Edman degradation** is the reaction of the terminal amino group of the peptide with phenyl isothiocyanate to give a thiourea. On treatment with anhydrous acid, cyclization occurs between the thiourea group and the carbonyl group of the first peptide bond. The result is cleavage to give a heterocyclic derivative of the *N*-terminal amino acid and a new peptide which is ready to undergo a second cycle. The reaction is highly selective for the terminal peptide bond because it is an internal nucleophilic addition to the amide carbonyl group, and not a hydrolytic process. The phenyl-thiohydantoin of the first amino acid is identified by comparison with reference derivatives of known amino acids.

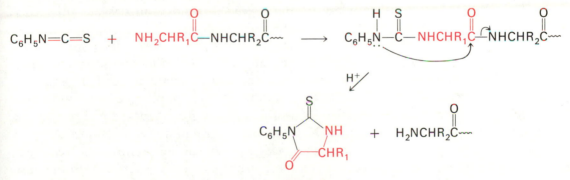

phenylthiohydantoin

If the peptide to be degraded is attached to an insoluble polymer, the steps of reaction with $C_6H_5N{=}C{=}S$, cleavage with acid, and extraction of the thiohydantoin can be carried out automatically with a very small amount of sample, and a sequence of 40 or more amino acids can be removed one by one and identified. For sequencing of proteins with hundreds of amino acids, the chain is cleaved into peptides of manageable size by methods that provide overlapping fragments. After sequencing, these segments of 15 to 30 amino acids are fitted together as in the example given earlier.

14.5   Synthesis of Peptides

Side by side with amino acid analysis and sequence determination, methods for the synthesis of polypeptide chains have been developed. Synthetic methods are needed to provide reference peptides of known sequence and, potentially, sources of scarce peptide hormones. The general approach to synthesis involves coupling of two amino acids, or of two peptide segments, through a new peptide bond, as outlined in Figure 14.3.

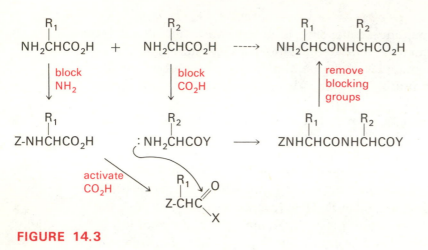

FIGURE 14.3

General scheme for peptide synthesis.

To effect the coupling in the desired direction and prevent side reactions, the NH_2 group in one amino acid and the CO_2H group in the other must be "protected" with groups that block reaction during the peptide bond formation and which can then be removed under mild conditions. In addition, some type of "activation" for the CO_2H group is needed, since the formation of an amide directly from the free acid requires severe conditions.

One of the most generally useful blocking groups for the NH_2 end is the *t*-butoxycarbonyl group **(BOC)**. The protected amino group is unreactive in the peptide coupling, and the blocking group can be removed by treatment with anhydrous acid at low temperature; this reaction depends on the ease with which the *t*-butyl ester undergoes elimination. The by-products are a low boiling alkene and CO_2, both of which are innocuous and easily removed.

Formation of protected amino acid

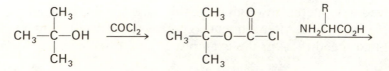

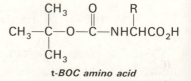

t-BOC amino acid

Removal of BOC group

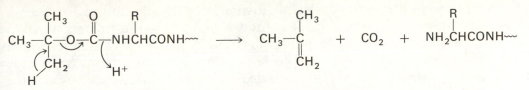

For CO_2H activation and amide formation, reagents have been found that permit a one-step intermolecular dehydration. Among the best of these are **carbodiimides,** highly reactive compounds in which both oxygens of CO_2 are replaced by $RN=$ groups. Reaction of the *N*-protected acid with the reagent gives an imino anhydride, and this acylates the amine.

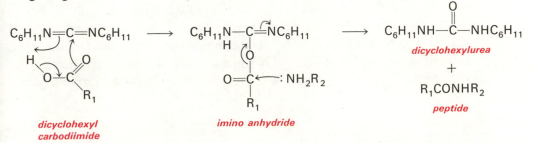

These selective protecting groups and coupling reagents have made it possible to carry out peptide coupling between two amino acids in high yields, but the labor of preparing even a small peptide in this way is enormous. A major task is the separation and purification of the peptides at each stage.

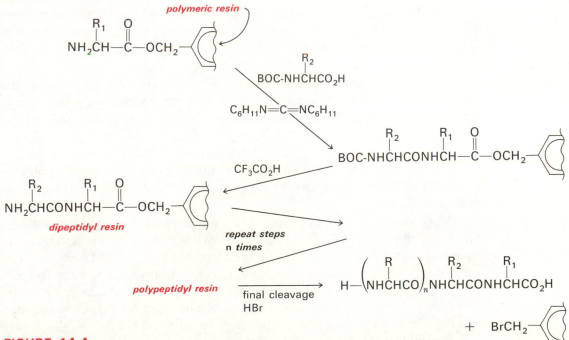

FIGURE 14.4

Solid phase peptide synthesis.

To circumvent the repetitive and inefficient isolation steps, **solid phase** peptide synthesis was developed (Figure 14.4). This technique has been a revolutionary advance since it permits a *fully automated* process. In solid phase synthesis, the CO_2H group is blocked by formation of an ester with a benzyl group attached to a polymer support. From this point on, all the steps—acylation with a BOC-protected amino acid, selective removal of the BOC group (leaving protective groups in the side chains intact), and repetition of the cycle—are performed with the peptide attached to the insoluble polymer. After each stage, extensive extraction and washing can be carried out to remove by-products.

The entire operation can be carried out automatically by connecting the reaction vessel to reservoirs containing solvents, reagents, and solutions of the various protected amino acids. At the end of the synthesis, the polypeptide is cleaved from the resin in a final step by somewhat more vigorous acid treatment which also removes the protecting groups in the side chains. The protein ribonuclease, containing 124 amino acids, has been synthesized by this procedure, using 369 separate reactions and 12,000 automated steps.

PROTEINS AND ENZYMES

Proteins range in size from single polypeptide chains of about 150 amino acids to giant polymers of 1800 units, corresponding to a molecular weight of about 200,000. The frequency of occurrence and the sequence of individual amino acids in proteins is determined by nucleic acid coding (Chapter 16), and varies according to the type and function of the protein. The protein fraction isolated from a sample of tissue is often a complex mixture, but it is a mixture of *specific compounds,* and not a random assortment of polypeptide chains.

The physical properties and macromolecular structure of proteins are functions of the amino acid composition. All proteins contain some amino acids with acidic CO_2H and basic NH_2 or $C(=NH)NH_2$ groups in the side chains. According to the number of these ionizable groups, the protein can be positively or negatively charged, and the properties therefore depend on the pH of the solution. Like amino acids, proteins have an isoelectric point at which the acidic and basic groups are exactly balanced to give an electrically neutral molecule.

Thiol groups in the chain are often present in the form of disulfides, which provide cross-links between different segments of the chain. Two separate polypeptide chains can be linked together by a disulfide bridge, as in the hormone insulin. One potential complication is absent, however. In spite of the extra CO_2H and NH_2 groups in side chains, proteins do not contain branches with peptide bonds extending from a pendant NH_2 or CO_2H group; a living organism requires fibers but not cross-linked resins.

Protein structure can best be described in terms of successive levels of organization: (1) **Primary** structure is the order of amino acids and disulfide bridges in the polypeptide chain, as determined by sequencing procedures. (2) **Secondary** structure refers to the conformation of the chain, specifically the spatial relationship of successive amino acid units. (3) **Tertiary** structure involves the overall conformation of the molecule, determined by *folding* of the chain. (4) **Quaternary** structure is the association of separate protein molecules into a larger aggregate. Information on the higher levels of structural organization comes from x-ray crystallography; remarkable progress has been made in visualizing the actual conformation of the chain.

14.6   Conformation of Polypeptide Chains

Because of resonance interaction in amides (Section 10.10), the —CONH— group in the peptide bond is *planar,* with the CHR groups *trans* to each other (Figure 14.5). Rotation can occur around the NH—CHR and CHR—CO bonds, however, and the polypeptide chain twists at these points to place the successive —CONH— planes at the proper angles for maximum stability. These angles are limited to a few combinations in order to avoid

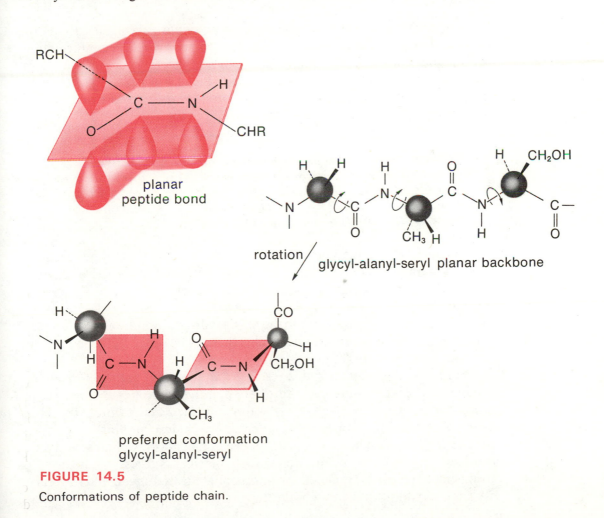

planar
peptide bond

glycyl-alanyl-seryl planar backbone

rotation

preferred conformation
glycyl-alanyl-seryl

FIGURE 14.5

Conformations of peptide chain.

steric interaction of R groups on the amino acid units with C=O groups and with each other.

A major factor in determining the conformation of the polypeptide chain is **hydrogen bonding.** The —CONH group acts as both acceptor and donor, and certain arrangements of the chain are stabilized by extensive hydrogen bonding between the C=O group of one amino acid and the NH of another unit. Two important types of conformation are a **helix,** which is stabilized by hydrogen bonds *within* a chain, and a **pleated sheet** arrangement with hydrogen bonding *between* adjacent polypeptide chains.

The most common helical structure is a right-handed spiral in which every C=O group is perfectly positioned for hydrogen bonding with the N—H group three units away and above it in the spiral. This conformation, called the **α-helix,** is shown in Plate 10. A complete turn around the axis contains 3.6 amino acid units. The C=O and N—H bonds, and the hydrogen bonds between them, lie along the surface of the "barber pole," almost parallel to the axis; the R groups are tipped outward, with minimum interactions. The α-helix provides a readily extensible, elastic chain; wool fiber and also the fibrous contractile protein of muscle are largely α-helix structures.

It is the ability of the α-helix to accommodate varying R groups that stabilizes this secondary structure relative to many other conformations. Glycine units, with no R group, can, of course, also be accommodated, but their presence lessens the advantage of the α-helix compared to other possible arrangements. The cyclic amino acid proline, with a rigid ring between the N and α-carbon, cannot be accommodated in the α-helix and disrupts the helical structure wherever it occurs in the chain.

Problem 14.6   Sketch the backbone conformation structure of a tripeptide segment like that at the top of Figure 14.5, with the amino acid proline taking the place of the center alanine unit, *i.e.,* glycyl-prolyl-serine. What specific effect does the presence of the proline unit have on the secondary structure of the polypeptide chain?

Collagen, the tough protein present in hide, tendon, and cartilage, contains mainly glycine, proline, and hydroxyproline arranged in more or less regular segments: gly-X-pro, gly-X-hypro, gly-pro-hypro. This rather unusual chain fits a special helical arrangement with the C=O and N—H groups extending perpendicular to the axis. Hydrogen bonding holds together three chains, which wrap around each other in a "super helix" to give an extremely strong fiber (Figure 14.6).

Silk is another protein with a restricted amino acid composition. Half of the units are glycine, about one-third are alanine, and one-sixth are serine, with the sequence (gly-ser-gly-ala-gly-ala)$_n$ repeated many times. The chain is an extended, flat ribbon with alternating chains running in opposite directions. Hydrogen bonding occurs between neighboring chains, and the R groups of alanine and serine are nestled between successive sheets. This arrangement (Plate 11) is called the **antiparallel pleated sheet,** and confers on silk high flexibility but limited elasticity.

FIGURE 14.6

Collagen helix.

14.7 Globular Proteins

The proteins discussed so far are fibrous structural polymers. A second major group are the **globular proteins** which occur in all tissues and carry out important functions as enzymes, oxygen or electron transport agents, and antibodies. These proteins are compact, roughly spherical molecules in which the chains are folded and convoluted into a complex tertiary structure.

The folding of a protein chain is not a random process like the snarling of a piece of yarn. Globular proteins are coiled in a specific pattern that results from the same driving force responsible for all chemical processes— the molecule seeks the lowest total energy. The conformation in which the protein is folded is actually determined by the primary amino acid sequence. Detailed structural analysis shows that the folding pattern of proteins places the alkyl groups and aromatic rings on the inside surface of the coiled chain. Polar groups such as CO_2H, $CONH_2$ and $C(=NH)NH_2$ are exposed on the outer surfaces of the chains. Hydrophobic (water-repelling) interactions between internal alkyl groups and hydrogen bonding of the external polar groups with water molecules both act to stabilize a specific folded structure.

One of the most thoroughly studied globular proteins is **myoglobin,** which serves as a reservoir for molecular oxygen in animal tissues. The three-dimensional structure of myoglobin (Figure 14.7) has been determined

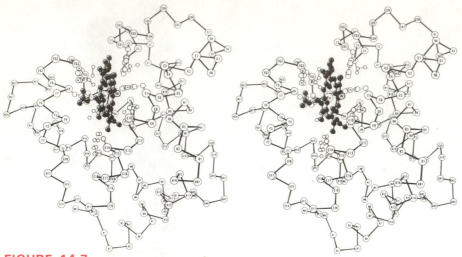

FIGURE 14.7

Myoglobin. This figure is a stereoview of the myoglobin structure. With a viewer, the structure can be seen in three-dimensional perspective.

in great detail by x-ray crystallography, and the complete primary structure, containing 153 amino acid units, is also known. Myoglobin contains a molecule of the porphyrin heme (Section 13.1), and helical segments of the chain are folded so that the porphyrin ring fits snugly into a hydrophobic pocket. A histidine group is positioned to provide an unshared electron pair for coordination with the iron atom.

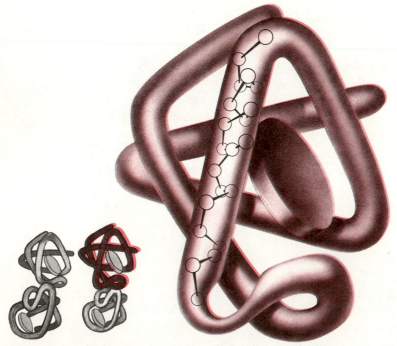

FIGURE 14.8

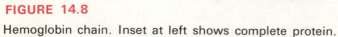

Hemoglobin chain. Inset at left shows complete protein.

Hemoglobin, the protein that transports oxygen from the lungs to cellular tissue where it is transferred to myoglobin, contains four chains. Each subunit contains one heme group and closely resembles the myoglobin chain. The association of the four chains into the functional protein cluster illustrates quaternary structure (Figure 14.8). Complete amino acid analysis of myoglobin and hemoglobin subchains from various animal species reveals many similarities, and suggests the course of evolutionary development of these proteins from an ancestral globulin.

14.8 Enzymes

Enzymes are proteins that are catalysts in biochemical reactions. As discussed in the following chapter, the metabolic processes of animal and plant cells are complex multi-reaction sequences. Each step is controlled by a specific enzyme, and over 1000 enzymes have been identified and fully or partially purified. Enzymatic activity is found in widely different types of proteins, ranging from relatively small polypeptide chains to clusters containing from 2 to 60 subunits.

Reactions catalyzed by enzymes run the gamut of organic chemistry from oxidation of ethanol to the hydrolysis of cellulose. Enzymes exert their catalytic role by bringing together the reactant molecules and providing an ideal environment for bond-making or bond-breaking. The same type of effect is seen in the much greater rate of intramolecular lactone formation of a hydroxy acid (Section 10.20) compared to the intermolecular esterification of an acid and an alcohol.

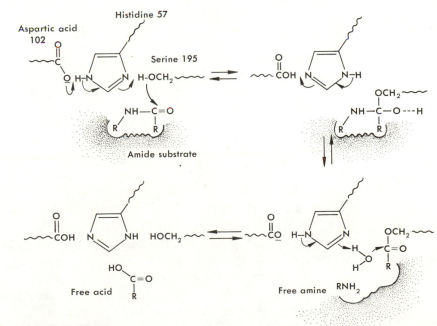

FIGURE 14.9
Hydrolysis by chymotrypsin.

Among the most thoroughly studied enzymatic reactions is the hydrolysis of amides or esters which occurs, for example, in the digestion of food. Several hydrolytic enzymes have been completely purified and crystallized, and the amino acid sequence and tertiary structure are known. The active site in these enzymes, where the hydrolysis takes place, has been probed by treating the protein with reagents that combine selectively with certain groups. If a given treatment inactivates the enzyme as a catalyst, and degradation shows that a specific functional group in one amino acid has been changed by the deactivating treatment, that group is implicated as part of the catalytic site.

In the digestive enzyme **chymotrypsin,** a serine OH group, an aspartic acid CO_2H group, and a histidine ring are involved in the catalytic mechanism. The three amino acids are widely separated in the polypeptide chain, occupying positions 195, 102 and 57, but they are brought together at a hydrophobic depression in the folded tertiary structure. Figure 14.9 shows how the functional groups can orchestrate the hydrolysis of an amide. Transfer of the acyl group from the amide to the serine OH group is the first stage. In solution, the reaction of an amide with an alcohol requires drastic conditions, with strong acid or base. In the hydrophobic pocket of the enzyme, however, the amide is held at just the right position for attack by the OH group, and the functional groups are free from interfering shells of solvent molecules.

SUMMARY AND HIGHLIGHTS

1. **α-Amino acids** are the monomer units in peptides and proteins. Amino acids exist as inner salts, $^+NH_3$—CH—CO_2^-—, with R substituent.

2. **Peptides,** NH_2—CH—CONH—CH—CO_2H, are polyamides formed from amino acids; several hormones are peptides with 8 to 40 amino acids.

3. The **sequence** of amino acids in a peptide is determined by partial hydrolysis to smaller fragments and step-by-step removal of the N-terminal amino acids as thiohydantoins.

4. Peptide **synthesis** involves protection of one reactive end of an amino acid by a removable blocking group and then coupling free NH_2 and CO_2H groups by reaction with a carbodiimide. In **solid-phase** synthesis, the growing peptide chain is attached to a polymer.

5. **Proteins** are polypeptides with 150 to 1800 amino acids. Protein structure depends on amino acid sequence, conformation of the chain, folding or looping of the chain, and association.

6. **Helical** structures, stabilized by hydrogen bonds, are present in many fibrous proteins.

7. In **globular** proteins, the chain is folded and looped in a specific pattern, with alkyl and aryl groups in the interior and polar groups on the outer surface.

8. **Enzymes** are globular proteins which have catalytic functions in biochemical reactions, such as hydrolysis.

ADDITIONAL PROBLEMS

14.7 A pentapeptide was found by amino acid analysis to contain one mole each of alanine (ala), cysteine (cys), glycine (gly), tyrosine (tyr), and valine (val). Reaction of the peptide with dansyl chloride followed by complete hydrolysis gave N-dansylglycine plus ala, cys, tyr, and val. Partial hydrolysis of the pentapeptide gave individual amino acids plus two dipeptides, A and B, and a tripeptide C. Further hydrolysis of these peptides individually gave the following amino acid compositions:

$$A = gly + cys$$
$$B = tyr + cys$$
$$C = tyr + cys + ala$$

Deduce the sequence of the amino acids in the pentapeptide.

14.8 A dipeptide was treated with phenyl isothiocyanate, followed by acid, and the following products were isolated:

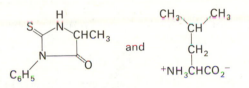

Write the structure and the name of the dipeptide.

14.9 Show the reactions involved in one cycle of solid phase peptide synthesis in which a leucine unit is added to the chain, beginning with the t-butoxycarbonyl amino acid and the peptidyl polymer.

14.10 The dipeptide glycylalanine, on treatment with dicyclohexylcarbodiimide, gave a compound $C_5H_8N_2O_2$. Suggest the structure of the product and show in detail how it is formed.

14.11 Arrange the following peptides in order of *increasing* isoelectric point (the pH at which each exists as an electrically neutral molecule): a) alanyl-seryl-lysine, b) alanyl-leucyl-serine, c) glycyl-glycyl-arginine, and d) alanyl-aspartyl-phenylalanine.

14.12 Arginine is the most strongly basic of all protein amino acids; the high basicity is due to the $HN=C(NH_2)NH-$ (guanidine) group. a) Write the structures of the principal ionic forms of arginine at pH $= 1$, 5, and 11. b) Suggest an explanation for the high basicity of the guanidine group.

15

ORGANIC REACTIONS IN LIVING SYSTEMS

The chemical reactions that occur in living organisms accomplish two basic purposes: (1) synthesis of the compounds that make up the organism, and (2) storage and utilization of energy to support physiological needs. The requirements that must be fulfilled by chemical systems in living cells range from growth and reproduction in primitive organisms to locomotion and mental activity in animals. One of the most remarkable facts of natural science is that most of these diverse requirements are met by the *same basic chemical mechanisms* in all living forms on earth.

In this chapter we will examine these universal biochemical processes and some specialized reactions that occur only in plants. The reactions and functional groups are those seen in earlier chapters. Condensations of nucleophiles (such as OH, NH_2, and SH groups, enols, and enamines) with $>C=O$, $-CO_2H$, and CO_2 as electrophiles are the main synthetic reactions. These reactions occur in cells that are highly organized, and they must be suitable for the requirements of the organism. As starting materials, man and other animals have available a wide variety of compounds from food, whereas plants must be able to manage with CO_2, nitrogen, water, and minerals.

15.1 Primary Metabolic Processes

Metabolism is the sum of the chemical reactions in living cells leading to the synthesis and breakdown of organic compounds. The range of conditions that prevail for metabolic processes is extremely limited. In contrast to the high temperatures, anhydrous solvents, and strong acids or bases that can be used in laboratory operations, biochemical reactions must take place in dilute aqueous solutions, usually between 20 and 40°C, and in a pH range of about 6 to 8. Metabolism goes on under these conditions because of the tremendous catalytic effect of **enzymes** (Chapter 14). Every biochemical reaction is controlled by an enzyme, and metabolic processes are described by biochemists primarily in terms of the enzymes that are responsible. A complex metabolic pathway may involve a number of steps in which the same type of reaction occurs with different substrates, but each step requires a different enzyme.

Enzymes, like all catalysts, affect only the *rate* of reactions; they do not change the position of an equilibrium, or cause an energetically unfavorable reaction to occur. Metabolic processes also require a *driving force*, which must be furnished by an overall decrease in energy. If a reaction A ⟶ B is exothermic and is used by the cell to *produce* energy, the reverse reaction B ⟶ A requires the *expenditure* of energy. To provide the driving force for the endothermic reverse reaction, it must be coupled, directly or indirectly, with another reaction that is exothermic. Metabolic pathways are therefore a complex network of synthesis, degradation, and energy transfer steps. To fully describe a metabolic process, the individual enzymes and energy changes at each step must be specified. Such descriptions are among the central elements of biochemistry. We will be concerned in this chapter only with the general nature of enzymatic reactions and with the overall changes of organic compounds in a few metabolic sequences.

15.2 Energy Transfer

The main process for energy transfer in biochemical reactions is the formation and hydrolysis of phosphate esters. Phosphoric acid or alkyl phosphate esters can form anhydrides containing P—O—P bonds. In the evolution of living organisms, one of these anhydrides, adenosine triphosphate (ATP), assumed a major role in energy transfer. Hydrolysis of the triphosphate group to the diphosphate (ADP) is a highly exothermic reaction,

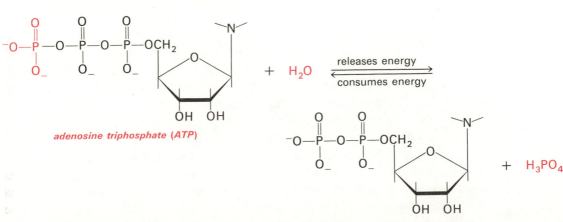

adenosine triphosphate (ATP)

adenosine diphosphate (ADP)

and this "energy-rich" compound is present in all cells. In metabolic cycles, ATP releases energy by transfer of a phosphate group, and is then regenerated in another step, in which energy from an exothermic reaction is stored by rephosphorylation of ADP. The phosphate OH groups in these esters are ionized at pH 7, and in many reactions the oxygens are complexed with magnesium ion.

15.3 Chemical Reactions in Metabolism

Chemical transformations that occur in metabolism are of the same type, and follow the same basic rules and electron shifts, that we have seen in earlier chapters in non-enzymatic reactions. Addition and elimination of water, formation and hydrolysis of esters and amides, and oxidation-reduction of $\diagup$CH—CH$\diagdown$ $\rightleftarrows$ $\diagup$C=C$\diagdown$ and $\diagup$CHOH $\rightleftarrows$ $\diagup$C=O systems are all encountered in metabolic pathways. Carbon-carbon bonds are formed by combination of nucleophilic and electrophilic carbons at all oxidation levels. The electrophilic carbon species are $ROPO_3H$, $R_2C=O$, $RCOSR$, and CO_2.

The carbon nucleophile can be an enolizable $\diagup$CH—$\overset{|}{C}$=O group or a derivative of the enol form. The carboxylation of pyruvic acid or its enol phosphate ester to oxaloacetic acid can be accomplished by at least four different enzymes. One of these reactions is shown below; the driving force is provided by hydrolysis of the high-energy enol phosphate and formation of a carbonyl group.

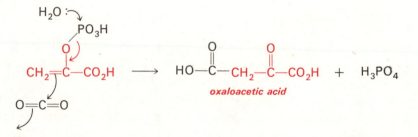

All these reactions are subject to the limitations imposed by the very mild conditions prevailing in the cell and by the necessity for balancing energy requirements. The enzymatic process may therefore involve several steps to accomplish a reaction that would be a single operation in a laboratory synthesis. Each step makes "chemical sense," however, and can be understood in terms of basic organic chemistry.

It must be kept in mind that enzymatic reactions involve a direct interaction of the reactants with the protein molecule. The enzyme-substrate interaction may be hydrogen-bonding or a covalent bond. For example, in many enzymatic aldol condensations a free NH_2 group in the protein chain forms an addition product with the C=O group to facilitate enolization (Section 8.18).

$$-\overset{|}{\underset{|}{C}}-\overset{H}{\underset{|}{C}}=O \;+\; :NH_2R \;\rightleftharpoons\; -\overset{|}{\underset{|}{C}}-\overset{H}{\underset{|}{C}}=\overset{+}{N}HR \;\rightleftharpoons\; -\overset{|}{\underset{|}{C}}=\overset{|}{\underset{|}{C}}-\overset{..}{N}HR$$

15.4   Coenzymes

In many enzymatic reactions, the functional group responsible for catalyzing the bond changes is part of a separate molecule called a **coenzyme.** Both coenzyme and substrate molecules are bound to the protein during the reaction; the complex of protein, coenzyme, and product then dissociates. The same coenzyme can be associated with a number of different proteins, each of which catalyzes a given reaction with a specific substrate. Coenzymes are complex molecules containing the catalytic functional group and usually also a phosphate or nucleotide unit. The functional group unit in each case is one of the B vitamins, discussed later in this chapter. In the structures following, the B vitamin part of the coenzyme is shown in color. About 10 coenzymes are known, but we will discuss only a few of the simpler ones.

Coenzyme A (HSCoA)

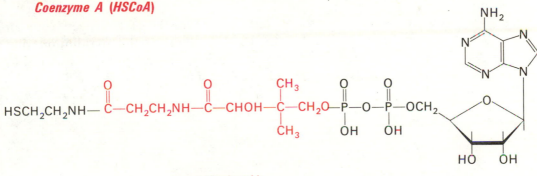

pantothenic acid

The key functional group in coenzyme A is the thiol or SH group, and the coenzyme function is carried out by the formation of thioesters. The most important of these is the ester of acetic acid, called acetyl coenzyme A. The CH_3CO group in acetyl CoA has a reactivity more like that of an acid anhydride than an oxyester. The thioester undergoes enzyme-catalyzed ester condensations like those in the synthesis of fatty acids (Section 10.18). Moreover, hydrolysis of the thioester bond is an energy-yielding step, and energy for metabolic reactions is "stored" when a molecule of acetyl CoA is formed.

$$CH_3\overset{O}{\overset{||}{C}}OH + HSCoA \rightleftharpoons CH_3\overset{O}{\overset{||}{C}}SCoA + H_2O$$

acetyl
coenzyme A

Nicotinamide Adenine Dinucleotide (NAD⁺)

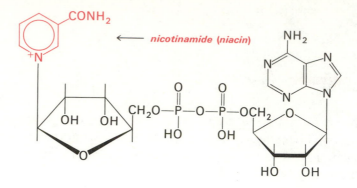

nicotinamide (niacin)

The pyridinium ring in NAD^+ is an *oxidizing agent* in many enzymatic reactions. The coenzyme is reduced to the 1,4-dihydropyridine, called NADH; a typical reaction is the oxidation of an alcohol:

$$CH_3CH_2OH + NAD^+ \rightleftharpoons CH_3-\overset{O}{\overset{\|}{C}}-H + NADH + H^+$$

The oxidation-reduction process occurs by transfer of hydrogen with an electron pair from the alcohol to the 4 position of the pyridinium ring (Section 13.2). The reverse reaction, in which a carbonyl group is reduced to the alcohol by NADH, has already been discussed in Section 8.8.

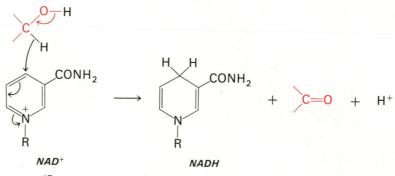

(**R represents the remainder of the NAD structure shown above**)

Oxidation of NADH to regenerate NAD^+ is accomplished by molecular oxygen in a series of steps called an **electron transport chain.** The enzymes in this chain are called cytochromes, and contain porphyrin nuclei bonded to proteins; the electron transfer occurs by reversible oxidation of the iron atom (Section 13.1). Energy liberated by the overall oxidation is stored by coupling the oxidation with conversion of ADP to ATP. The energy released in the oxidation of NADH is sufficient to produce three moles of ATP.

$$NADH + H^+ + \tfrac{1}{2}O_2 + 3\ ADP + 3\ H_3PO_4 \longrightarrow NAD^+ + 3\ ATP + 4\ H_2O$$

Flavin Adenine Dinucleotide (FAD)

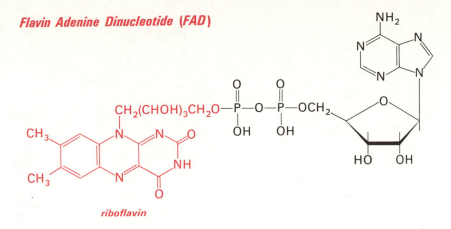

riboflavin

The flavin coenzyme system FAD-FADH$_2$ provides a second means of oxidation-reduction for enzymatic reactions. FAD is a less powerful oxidizing agent than NAD$^+$, and is often linked with the NADH-NADH$^+$ system in electron transport. The riboflavin ring is the equivalent of a quinone, and forms a relatively stable free radical; reduction can therefore occur by one-electron steps.

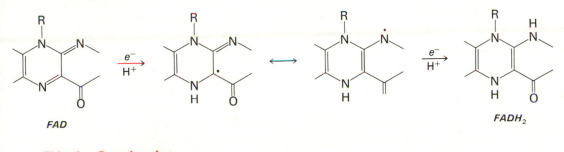

FAD   **FADH$_2$**

Thiamine Pyrophosphate

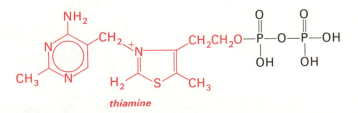

thiamine

One of several important reactions involving thiamine is the decarboxylation and oxidation of an α-keto acid. This sequence is a major step in the breakdown of carbohydrates to release energy. The chemical reaction is a

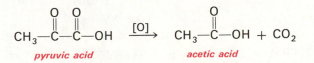

pyruvic acid   *acetic acid*

simple one which could be carried out under mild conditions in the laboratory with a variety of oxidizing agents. In the cell, however, the important

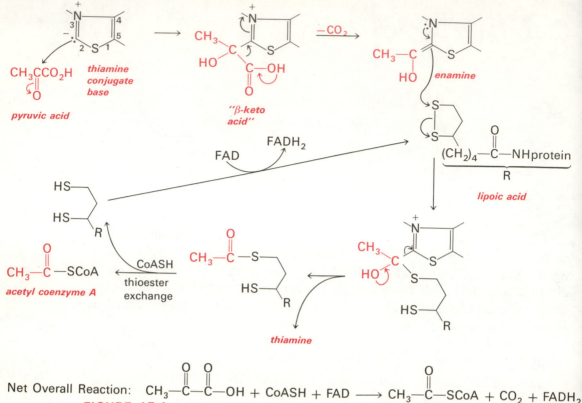

Net Overall Reaction:

$$CH_3-\overset{O}{\underset{}{C}}-\overset{O}{\underset{}{C}}-OH + CoASH + FAD \longrightarrow CH_3-\overset{O}{\underset{}{C}}-SCoA + CO_2 + FADH_2$$

FIGURE 15.1

Oxidative decarboxylation.

aspect is transfer of the energy released to some form in which it can be stored or used to do muscular work. This is accomplished by a series of steps that requires thiamine and three other coenzymes (Figure 15.1). **Oxidative decarboxylation** is an excellent illustration of the type of complex chain by which enzymatic processes occur.

The conversion of pyruvic acid to acetic acid and CO_2 cannot occur by direct decarboxylation of the α-keto acid, since a carbonyl group in the β-position of an acid is required for easy loss of CO_2 (Section 10.18). The cell calls on the thiamine coenzyme to overcome this problem. The catalytic role of thiamine depends on the high acidity of the proton in the 2-position of the thiazolinium ring. The conjugate base is stabilized by the adjacent positively charged nitrogen and also by the sulfur atom. Addition of the base to the carbonyl group of an α-keto acid gives an intermediate in which the original $\alpha \diagdown C{=}O$ group is converted to a $\beta \diagdown C{=}N^+$ group, which is part of the thiazole ring. The "temporary" β-keto acid readily undergoes decarboxylation to give an enamine, which can then react as a nucleophile.

The next two steps accomplish the actual oxidation. Another coenzyme, lipoic acid, which is a cyclic disulfide, is reduced by attack of an electron pair from the enamine on the S—S bond. Lipoic acid emerges as a thioester of the dithiol form, and the original thiamine is recovered. Coenzyme A then picks up the acetyl group by thioester exchange, and the reduced lipoic acid is reoxidized to the cyclic disulfide.

15.5 Carbohydrate Metabolism and Photosynthesis

Glucose is the key link between the metabolic pathways of the plant and animal kingdoms. Glucose is ingested by animals in the form of starch and sucrose and is burned in the "metabolic furnace" to provide energy for cellular processes and muscular work. The CO_2 produced by combustion is then recycled by photosynthesis in plants, which leads back to glucose.

$$C_6H_{12}O_6 \quad + \quad O_2 \quad \underset{photosynthesis}{\overset{metabolic\ oxidation}{\rightleftharpoons}} \quad CO_2 \quad + \quad H_2O$$

The metabolic breakdown of glucose occurs in two stages: (1) **anaerobic glycolysis** to two three-carbon units and (2) **oxidation** of these units to CO_2. In the first stage (see *top* of Figure 15.2), cleavage of the six-carbon chain occurs by a reverse aldol condensation of fructose diphosphate. Two moles of lactic acid, $CH_3CHOHCO_2H$, are eventually formed in a number of further steps. Only a few of the intermediates are shown in Figure 15.2.

This anaerobic stage of the metabolic sequence involves no overall oxidation. About 10 per cent of the total energy available in glucose is converted to high-energy ATP, and this is the primary source of energy for muscular work. During vigorous exercise, the lactic acid concentration of blood increases several fold. Accumulation of lactic acid ultimately leads to fatigue and muscular rigor until balance is restored by oxidation. In yeast, the anaerobic pathway takes a sidepath in which pyruvic acid undergoes decarboxylation to acetaldehyde. Reduction then gives ethanol, and the overall process in this case is called **fermentation.**

In the second phase of glucose metabolism, pyruvic acid from the glycolysis sequence is oxidized to CO_2 by the oxidative decarboxylation shown in Figure 15.1. Acetyl coenzyme A then condenses with oxaloacetic acid and enters a series of reactions called the **citric acid cycle.** Subsequent steps accomplish the successive loss of two moles of CO_2 and lead back to oxaloacetic acid. A few of the steps are shown in the bottom part of Figure 15.2. The energy liberated in the various oxidative steps is taken up by converting NAD^+ to NADH, and is eventually stored as ATP.

Photosynthesis is the path by which chemical energy from sunlight and carbon from CO_2 enter living systems. Photosynthesis occurs in special particles called chloroplasts in the leaf cells of green plants and in green algae. Energy from sunlight is absorbed by the pigment chlorophyll, which makes up about 10 per cent of the organic material in chloroplasts. In a complex series of reactions, energy from the photo-excited chlorophyll is used to transfer hydrogen from water to a pyridinium coenzyme.* The

---

*The coenzyme is NAD phosphate, with an additional phosphate group in the 2-position of the adenosine unit.

Anaerobic Glycolysis: $C_6H_{12}O_6 + 2\ ADP + 2\ H_3PO_4 \longrightarrow 2\ CH_3CHOHCO_2H + 2\ ATP + 2\ H_2O$

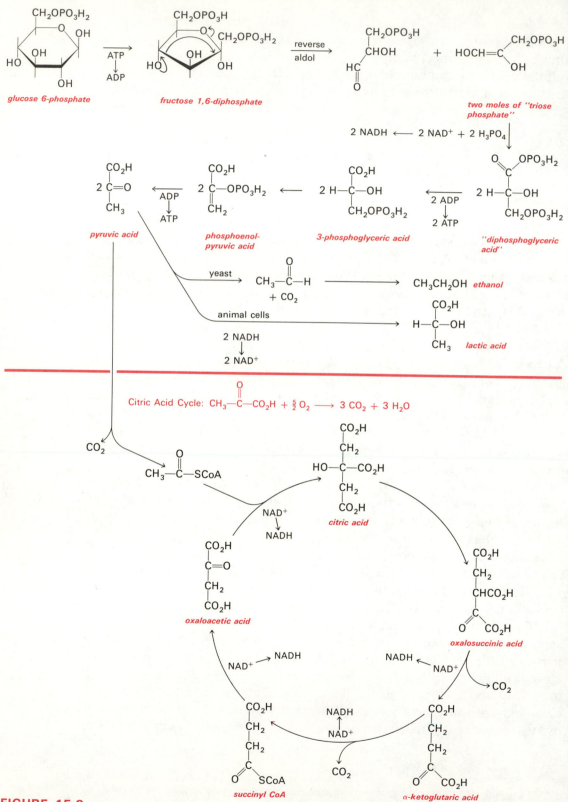

FIGURE 15.2

Abbreviated scheme of the major steps in the two phases of carbohydrate metabolism.

products are oxygen, the energy-rich reduced coenzyme, and high-energy ATP.

$$2\ NADP^+ + H_2O + 2\ ADP + 2\ H_3PO_4 \longrightarrow 2\ NADPH + 2\ H^+ + O_2 + 2\ ATP$$

The CO_2 "fixation" steps in photosynthesis utilize the energy trapped by the photoreaction. In the overall process, six carbons from CO_2 are incorporated into glucose. The first step is nucleophilic addition of an enolized five-carbon pentose to carbon dioxide; this is followed by cleavage to two moles of phosphoglyceric acid.

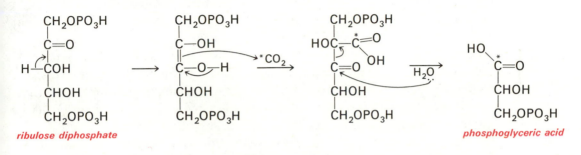

In order to build up hexose molecules entirely from CO_2, all the C—C bonds must be broken and remade in different ways so that each carbon in the chain can be supplied by a C-1 unit. This scrambling is accomplished by a neat cycle of condensations and cleavages that proceeds through carbon

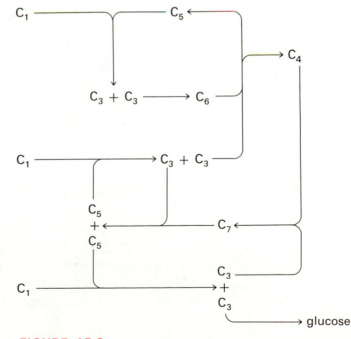

FIGURE 15.3

Schematic diagram of chain scrambling in photosynthetic carbon cycle.

chains of various lengths. Figure 15.3 shows the net formation of a C-3 sugar from three C-1 fragments. All the intermediates with four, five, six and seven carbons are continuously regenerated, and C-1 units from CO_2 are fed into every position.

15.6   Vitamins

As noted in the previous section, the functional group units in coenzymes are B vitamins. Before their biochemical role was known, these compounds were recognized as *nutritional* entities. Vitamins are compounds that are required in small amounts for the growth and function of an animal and which must be supplied in the diet.* Absence of a given vitamin in the diet can lead to a deficiency disease such as scurvy, which was common in the seventeenth and eighteenth centuries among sailors on long sea voyages, when the diet for months consisted of preserved meat, hardtack, and rum.

It was found that scurvy was prevented by addition of citrus fruits or green vegetables to the diet, and this observation eventually led to the isolation of a compound which was named **ascorbic acid,** or vitamin C. Ascorbic acid is an enolic lactone derived from glucose in plants and some animals. Primates and a few other mammals lack the enzyme required to produce the vitamin from glucose. The enediol grouping in ascorbic acid is a strong reducing agent, and the biochemical role is associated with the reducing properties of the compound.

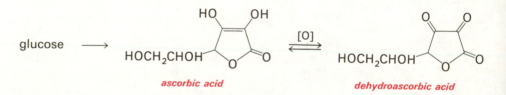

ascorbic acid   *dehydroascorbic acid*

Other vitamins were similarly recognized because of deficiency symptoms associated with a restricted diet in man or experimental animals. About 15 compounds, most of them designated by letters such as A, B_1, B_6, B_{12}, C, D, E, and so on are known to be vitamins for man. Vitamins A (Figure 3.9) and D are required in color vision and calcium metabolism, respectively. Each vitamin has an essential biochemical role; the most thoroughly understood are the coenzyme functions of the B vitamins in general metabolism. Clear-cut deficiency conditions have not been established in every case, however. Even in restricted diets, an adequate amount of the vitamin may be supplied by the bacterial flora present in the digestive tract of all animals. A balanced diet supplies all the vitamin requirements of a healthy person. Several of the vitamins are simple compounds that can be manufactured very

---

*In addition to vitamins, certain amino acids and fatty acids are also essential in the nutrition of man and other mammals; these compounds are not considered as vitamins since they are building blocks rather than part of the metabolic apparatus.

cheaply, and they are frequently added to prepared foods without any direct relationship to the likelihood of a dietary deficiency.

15.7   Secondary Biochemical Processes

The fact that vitamins and certain amino acids are required in the diet of mammals is significant from the standpoint of the relative biochemical capabilities of different organisms. At the beginning of the chapter we stressed the uniformity of biochemistry as reflected in the biopolymers and enzymatic reactions common to all living forms. In addition to these universal metabolic pathways, however, some other very important processes take place only in certain organisms. Foremost among these is photosynthesis, which occurs only in green plants and algae.

The scope and diversity of metabolic activity in plants are far greater than those in animals. The reason for this is simple—plants must carry out all the biochemical reactions needed for growth and reproduction with CO_2, inorganic nitrogen, and salts as the only starting materials; the biochemical apparatus must be complete and all-encompassing. On the other hand, animal forms in the course of evolution seem to have "lost" the enzymes required to produce a number of basic compounds that can be obtained from food.

Enzyme systems in the animal kingdom are limited to the primary reactions of energy production and the synthesis and condensation of simple aliphatic compounds and heterocyclic bases. The relatively straightforward condensations required for pyrimidines can be carried out in animal metabolism, but benzene rings cannot be synthesized in mammalian tissues. Essential phenyl-substituted amino acids must be supplied from food derived ultimately from plants or bacteria which can carry out the condensation of carbohydrates to benzenoid compounds.

Beyond the basic metabolic processes central to vital functions, some additional biochemical activities of bacteria and higher plants are termed **"secondary metabolism."** These processes lead to a variety of complex compounds, including alkaloids and antibiotics (Chapter 13), lignans (Chapter 12), and terpenes (Chapter 3), collectively called **natural products.** The distribution of these compounds in plants is often quite restricted, sometimes to a single species. Chemical mechanisms by which these natural products are formed can be traced from simple precursors by ^{14}C labeling, as illustrated for the alkaloids (Section 13.5), but relatively little is known about the enzymes involved.

The function of these secondary metabolites in the microorganism or plant is obscure. The quinones (Section 12.8) produced by certain plants are toxic to wood-rotting fungi, and some alkaloids are very bitter or toxic, and deter attack by animals. However, other plant species survive quite well without producing detectable amounts of these compounds. Regardless of their role in the ecology of the plant or mold, however, the importance of alkaloids and antibiotics as drugs has made natural products a major area of organic chemistry. The structural complexity of these compounds has provided a great stimulus in the development of spectroscopic methods for structure analysis and also synthetic reactions.

PREBIOLOGICAL SYNTHESIS OF ORGANIC COMPOUNDS

Having seen the complex structures of compounds produced by enzymes, and the elaborate processes by which they are formed and transformed, we must inevitably ask how these compounds and pathways have evolved from the inorganic materials present in the newly created planet about 4.5 billion years ago. Several lines of evidence suggest that the primitive earth possessed a reducing atmosphere, containing H_2, N_2, H_2O, NH_3, CH_4, and H_2S, and probably little or no O_2 and CO_2. These gases plus liquid water, salts, and minerals were presumably the raw materials for the earliest abiotic synthesis of organic compounds which must have occurred under the flux of solar energy and cosmic radiation prior to the existence of enzymes.

Experiments have shown that a number of compounds are produced by exposing mixtures of these ingredients to electric discharges, ultraviolet radiation, or high temperatures. It is impossible to know which sets of conditions most closely simulate the primitive environment. There is no question, however, that substances which are elaborated in living cells arise spontaneously under conditions resembling those that existed about four billion years ago.

Extensive work has been done with mixtures of CH_4, NH_3, H_2, and H_2O with a variety of energy sources. In an electric discharge, about six α-amino acids and several α-hydroxy acids have been produced. Compounds such as HCN and HCHO can also be detected, and addition of these to the reaction mixtures leads to additional products. Heating mixtures of CH_4, NH_3, and H_2O to 900°C in the presence of quartz gives 12 of the 20 α-amino acids found in proteins.

These reactions are initiated by the formation of free radicals and other highly reactive intermediates. Formation of α-amino acids probably occurs by condensation of an aldehyde, HCN, and ammonia, a well-known laboratory synthesis.

$$CH_4 \xrightarrow{h\nu} CH_3\cdot \xrightarrow{H_2O} CH_2O \xrightarrow[HCN]{NH_3} NH_2CH_2CN \longrightarrow NH_2CH_2CO_2H$$

Aspartic acid is formed in relatively high yield in some of these experiments. This product is thought to arise by way of cyanoacetylene.

$$NH_3 + CH_4 \longrightarrow \underset{\textit{cyanoacetylene}}{HC\equiv C-CN} \xrightarrow{NH_3} \overset{NH_2}{\underset{}{H-C=CHCN}} \Big\downarrow HCN$$

$$\underset{\textit{aspartic acid}}{HO_2C-\overset{NH_2}{\underset{}{CH}}-CH_2CO_2H} \xleftarrow{H_2O} NC-\overset{NH_2}{\underset{}{CH}}-CH_2CN$$

An important finding in these experiments is the formation of the purines and pyrimidines that are present in nucleic acids. Adenine is obtained by simply heating solutions of HCN and NH_3 at 90°C, and it can also be isolated in high-energy reactions of CH_4 and NH_3. Pyrimidines are readily obtained under mild conditions from simple precursors, and these are also formed in simulation experiments.

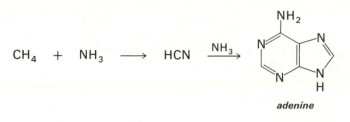

$$CH_4 \ + \ NH_3 \ \longrightarrow \ HCN \ \xrightarrow{NH_3}$$

adenine

Other experiments with simple molecules that are products of the primary reactions lead to ribose, glucose, fatty acids, and porphyrins. The next step in reconstructing abiotic synthesis is condensation of monomer units to give polymers. The conditions in which the original biopolymers were formed are subjects of controversy. Relatively high temperature reactions and catalytic condensations on the surface of clay minerals are two of the possibilities that have been explored. Polypeptides are produced from amino acid mixtures in these experiments, and the amino acids are incorporated in a nonrandom way. The preference for certain dipeptide sequences in these products is in the same direction as that seen in contemporary proteins. Finally, the proteinoid materials formed in high-temperature condensations show a tendency to form cell-like spherical aggregates in water.

Overall, these simulation experiments have shown that a number of biochemically significant compounds can be produced under conditions that are thought to have prevailed prior to the existence of living organisms. A great many other compounds are also found in the experiments, and attention has naturally been focused on those of biochemical interest. Despite the limitations and assumptions of these experiments, they provide a plausible concept of chemical evolution. As compounds became more complex, the degree of selection of structures and reactions would increase. At some point, a polymer could serve as a template for the organization of smaller precursor molecules, marking the beginning of a living system.

SUMMARY AND HIGHLIGHTS

1. The same basic chemical processes occur in the cells of all living organisms.

2. **Metabolic reactions** are enzyme-catalyzed; the reactions are fundamentally the same as those of organic compounds in laboratory glassware, but they occur under a narrow range of conditions, and the processes may require many steps.

3. Chemical energy is stored and transferred by forming and breaking **phosphate esters.**

4. **Coenzymes** are compounds that contain the catalytic functional group in many enzymatic reactions; the same coenzyme, in combination with different enzymes, can catalyze a reaction with a variety of substrates.

5. **Coenzyme A** contains an —SH group; acyl groups are transferred as thioesters, $R-S-\overset{\overset{\displaystyle O}{\|}}{C}-R'$.

6. **NAD$^+$-NADH** and **FAD-FADH$_2$** contain pyridinium and flavin rings, respectively, and function in electron-transfer reactions.

7. **Thiamine phosphate** catalyzes reactions by reversibly converting a $C=O$ group to $\overset{\overset{\displaystyle OH}{|}}{C}-\overset{+}{C}=N$, thus permitting reactions such as decarboxylation of an α-keto acid.

8. **Breakdown of carbohydrates** is the main source of energy in cells. Glucose is first converted without oxidation to pyruvic acid and is then broken down to CO_2 with release of energy.

9. The reverse process, **photosynthesis,** is carried out in green plants. Energy from sunlight is used to convert H_2O to oxygen plus high-energy reaction intermediates, which provide the energy needed to build up glucose from CO_2.

10. **Vitamins** are compounds required for biochemical functions, usually as coenzymes, that are not synthesized by animals and must be supplied in the diet.

11. **Secondary metabolic processes** in plants and lower organisms produce alkaloids, antibiotics, terpenes, and other substances which have no direct role in metabolism of the organism.

12. The origin of metabolic processes is thought to be the slow condensation of CH_4, NH_3, H_2, and H_2O under conditions of high light energy, lightning, and heat, prior to the existence of living systems. Reactions resembling those that could have occurred lead to amino acids, hydroxy acids, and purine bases; further condensation of these compounds leads to more complex molecules.

PROBLEMS

15.1 For each of the following reactions, complete and balance the equation and identify the "energy storing" and "energy releasing" components of the overall process; *i.e.*, which reactant absorbs, and which releases, energy?

a) $2 \text{ NADH} + 2 \text{ H}^+ + O_2 + 6 \text{ ADP} + 3 \text{ H}_3PO_4 \longrightarrow$

b) $CH_3COCO_2H + CoASH + FAD \longrightarrow$

c) $C_6H_{12}O_6 + 6 O_2 + 38 \text{ ADP} + 38 \text{ H}_3PO_4 \longrightarrow$

15.2 Distinguish between the terms *B vitamin* and *coenzyme*.

15.3 The decarboxylation of pyruvic acid can be catalyzed by cyanide ion, CN^-, in a process that parallels the action of thiamine. (The reaction with CN^- was a key clue in determining the mechanism of thiamine catalysis.) Write steps which would account for the catalysis by CN^- (the first step, as with thiamine, is nucleophilic addition to the ketone carbonyl group of pyruvic acid).

15.4 The last step in anaerobic glycolysis (Figure 15.2) is the reduction of pyruvic acid by NADH. (a) Using a partial structural formula for NADH, show the reaction by which lactic acid is produced. (b) Does this reaction liberate or consume energy? (c) Pyruvic acid is an achiral (symmetrical) compound; lactic acid is asymmetric. Is the lactic acid produced by NADH optically active? (d) Suggest a chemical reducing agent that resembles NADH (*i.e.*, one which could bring about the same reaction in aqueous solution).

15.5 The average CO_2 concentration in the atmosphere is about 320 parts per million. In the Northern Hemisphere, the value fluctuates annually over a range of about $\pm3\%$. At what times of the year would you expect CO_2 concentration to be maximal or minimal?

16
NUCLEIC ACIDS

Nucleic acids are polymers with **nucleotides** as the repeating units. These biopolymers carry the information, or instructions, necessary for an organism to develop and reproduce itself. The information, in a "language" called the **genetic code,** is contained in the *sequence of the nucleotides* in the polymer. This sequence in turn determines the *sequence of amino acids* in proteins, including the enzymes that control metabolism.

Two types of polynucleotide are involved in this complex process; one contains the sugar **2-deoxyribose** and the other, **ribose.** Deoxyribonucleic acid **(DNA)** occurs in the chromosomes of cell nuclei, and is the master template on which genetic traits are preserved and transmitted. Ribonucleic acid **(RNA)** is formed on the DNA pattern and translates the genetic code from DNA to the amino acid sequence of proteins.

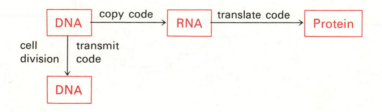

16.1 Deoxyribonucleic Acid (DNA)

As described in Chapter 11, a nucleotide is a glycoside containing a **base,** a **pentose sugar,** and a **phosphate ester** group. The nucleotides in DNA contain one of four different bases attached to C-1' of 2'-deoxyribose (the primes indicate numbering of the sugar unit). The nucleotides are linked together in the polymer chain as a phosphate diester through the 3' OH group of one sugar unit and the 5' OH of the sugar in the next nucleotide (see Figure 16.1).

Analysis of the nucleotide mixtures obtained by enzymatic hydrolysis of DNA reveals a very important relationship among the bases. The amounts of guanine and cytosine are almost exactly the same (G/C = 1), and the amounts of adenine and thymine are the same (A/T = 1). These ratios are a

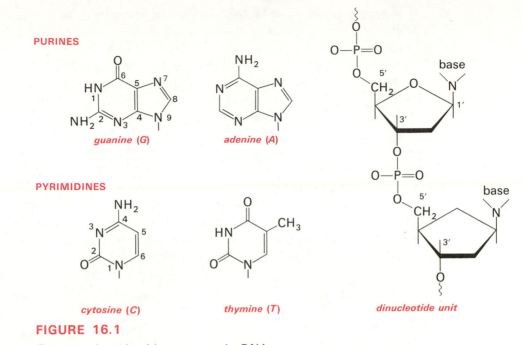

FIGURE 16.1

Bases and nucleotide structure in DNA.

consequence of complementary *pairing* of a purine with a pyrimidine in the nucleic acid structure.

Base-pairing in nucleic acids is explained by the possibilities for multiple hydrogen-bonding between G—C and A—T combinations. Models of polynucleotides show that when the bases of two chains are lined up back-to-back, guanine can form *three* hydrogen bonds to a cytosine unit in the complementary chain, and adenine can form *two* hydrogen bonds to a thymine unit. *Any other combinations result in fewer hydrogen bonds between base pairs.*

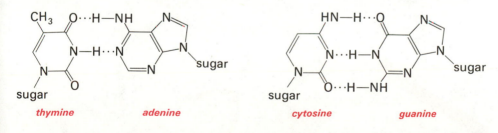

Structure and Replication of DNA.   DNA is an extremely large polymer, with molecules containing more than a million nucleotide units. The polymer is isolated from cells as a fibrous mass; because of the huge size, samples containing a single molecular species with definite chain length and structure cannot be obtained, as is possible with proteins. X-ray diffraction patterns of DNA indicate a regular spiral structure with the chain of phosphate and sugar units twisted around the long axis of the polymer. The position and sequence of individual bases cannot be defined, however.

A crucial property of DNA that must be accounted for in any structural model is the ability of the polymer to be *duplicated* in the process of cell division so that each new cell contains an exact copy of all DNA molecules, with every nucleotide in place. Putting this fact together with the x-ray measurements and the phenomenon of base-pairing, a structure emerged in which two strands of polynucleotide are held together by hydrogen-bonding of the bases in a **double helix** (Figure 16.2 and Color Plate 12).

In this model, the sugar-phosphate-sugar backbones of the two chains form the outer surface of the polymer and run in opposite directions. In one chain, the deoxyribose units are connected $5' \longrightarrow 3' \longrightarrow 5'$; the complementary chain has the reverse sequence, $3' \longrightarrow 5' \longrightarrow 3'$. The pairs of bases fit neatly between the deoxyribose phosphate chains, with a purine pointing toward a pyrimidine and vice versa across the central axis of the helix. Successive base pairs, which can come in any sequence, are thus stacked on one another, about 3.4 Å apart.

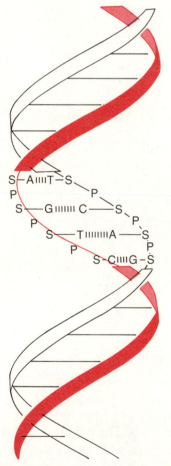

FIGURE 16.2

Double helix structure of DNA. (From Jones, Netterville, Johnston and Wood: Chemistry, Man and Society. Philadelphia, W. B. Saunders Company, 1972.)

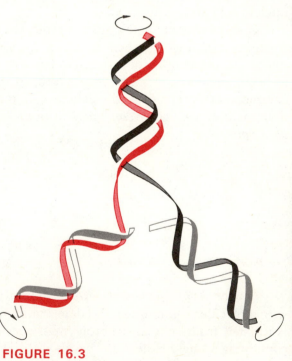

FIGURE 16.3

Replication of DNA.

Replication of the DNA chain begins with the unwinding of the double helix. As the core of bases is exposed, nucleotides from the cell fluid are directed by an enzyme to take the place of those that were present in the original opposite strand; their sequence is determined by base-pairing. New 3'-phosphate bonds are then formed, and the result is two new double chains, each containing one strand of the original. The duplication process is shown schematically in Figure 16.3.

16.2   Ribonucleic Acid (RNA)

RNA has the same 3',5'-phosphate diester repeating unit as DNA, with ribose instead of deoxyribose as the sugar. Three of the four bases in RNA are G, C, and A, as in DNA; the fourth base is uracil (U), which lacks the 5-methyl group of thymine. The tautomeric structures of uracil and thymine are the same, and U is complementary with A.

Information stored in the base sequence of DNA is transmitted to the cell by RNA. In essentially the same way that deoxynucleotides form a new strand of DNA, *ribonucleotides* of guanine, cytosine, adenine, and uracil pair with the complementary bases (G—C and A—U) on an uncoiled DNA strand to form an RNA strand. RNA is a much smaller polymer than DNA, and does not have a double-stranded structure. However, the RNA chain can fold and form helical loops that are stabilized by hydrogen bonds between complementary base pairs.

Most of the RNA in a cell is localized in particles called ribosomes, which are rich in **ribosomal RNA** and are the sites of protein synthesis. Two special types of RNA play major roles in translating the genetic code from the nucleotide language of DNA to the amino acid language of proteins. **Messenger RNA** (*m*RNA) carries the code from the DNA molecule in the nucleus to the ribosome. A group of **transfer RNA** (*t*RNA) molecules perform the ordering of the individual amino acids. About 80 *t*RNA's have been isolated—from one to four for each of the 20 amino acids in proteins. *t*RNA's are relatively small polymers with 70 to 90 nucleotides; they contain several modified bases in addition to G, C, A, and U. The complete sequence of bases is known for many *t*RNA molecules; one of these is shown in Figure 16.4*A*. The "cloverleaf" folding pattern is a schematic representation, which shows the possibility for base-pairing and the existence of exposed loops. By means of x-ray diffraction, the three-dimensional structure of a crystalline *t*RNA has been determined (Figure 16.4*B*), and the loops can be seen to be a physical reality.

16.3   The Genetic Code

One of the most striking developments in chemistry in the 1960's was the "cracking" of the genetic code—the chemical mechanisms by which genetic traits are passed from parent to progeny, and which determine that one fertilized egg develops into a fruit fly and another into a human. Biologists have recognized for many years that the key to these mysterious processes lay in the **chromosomes** of the cell nucleus. Structural and functional characters of the organism are determined by **genes** within the chromosomes.

In a simplified view, a chromosome is a collection of DNA molecules; duplication of the double helix provides for the preservation of hereditary characteristics. Since all aspects of growth and development are determined by the enzymes available to the cell, a gene represents the *information* required to produce a given enzyme. A bacterial cell may contain a thousand genes, and a mammalian cell, a million. Human genes are divided among 46 chromosomes, and may be repeated several times. In chemical terms, a gene is that portion of a DNA template on which a specific *m*RNA chain is constructed by base-pairing of ribonucleotides and polymerization.

Protein Synthesis. For the synthesis of a polypeptide chain, the amino acids are first attached to *t*RNA. Each amino acid requires a specific enzyme for this reaction, and one of several specific *t*RNA molecules. The amino acids are activated as mixed anhydrides with ATP and are then attached as esters with the 2′ or 3′ OH group in the terminal adenosine unit of the *t*RNA. (All *t*RNA's contain the sequence C—C—A at the end of the chain.)

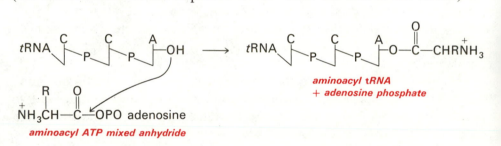

aminoacyl tRNA
+ adenosine phosphate

aminoacyl ATP mixed anhydride

We now come to the question of how the aminoacyl *t*RNA molecules are ordered to produce the correct sequence in the polypeptide chain. It has been shown by an ingenious experiment that recognition of the correct aminoacyl *t*RNA by the *m*RNA-ribosome complex depends on some feature in the nucleotide chain of the *t*RNA, and not on the amino acid. The *t*RNA for cysteine with ^{14}C cysteine attached was isolated from a cell and then was hydrogenated to desulfurize the cysteine to alanine. When the modified tRNA$_{cys}$ carrying ^{14}C-alanine was added to protein-synthesizing cells, the ^{14}C alanine was incorporated in the chain where the cysteine should have been located.

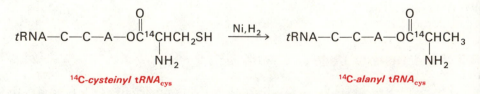

^{14}C-cysteinyl tRNA$_{cys}$ ^{14}C-alanyl tRNA$_{cys}$

The recognition site in each aminoacyl *t*RNA is located on an exposed loop of the nucleotide chain in a segment called the **anticodon** (Figure 16.4).

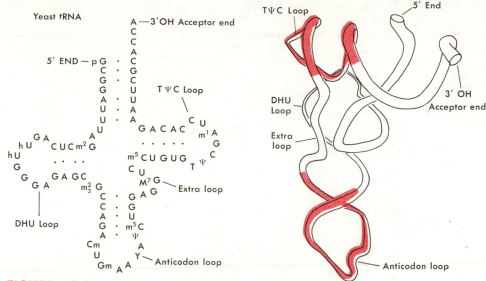

FIGURE 16.4

*t*RNA. *A.* Cloverleaf representation. *B.* Three-dimensional looping of chain. (From Kim *et al.,* Science *179:*285, 1973.)

This segment is complementary to a **codon** in the *m*RNA. Base-pairing between codon and anticodon provides the translation from nucleotide sequence to amino acid sequence.

The Triplet Code.   Having seen the process of replication of DNA, and transcription from DNA nucleotide sequence to *m*RNA sequence to amino-acyl *t*RNA sequence, we can now ask what message this elaborate machinery transmits. What type of code is needed to provide for the sequence of 20 amino acids in the protein, using as a language the sequence of four bases in the nucleic acid?

Direct evidence that nucleotide sequence does determine the incorporation of amino acids into proteins was obtained by using a synthetic polyribonucleotide containing a single base, uracil. The synthetic poly-U can take the place of the natural *m*RNA in a cell-free system. A mixture of amino acids was added, and only *one* of them, phenylalanine (phe), was incorporated. This experiment demonstrated that some combination of the repeating sequence U—U—U—U codes specifically for this amino acid.

A code based on a sequence of two nucleotides, such as U—U, would provide 4^2 or 16 combinations, which is inadequate to specify 20 different amino acids. An alphabet using the sequence of three nucleotides contains 4^3 or 64 "letters," and this has been found to be the code. The excess number of combinations means that several triplets are available for each amino acid, and this redundancy is insurance that the message will go through even if there is some uncertainty in the third digit.

With the concept of a code based on triplet sequences in mind, synthetic polynucleotides with known repeating sequences were used to replace RNA. Certain amino acids were found to correspond to specific nucleotide triplets.

TABLE 16.1 **THE DICTIONARY OF THE
GENETIC CODE**

First Digit	Second Digit				Third Digit
(5′ end of mRNA)	U	C	A	G	(3′ end of mRNA)
	Phe	Ser	Tyr	Cys	U
U	Phe	Ser	Tyr	Cys	C
	Leu	Ser	+	+	A
	Leu	Ser	+	Try	G
	Leu	Pro	His	Arg	U
C	Leu	Pro	His	Arg	C
	Leu	Pro	Gln	Arg	A
	Leu	Pro	Gln	Arg	G
	Ileu	Thr	Asn	Ser	U
A	Ileu	Thr	Asn	Ser	C
	Ileu	Thr	Lys	Arg	A
	Met	Thr	Lys	Arg	G
	Val	Ala	Asp	Gly	U
G	Val	Ala	Asp	Gly	C
	Val	Ala	Glu	Gly	A
	Val	Ala	Glu	Gly	G

+ = termination.

In a few years' time the complete code, with the amino acids corresponding to all the triplets, was worked out (Table 16.1). The code has no overlap and no gaps; a sequence of 21 nucleotides specifies seven amino acids—no more and no less. In most of the triplets, the third base is flexible, meaning that the first two bases define the amino acid. Three codons, UAA, UAG, and UGA, are used to stop a "sentence" in the code so that a second polypeptide chain can be started on the same mRNA chain.

The tRNA anticodons which read the mRNA "tape" are nucleotide triplets in which the first two bases are usually the standard ones, U, A, G, and C. These bases pair only with their complementary partners—U-A, A-U, G-C, and C-G. The third digit in the tRNA anticodon is frequently the modified base inosine, which is less specific and more flexible in base-pairing. Inosine can form two hydrogen bonds to C, U, or A, and can pair with any of these when they are the last digit in the mRNA codon.

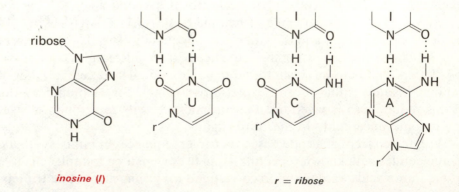

inosine (I)   r = ribose

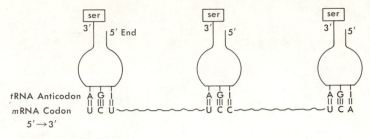

FIGURE 16.5

Base-pairing of anticodon loops.

The translation process is illustrated in Figure 16.5. Three different serine *t*RNA's have the same triplet, AGI, in the anticodon loop. In reading, the anticodon sequence runs in the 3′ ⟶ 5′ direction and the *m*RNA codon sequence is the opposite, 5′ ⟶ 3′. As seen in Table 16.1, three of the codons for serine are UCU, UCC, and UCA. The other three serine codons require different anticodons.

16.4 Mutations and Genetic Damage

The intricate system of DNA replication, RNA transcription, and protein synthesis provides a marvelous mechanism for reproduction of a species, but it also contains the potential for change. Genetic **mutations** occur in the course of natural events, and they can also be induced in a variety of ways. A mutation is an abrupt change in a genetic character; it may be beneficial, but more frequently it is lethal and is eliminated by natural selection.

In terms of the molecular picture of genetics that we have seen in this chapter, a mutation is a change in the sequence of bases in the DNA template. A new base pair may become inserted in the chain, or an existing pair modified or lost. The result is disruption of the chemistry that depends on the affected portion of the DNA chain; the change may be manifested in loss of a certain enzyme. The relative distance between mutations on the chain (the genetic map) and alterations in the amino acid sequence of an enzyme can be correlated, and studies of this kind have greatly amplified understanding of nucleic acid chemistry.

Mutations can be brought about by high-energy radiation or by chemical reagents. The resulting changes in polynucleotide structure are known in several cases. Damage from x-rays is due to a break in the DNA chain. Mutant bacteria that are deficient in the enzymes needed for the repair of DNA chains are particularly vulnerable to radiation. Ultraviolet radiation can cause the photochemical dimerization of adjacent thymine units to give a cyclobutane. This reaction forms an extra link in the strand and interferes with replication. Alkylating agents, such as nitrogen mustards (Section 13.7), similarly form a cross-link between guanine units, accounting for the toxicity of these compounds in rapidly proliferating neoplastic tissue.

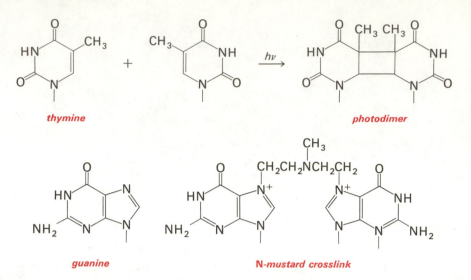

thymine + thymine → photodimer

guanine   N-mustard crosslink

SUMMARY AND HIGHLIGHTS

1.   **Nucleic acids** are polymers of nucleotide units; the nucleotides contain a base, a pentose sugar, and a phosphate group. The units are linked as phosphate diesters between the 3-hydroxy group of one sugar and the 5-hydroxy group of the sugar in the next unit.

2.   **Deoxyribonucleic acid** or **DNA** contains deoxyribose and four bases, the purines **guanine (G)** and **adenine (A),** and the pyrimidines **cytosine (C)** and **thymine (T).**

3.   The bases in DNA are **paired,** G with C and A with T, to obtain maximum hydrogen bonding.

4.   The DNA polymer is a **double helix** with bases on the inside of the sugar-phosphate backbone. The sequence of bases in the chains is complementary, with G in one chain paired to C in the other, etc.

5.   **Replication** of the DNA molecule occurs by untwisting; nucleotides then line up on the single strands to form new complementary chains with the bases paired.

6.   **Ribonucleic acid** or **RNA** contains ribose as the sugar and **uridine (U)** in place of thymine. RNA chains are built up by base-pairing with a DNA chain, preserving and transcribing the base sequence.

7.   Protein synthesis occurs with an RNA chain as a template.

8.   Transfer RNA (*t*RNA) molecules, each carrying a specific amino acid, line up on a messenger RNA (*m*RNA) chain and deliver the amino acids in the order that is determined by the base sequence.

9.   The *t*RNA chains contain a loop in which a group of three nucleotides called the **anticodon** matches a complementary triplet **(codon)** on the *m*RNA chain.

APPENDIX

DETERMINATION OF MOLECULAR FORMULAS

The molecular formula of a compound, *e.g.*, C_3H_8O, gives the number of each type of atom in the molecule. The calculation of formulas from percentage composition is covered in nearly all general chemistry courses, and this type of calculation is not crucial to an understanding of organic chemistry. However, the analysis of compounds and derivation of molecular formulas are essential parts of organic chemistry, and these topics are briefly outlined in this appendix.

Analysis. The presence of elements other than carbon, hydrogen, and oxygen in an organic compound can be detected qualitatively by complete decomposition of the compound with molten sodium metal. This is a drastic reaction in which carbon is largely converted to the element (carbon black). Halogen, oxygen, or sulfur are liberated as simple anions, and nitrogen is converted to cyanide ion (CN^-).

$$(C,H,O,X,N,S) \xrightarrow{\text{Na}} C, H^+, O^{-2}, X^-, CN^-, S^{-2}$$
$$\downarrow H_2O$$
$$2\ OH^-$$

To determine the percentages of carbon and hydrogen in a compound, an accurately weighed sample is burned in a stream of oxygen to convert all the carbon in the sample to CO_2 and all the hydrogen to water. These products are absorbed and weighed, and the amounts of carbon and hydrogen in the sample are calculated. This determination is carried out routinely with a few milligrams of sample. When only carbon, hydrogen, and oxygen are present, the percentage of oxygen can be determined by difference. If nitrogen or other elements are present, separate determinations for these elements must be carried out. Finally, the atomic ratios are calculated from the weight percentages of the elements.

Calculation of Formulas. The formula derived from combustion analysis is *empirical; i.e.*, it is the simplest formula with whole-number atomic ratios that can be calculated from the percentage composition. The *molecular formula* is a whole-number multiple of the empirical formula.

$$\text{empirical formula} \times n = \text{molecular formula}$$
$$n = 1, 2, 3 \ldots$$

To determine the molecular formula, the molecular weight must be known. The most rapid and accurate means of measuring molecular weight is mass spectrometry (see Section 9.6).

EXAMPLE 1

An unknown compound contains carbon, hydrogen, and possibly oxygen. Combustion of 4.55 mg of the compound gave 9.32 mg of CO_2 and 2.86 mg of H_2O. The molecular weight, from the mass spectrum, is 172. Calculate the empirical and molecular formulas.

Step 1. Percentage composition from combustion products:

To determine the percentages of carbon and hydrogen in the sample, the weight of carbon in the CO_2 and the weight of hydrogen in the water are calculated from the weights of CO_2 and H_2O:

$$\underset{12}{C} \longrightarrow \underset{44}{CO_2} \quad \text{Let } X = \text{mg of C in sample; 9.32 mg } CO_2 \text{ formed.}$$

$$\frac{X}{12} \text{ mg C} = \frac{9.32}{44} \text{ mg CO}_2; \quad X = \frac{12 \times 9.32}{44} = 2.54 \text{ mg C}$$

$$\underset{2 \times 1 = 2}{2\,H} \longrightarrow \underset{18}{H_2O} \quad \text{Let } Y = \text{mg of H in sample; 2.86 mg of } H_2O \text{ formed.}$$

$$\frac{Y}{2} \text{ mg H} = \frac{2.86}{18} \text{ mg H}_2O; \quad Y = \frac{2.86}{9} = 0.318 \text{ mg H}$$

% C in sample: $\dfrac{2.54 \text{ mg C}}{4.55 \text{ mg sample}} \times 100 = 55.8\% \text{ C}$

% H in sample: $\dfrac{0.318 \text{ mg C}}{4.55 \text{ mg sample}} \times 100 = 7.00\% \text{ H}$

% O by difference: $100 - (55.8 + 7.00) = 37.2\% \text{ O}$

Step 2. Atomic ratio from weight per cent:

The atomic ratios of each element in the compound are obtained by dividing weight per cent by atomic weight:

$$\text{C:} \quad \frac{55.8}{12} = 4.6; \qquad \text{H:} \quad \frac{7.0}{1} = 7.0; \qquad \text{O:} \quad \frac{37.2}{16} = 2.3$$

Step 3. Convert to whole-number ratio:

The formula $C_{4.6}H_{7.0}O_{2.3}$ is converted to a whole-number ratio by dividing through by the smallest subscript; *i.e.*, assuming one oxygen:

$$C: \frac{4.6}{2.3} = 2.0; \qquad H: \frac{7.0}{2.3} = 3.05; \qquad O: \frac{2.3}{2.3} = 1$$

Rounding off, the empirical formula from analysis is C_2H_3O. The formula weight for C_2H_3O is $2 \times 12 + 3 + 16 = 43$. The molecular weight of the compound is 172, and $172/43 = 4$. The molecular formula is therefore C_2H_3O multiplied by 4, or $C_8H_{12}O_4$.

EXAMPLE 2

An unknown compound contains carbon, hydrogen, nitrogen, and possibly oxygen. The percentage composition from analysis is C, 68.3; H, 7.3; N, 11.2. The molecular weight is 123 ± 1. What is the molecular formula?

1. In this case the percentage composition is given for the three elements that are determined directly. The percentage of oxygen by difference is $100 - (68.3 + 7.3 + 11.2) = 13.2$ per cent.

2. Atomic ratios:

$$C: \frac{68.3}{12} = 5.7; \quad H: \frac{7.3}{1} = 7.3; \quad N: \frac{11.2}{14} = 0.8; \quad O: \frac{13.2}{16} = 0.8$$

3. Dividing through by 0.8, we have C, 7.1; H, 9.1; N, 1; O, 1. Rounding off, the empirical formula is C_7H_9NO. The formula weight is 123, and the molecular formula is therefore the same as the empirical formula.

A very useful point to be noted about molecular formulas is that carbon has an *even* number (4) of bonding electrons. In any neutral compound with a complete shell of eight electrons, each carbon has a total of four covalent bonds:

Moreover, oxygen also has an even number of bonding electrons, and an oxygen atom in a neutral compound with completed electron shells has two bonds.

It follows that any stable neutral compound containing only carbon and hydrogen or carbon, hydrogen, and oxygen will always have an *even* number of hydrogen atoms, *e.g.*, CH_4, C_2H_6, C_2H_4, C_2H_4O, $C_6H_8O_2$ and so on. In Example 1, the empirical formula C_2H_3O cannot represent a stable compound; it would be an odd-electron molecule (*i.e.*, a free radical). By inspection, therefore, we know that the molecular formula must be some *even* multiple—$C_4H_6O_2$, $C_8H_{12}O_4$, and so forth.

Halogens and nitrogen have *odd* numbers of bonding electrons, however. Therefore, if a compound contains one chlorine or one nitrogen, or any *odd* combination of these elements, the number of hydrogens in the molecular formula must be *odd* (e.g., CH_3Cl, C_2H_7N, $C_3H_3Cl_3$). With an *even* number of halogen or nitrogen atoms, the number of hydrogens is *even*.

PROBLEMS

1. A compound contains only carbon, hydrogen, and oxygen. Combustion of 4.56 mg of a compound gave 12.52 mg of CO_2 and 2.86 mg of H_2O. Calculate the empirical formula.

2. A compound contains 79.5 per cent C, 12.8 per cent H, and 7.7 per cent N. The molecular weight is known approximately from the boiling point to be no greater than 200. Calculate the molecular formula.

3. A compound contains carbon, hydrogen, chlorine, and possibly oxygen. Combustion of a 5.11 mg sample gave 8.49 mg of CO_2 and 2.62 mg of H_2O. A separate analysis for chlorine gave a value of 33.6 per cent Cl. Calculate the empirical formula. What is the simplest possible molecular formula?

GLOSSARY
OF
TERMS

Absorption Spectrum. A chart showing the characteristic absorption of differing frequencies or wavelengths of radiation by a compound.

Acetal. The condensation product $RCH(OR')_2$ of an aldehyde with two moles of alcohol, with loss of water.

Activating Group. In aromatic substitution reactions, an electron-donating group that causes the reactivity of the ring to be greater than that of benzene.

Activation Energy. The extra potential energy required to enable reactants to reach the transition state of a reaction; *i.e.*, the potential energy difference between reactants and the transition state.

Acylation. The replacement of H by an acyl group, $R\overset{\displaystyle O}{\overset{\|}{C}}$.

Alcohol. A compound in which a hydroxyl (OH) group attached to an alkyl chain is the major functional group.

Aldehyde. A compound in which the main functional group is a carbonyl bonded to hydrogen, *i.e.*, $R\overset{\displaystyle O}{\overset{\|}{C}}H$.

Aliphatic. Denotes a saturated non-cyclic molecule.

Alkane. An open chain compound containing only C—C and C—H σ bonds, with no π bonds or functional groups, exemplified by the series methane, ethane. . . .

Alkene. A compound in which a carbon–carbon double bond, C=C, is the principal functional group; also called an olefin.

Alkoxide. The anion RO^- obtained by removal of a proton from an alcohol, *i.e.*, the conjugate base of the alcohol.

Alkylation. A reaction in which an alkyl group R replaces a hydrogen.

alpha-. In names of compounds, alpha refers to the position in the chain adjacent to a functional group such as C=O.

Amide. The condensation product of an acid and an amine with loss of water, *i.e.*, $R\overset{\displaystyle O}{\overset{\|}{C}}NH_2$, $R\overset{\displaystyle O}{\overset{\|}{C}}NHR$, $RCNR_2$.

Amine. A compound having the general structure RNH_2, R_2NH, or R_3N.

Ammonium Ion. A compound containing the group $R\overset{+}{N}H_3$, $R\overset{+}{N}H_2R$, or $R\overset{+}{N}HR_2$.

Anhydride. A compound obtained by removal of water between two molecules of acid, *e.g.*, $R\overset{\displaystyle O}{\overset{\|}{C}}-O-\overset{\displaystyle O}{\overset{\|}{C}}R$.

Anion. A negatively charged particle.

Antioxidant. A compound, such as a phenol, which can act as a trap for free radicals and reacts readily with oxidizing species to prevent the build-up of peroxides and further oxidation products.

Aromatic Compound. A compound containing a ring with $4n + 2$ π electrons in a continuous conjugated system, such as benzene and its derivatives.

Aryl. A general term denoting an aromatic group, for example an aryl halide, ArCl, with halogen attached to an aromatic ring.

Atomic Orbital. A region of space around an atom which can contain two electrons of a specified energy.

Carbanion. A negatively charged molecule with an unshared electron pair located on carbon (as in $R_3C^{\,:-}$).

Carbocation. A cation with the positive charge located on carbon, as in $(CH_3)_3C^+$; the carbon is electron-deficient, and is a good electrophile.

Carbohydrate. A naturally occurring compound, often with the general formula

$$H(CHOH)_n\overset{\displaystyle O}{\overset{\|}{C}}H;$$ carbohydrates range from simple 5 or 6 carbon sugars to polymers such as starch and cellulose.

Carbonyl Group. The group C=O; a carbon-oxygen double bond.

Carboxylic Acid. A compound containing the group $-\overset{\displaystyle O}{\overset{\|}{C}}OH$; the carboxylate ion is $-CO_2^-$.

Carcinogen. A substance that produces cancer on absorption or prolonged contact.

Catalyst. A substance that increases the rate of a reaction by lowering the activation energy; although the catalyst is involved in the reaction, it is unchanged at the end.

Cation. A positively charged particle.

Chain Process. A reaction in which one step is formation of a product plus a reactive intermediate, such as a free radical, which then starts the next step. An example is free radical chlorination of alkanes.

Chirality. The property of being left- or right-handed, as in the case of enantiomers.

-cide. An ending indicating lethal, life-terminating action; *e.g.*, insecticide, herbicide, and fungicide are compounds lethal to insects, plants, and fungi, respectively.

Coenzyme. A compound which combines with a protein to form an enzyme and provides the functional group that is involved in the catalytic role of the enzyme.

Condensation. A reaction in which two compounds combine with loss of a small by-product molecule such as water.

Configuration. The arrangement of the four different groups attached to an asymmetric (chiral) carbon atom; or, more generally, the arrangement of groups that distinguishes one stereoisomer from another.

Conformation. The conformations of a molecule are isomers that can be interconverted by rotation around a single bond, and in general cannot be separated from one another because of the rapid interconversion.

Conjugate Acid. The acid obtained by bonding of a given base with a proton.

Conjugate Base. The base obtained by removal of a proton from an acid.

Covalent Bond. The attractive force between two atoms owing to the presence of two electrons in a bonding molecular orbital.

Cracking. An industrial process, normally catalytic, for converting alkanes into smaller alkanes and alkenes.

Cyanide. A compound containing the $-C\equiv N$ or cyano group; the anion $C\equiv N^-$. An alkyl cyanide, RCN, can also be named as a nitrile.

Cyanohydrin. A compound containing the group $-CHOHCN$, formed by addition of HCN to an aldehyde.

Cyclization. A reaction in which intramolecular bond formation occurs to give a product containing a ring.

Deactivating Group. In aromatic substitution reactions, an electron-withdrawing group that causes the reactivity of the ring to be less than that of benzene.

Delocalization. The dispersal or redistribution of electron density or electric charge over several atoms.

Diastereoisomers. Stereoisomers which are not enantiomers, as in the case of two compounds, each with two asymmetric carbons, in which the configuration is the same at one carbon and opposite at the other; geometric isomers are another case of diastereoisomers.

Dissociation. In the case of an acid HA, dissociation refers to the reaction $HA \longrightarrow H^+ + A^-$. The dissociation constant K_A is the equilibrium constant for this reaction, $[H^+][A]/[HA]$; pK_A is the negative logarithm of K_A. Dissociation also refers more generally to cleavage of any bond.

Dye. A compound that has intense color (due to absorption of certain wavelengths of light) and which combines with a fiber.

Electronegativity. Tendency of an atom to attract electrons; fluorine is the most electronegative element.

Electrophile. Any molecule or cation which can accept a pair of electrons from a nucleophile.

Electrophilic Substitution. A reaction, usually of an aromatic ring, in which an electron-seeking reagent, E^+, replaces a proton, H^+; specific examples are nitration, halogenation, alkylation, and sulfonation.

Elimination. A reaction in which two groups on adjacent atoms are removed with formation of a double bond.

Enantiomers. Stereoisomers which are non-superimposable mirror images, with opposite chirality and equal and opposite optical rotations.

Endothermic Reaction. A chemical reaction in which the products are *less* stable than the reactants; thus, heat must be supplied.

Enolization. The conversion of a carbonyl compound containing an α-hydrogen,

$$\begin{matrix} & O & \\ & \parallel & \\ -CH-C- \end{matrix}, \text{ to an equilibrium mixture with the enol isomer, } \begin{matrix} OH \\ | \\ -C=C- \end{matrix}.$$

Enzyme. A protein which has a catalytic effect in bringing about a specific chemical reaction.

Ester. The condensation product of an acid and an alcohol with the loss of water, *e.g.*, a

$$\text{carboxylic ester } R\overset{\overset{\textstyle O}{\parallel}}{C}OR, \text{ or a phosphate ester } HOPO_2R.$$

Ether. A compound in which an oxygen atom with two alkyl or aryl groups attached (ROR or ROAr, etc.) is the principal functional group.

Exothermic Reaction. A chemical reaction in which the products are *more* stable than the reactants; thus, heat is produced.

Extraction. A laboratory process in which a compound is removed from a mixture or from a solution by contact with a solvent that forms a separate liquid phase.

Fat. A triester of glycerol ($HOCH_2CHOHCH_2OH$) with three acid molecules, usually with chain length between C_{14} and C_{18}.

Free Radical. A reactive molecule containing an odd or unpaired electron; for example, the methyl radical $\cdot CH_3$.

Glycoside. An acetal formed by condensation of the carbonyl group of a sugar with an —OH group in the sugar and the OH group of an alcohol or phenol.

Hemiacetal. The addition product $RCH(OH)(OR)$ of an alcohol and an aldehyde.

Heterocyclic Compound. Any cyclic compound in which a ring contains an atom other than carbon (N, O, S, etc.).

Hexose. A six-carbon sugar, $C_6H_{12}O_6$.

Hybrid Orbitals. Atomic orbitals produced by "mixing" two or more atomic orbitals to obtain new orbitals. For example, mixing of one s and two p atomic orbitals produces three sp^2 hybrid orbitals.

Hydrogenation. A reaction with hydrogen as one of the reactants, usually resulting in addition of hydrogen to a multiple bond, *e.g.*, $H_2C=CH_2 + H_2 \longrightarrow H_3C-CH_3$.

Hydrogen Bonding. Attractive force between molecules of water, alcohols, or amines due to electrostatic attraction of H attached to oxygen or nitrogen to a second O or N atom, as in RO—H···OR.

$$\begin{matrix} | \\ H \end{matrix}$$

Inductive Effect. The transmission of electronic effects through single bonds.

Intermolecular. Bonding or reaction occurring between two molecules.

Intramolecular. Bonding or reaction occurring within a single molecule.

Inversion. Reversal of stereochemical configuration, such as that occurring in an S_N2 reaction at a chiral center.

Isomers. Different compounds with the same numbers and kinds of atoms, *i.e.*, *n*-butane, $CH_3CH_2CH_2CH_3$, and isobutane, $(CH_3)_2CHCH_3$.

Ketone. A compound in which the major functional group is a carbonyl bonded to two R or Ar groups, *e.g.*, CH_3COCH_3 or $CH_3COC_6H_5$.

Leaving Group. In a substitution reaction, the group which is displaced; in nucleophilic substitutions, usually a water molecule or an inorganic anion such as Cl^-, Br^-, or $^-OSO_3H$.

Meso Isomers. Compounds with two asymmetric centers, having the same substituents, that are related by a molecular plane of symmetry. A *meso* isomer is therefore *not chiral*.

Methylene Group. The group $-CH_2-$, attached to two other atoms or groups.

Molecular Orbital. An orbital, used in bonding, which is formed by the overlap of atomic orbitals from the two bonded atoms.

Monomer. Any small molecule, usually an alkene, which serves as the repeating unit in the formation of a polymer.

Nucleophile. Any molecule or anion in which an unshared electron pair can form a bond to an electron-deficient center, especially carbon; a Lewis base.

Nucleophilic Substitution. A reaction in which an electronegative leaving group such as halogen or water is replaced by a nucleophile; see S_N1 and S_N2.

Nucleotide. A compound consisting of a purine or pyrimidine base, ribose or deoxyribose, and phosphoric acid; the basic building block of nucleic acids.

Oxidation. With organic compounds, a reaction in which electrons are removed from carbon, usually by replacing bonds to hydrogen or to another carbon by bonds to a more electronegative element such as oxygen, *e.g.*, $R-CH_2OH \longrightarrow R\overset{\overset{O}{\|}}{C}H$.

Oxonium Ion. A compound containing the group $-\overset{+}{O}H_2$, $-\overset{+}{O}R$, or $-\overset{+}{\underset{H}{O}}R_2$.

Ozonolysis. The process of cleaving the double bond of an alkene with ozone, O_3, to produce aldehydes and ketones.

Peptide. A substance containing amino acids linked together by amide bonds, referred to in this context as *peptide* bonds.

Phenol. A compound with a hydroxyl group attached to an aromatic ring.

Phenyl. The name of the group C_6H_5- formed by removing a hydrogen atom from benzene.

Pi (π) Bond. A covalent bond formed by overlapping parallel *p*-orbitals on adjacent atoms.

Polar Bond. A covalent bond between two atoms of different electronegativities, having an unequal distribution of bonding electrons.

Polymer. A very large molecule made up of repeating small units called monomers.

Protein. A high molecular weight substance made by condensation of α-amino acid units, $CHRCO_2H$, with loss of water.

Quaternary Ammonium Ion. A compound containing the group R_4N^+.

Racemic Mixture. A mixture of exactly equal amounts of enantiomers.

Rate Determining Step. The slowest reaction (that with the largest activation energy) in a sequence of several steps.

Reaction Mechanism. A detailed, step-by-step picture of exactly how a reaction proceeds.

Rearrangement. A change in the carbon skeleton or position of substitution in a chain, caused by migration of a group during a reaction.

Reduction. A reaction in which bonds from carbon to oxygen or some other electronegative element are replaced by bonds to hydrogen, or in which hydrogen is added across a multiple bond.

Resolution. The separation of a racemic mixture into the separate enantiomers.

Resonance. The situation in which a molecule is better represented as a hybrid of two or more electronic structures than by a simple structure.

-saccharide. An ending designating a carbohydrate made up of two (disaccharide), several (oligo-), or many (poly-) sugar units.

Sigma (σ) Bond. A covalent bond with cylindrical symmetry (like a hotdog).

S_N1. Nucleophilic substitution, unimolecular, in which formation of a carbocation is the initial and rate-determining step.

S_N2. Nucleophilic substitution, bimolecular, in which bonding of the nucleophile and departure of the leaving group occur in the same step.

Stereoisomers. Isomers which have the same atom-to-atom bonding sequence but differ in the spatial arrangement of these atoms; includes enantiomers, diastereoisomers, and conformers.

Steroid. A class of compounds occurring in plant and animal tissues and related to terpenes. Steroids contain a fused system of three six- and one five-membered ring plus hydroxyl and carbonyl groups.

Surfactant. A surface active agent; substance that promotes the spreading of water on a surface, aiding in the removal of grease or other water-insoluble material.

Terpenes. Naturally occurring compounds made up of two to six branched five-carbon isoprene units.

Thio-. A term indicating replacement of oxygen by sulfur, as in thiol (RSH), thiol acid (RCOSH), and thione ($R_2C=S$).

Transition State. The point of highest energy in a reaction, at which bond changes are not completed.

Wax. A substance with a long alkyl chain that provides a water-repellent coating; plant waxes are carboxylic esters RCO_2R', with R and R' totalling 40 to 50 carbons.

INDEX